Biotechnology in Agriculture and Forestry

Springer

Berlin
Heidelberg
New York
Barcelona
Hong Kong
London
Milan
Paris
Singapore
Tokyo

Biotechnology in Agriculture and Forestry 47

Transgenic Crops II

Edited by Y.P.S. Bajaj

With 96 Figures, 5 in Color, and 71 Tables

 Springer

Professor Dr. Y.P.S. Bajaj†
New Delhi, India

Agr.
SB
123
.57
.T724
2000
v.2

ISSN 0934-943-X
ISBN 3-540-67131-5 Springer-Verlag Berlin Heidelberg New York

Die Deutsche Bibliothek – CIP-Einheitsaufnahme

Transgenic crops / ed. by Y. P. S. Bajaj. – Berlin; Heidelberg; New York; Barcelona; Hong Kong; London; Milan; Paris; Singapore; Tokyo: Springer 2. – (2001)
(Biotechnology in agriculture and forestry; 47)
ISBN 3-540-67131-5

Library of Congress Cataloging-in-Publication Data applied for.

Springer-Verlag Berlin Heidelberg New York a member
of BertelsmannSpringer Science & Business media GmbH

© Springer-Verlag Berlin Heidelberg 2001
Printed in Germany

Production: PRO EDIT GmbH, Heidelberg

Cover design: *design & production* GmbH, Heidelberg

Typesetting: Best-set Typesetter Ltd., Hong Kong

Printed on acid-free paper   SPIN: 10691031   31/3130/SO 5 4 3 2 1 0

Dedicated to my dear and kind friends Bal Krishan Ghai and Sunita Ghai of Sun Valley, California, whose pleasant company brought me a lot of happiness and comfort.

Preface

Over the last decade there has been tremendous progress in the genetic transformation of plants, which has now become an established tool for the insertion of specific genes. Work has been conducted on more than 200 plant species of trees, cereals, legumes and oilseed crops, fruits and vegetables, medicinal, aromatic and ornamental plants etc. Transgenic plants have been field-tested in a number of countries, and some released to the farmers, and patented.

Taking in above-mentioned points into consideration, it appeared necessary to review the literature and state of the art on genetic transformation of plants. Thus 120 chapters contributed by experts from 31 countries (USA, Russia, Canada, France, Germany, England, The Netherlands, Belgium, Switzerland, Italy, Spain, Bulgaria, Yugoslavia, Denmark, Poland, Finland, Australia, New Zealand, South Africa, China, Japan, Korea, Singapore, Indonesia, India, Israel, Mexico, Brazil, Moroco, Senegal, Cuba, etc.) have been complied in a series composed of the following five books:

1. *Transgenic Trees* comprises 22 chapters on forest, fruit, and ornamental species such as *Allocasuarina verticillata*, *Casuarina glauca*, *Cerasus vulgaris*, *Citrus* spp., *Coffea* species, *Diospyros kaki*, *Eucalyptus* spp., *Fagara zanthoxyloides*, *Larix* spp., *Lawsonia inermis*, *Malus* × *domestica*, *Picea mariana*, *Pinus palustris*, *Pinus radiata*, *Poncirus trifoliata*, *Populus* spp., *Prunus* species, *Rhododendron*, *Robinia pseudoacacia*, *Solanum mauritianum*, *Taxus* spp., and *Verticordia grandis*.

2. *Transgenic Medicinal Plants* comprises 26 chapters on *Ajuga reptans*, *Anthemis nobilis*, *Astragalus* species, *Atropa belladonna*, *Catharanthus roseus*, *Datura* species, *Duboisia* species, *Fagopyrum* species, *Glycyrrhiza uralensis*, *Lobelia* species, *Papaver somniferum*, *Panax ginseng*, *Peganum harmala*, *Perezia* species, *Pimpinella anisum*, *Phyllanthus niruri*, *Salvia miltiorrhiza*, *Scoporia dulcis*, *Scutellaria baicalensis*, *Serratula tinctoria*, *Solanum aculeatissimum*, *S. commersonii*, *Swainsona galegifolia*, tobacco, and *Vinca minor*.

3. *Transgenic Crops I* comprises 25 chapters divided into 2 sections:

    Section I. Cereals and grasses, such as wheat, rice, maize, barley, sorghum, pearl millet, triticale, *Agrostis*, *Cenchrus*, *Dactylis*, *Festuca*, *Lolium*, and sugarcane.

    Section II. Legumes and Oilseed Crops. Arachis hypogaea, *Brassica juncea*, *Brassica napus*, *Cicer arietinum*, *Glycine max*, *Gossypium hirsutum*,

Helianthus annuus, Lens culinaris, Linum usitatissimum, Sinapis alba, Tri-folium repens, and *Vicia narbonensis.*

4. *Transgenic Crops II* comprises 21 chapters on fruits and vegetables, such as banana, beetroot, grapes, strawberry, kiwi, watermelon, cucumber, tomato, asparagus, carrot, cabbage, kale, turnip, rutabaga, broccoli, sweet pea, common bean, *Luffa, Amaranthus,* horseradish, sugarbeet, chicory, cassava, sweet potato, and potato.

5. *Transgenic Crops III* comprises 26 chapters arranged in 2 sections:

Section I. Ornamental, Aromatic and Medicinal Plants. Anthurium, Antirrhinum, Artemisia absinthium, Begonia, Campanula, Carnation, *Chrysanthemum, Dendrobium, Eustoma, Gentiana, Gerbera, Gladiolus, Hyoscyamus muticus, Hyssopus officinalis, Ipomoea, Leontopodium alpinum, Nierembergia, Phalaenopsis, Rudbeckia, Tagetes* and *Torenia.*

Section II. Miscellaneous Plants. Craterostigma plantagineum, Flaveria spp., *Moricandia arvensis, Solanum brevidens,* and freshwater wetland monocots.

These books will be of special interest to advanced students, teachers, and research workers in the field of molecular biology, genetic manipulation, tissue culture, and plant biotechnology in general.

New Delhi, April 1999

Professor Dr. Y.P.S. BAJAJ
Series Editor

Contents

List of Contributors

ABID, M., Laboratoire d'Amélioration et de Production Végétales, Faculté des Sciences, Université Mohammed 1er, Oujda, Morocco

ADAME-ALVAREZ, R.M., Department of Plant Genetic Engineering, Cinvestav-Irapuato. Apdo. Postal 629, 36500 Irapuato, Gto. México

AKIHAMA, T., Faculty of Agriculture, Meiji University, 1-1-1, Higashimita, Tamaku, Kawasaki 214-8571, Japan

ARAGÃO, F.J.L., Embrapa Recursos Genéticos e Biotecnologia, Parque Estação Biológica, Final Av. W3 Norte, 70.770-900 Brasília, DF, Brazil

BARG, R., Department of Plant Genetics, Institute of Field and Garden Crops, The Volcani Center, A.R.O., Bet-Dagan 50250, Israel

BELKNAP, W.R., United States Department of Agriculture, Agricultural Research Service, 800 Buchanan Street, Albany, California 94710, USA

BESTWICK, R.K. 15052, Fall River Drive, Bend, Oregon 97707, USA

BRAUN, R.H., New Zealand Institute for Crop & Food Research Limited, Private Bag 4704, Christchurch, New Zealand

CABRERA-PONCE, J.L., Department of Plant Genetic Engineering, Cinvestav-Irapuato. Apdo. Postal 629, 36500 Irapuato, Gto. México

CÁRCAMO, R., Akkadix Corporation, 11099 North Torrey Pines Road, Ste. 200, La Jolla, California 92037, USA

CAMMUE, B.P.A., FA Janssens Laboratory of Genetics, Catholic University of Leuven, Kardinaal Mercierlaan 92, 3001 Heverlee, Belgium

CHEE, P.P., Pharmacia & Upjohn Inc 301 Henrietta St. Kalamazoo, Michigan 49007, USA. Present address: PCCom.Inc 3305 Henrietta St. Kalamazoo, Michigan 49008, USA

CHMELEV, V., Laboratory of Biochemistry, All-Russian Plant Research
Institute, Isaakievskaya Sq. 2, 199000, St. Petersburg, Russia

CHOI, P.S., Plant Cell and Molecular Biology Research Unit, Korea Research
Institute of Bioscience and Biotechnology, KIST, P.O. Box 115, Yusong,
Taejon, 305-600, Korea

CHRISTEY, M.C., New Zealand Institute for Crop & Food Research Limited,
Private Bag 4704, Christchurch, New Zealand

DELBREIL, B., Laboratoire de Physiologie de la Différenciation Végétale
UPRES 2702. Université des Sciences et Technologies de Lille,
59655 Villeneuve D'Ascq CEDEX, France

FAUQUET, C.M., ILTAB/Donald Danforth Plant Science Center,
8001 Natural Bridge Road, St. Louis, Missouri 63121, USA

GALLIE, D.R., Department of Biochemistry, University of California,
Riverside, California 92521, USA

GONZÁLEZ, A.E., The Scripps Research Institute, Department of Cell
Biology, 10550 North Torrey Pines Road, La Jolla, California 92037, USA

HERRERA-ESTRELLA, L., Department of Plant Genetic Engineering,
Cinvestav-Irapuato. Apdo. Postal 629, 36500 Irapuato, Gto. México

HUSS, B., Physiologie et Génétique Moléculaire Végétales, Université des
Sciences et Technologies de Lille, 59655 Villeneuve d'Ascq cedex, France

JOFRE-GARFIAS, A.E., Department of Plant Genetic Engineering,
Cinvestav-Irapuato. Apdo. Postal 629, 36500 Irapuato, Gto. México

JULLIEN, M., UMR Biologie des Semences, INRA-INAP-G,
Route de Saint-Cyr, 78026 Versailles CEDEX, France

KIM, Y.S., Plant Cell and Molecular Biology Research Unit, Korea Research
Institute of Bioscience and Biotechnology, KIST, P.O. Box 115, Yusong,
Taejon, 305-600, Korea

KINO-OKA, M., Department of Chemical Science and Engineering,
Osaka University, Toyonaka, Osaka 560-8531, Japan

KISLIN, Y.N., Laboratory of Plant Growth Regulators, All-Russian Research
Institute for Plant Protection, Sh. Podbelskiy 3, 189620 Pushkin 8,
St. Petersburg, Russia

KRAVCHENKO, L.V., Laboratory of Biotechnology, All-Russian Research Institute for Agricultural Microbiology, Sh. Podbelskiy 3, 189620 Pushkin 8, St. Petersburg, Russia

LIMANTON-GREVET, A., Laboratoire "in vitro", J. Marionnet GFA, Route de Courmenin, 41230 Soings-en-Sologne, France
Present address:
UMR Biologie des Semences, INRA-INAP-G, Route de Saint-Cyr, 78026 Versailles CEDEX, France

LIU, J.R., Plant Cell and Molecular Biology Research Unit, Korea Research Institute of Bioscience and Biotechnology, KIST, P.O. Box 115, Yusong, Taejon, 305-600, Korea

LUTOVA, L.A., Department of Genetics, St. Petersburg State University, Universitetskaya Emb., 7/9, 199034 St. Petersburg, Russia

MACCREE, M.M., United States Department of Agriculture, Agricultural Research Service, 800 Buchanan Street, Albany, California 94710, USA

MALYSHEVA, N.V., Department of Genetics, St. Petersburg State University, Universitetskaya Emb., 7/9, 199034 St. Petersburg, Russia

MANDOLINO, G., Istituto Sperimentale per le Colture Industriali, via di Corticella 133, 40129 Bologna, Italy

MANO, Y., Department of Biological Science and Technology, Tokai University, 317 Nishino, Numazu, Shizuoka 410-0321, Japan

MARTINELLI, L., Laboratorio Biotecnologie, Istituto Agrario, 38010 San Michele all' Adige (TN), Italy

MASONA, V.M., University of Zimbabwe, Crop Science Department, MP 167, Harare, Zimbabwe

MATHEWS, H. Agritope Inc. 16160 SW Upper Boones Ferry Road, Portland, Oregon 97224-7744, USA

MATSUYAMA, T., Faculty of Agriculture, Meiji University, 1-1-1, Higashimita, Tamaku, Kawasaki 214-8571, Japan

METZ, T.D., Biology Department, Campbell University, P.O. Box 308, Buies Creek, North Carolina 27506, USA

OTANI, M., Research Institute of Agricultural Resources, Ishikawa Agricultural College, Nonoichi-machi, Ishikawa 921-8836, Japan

PAVLOVA, Z.B., Department of Genetics, St. Petersburg State University, Universitetskaya Emb., 7/9, 199034 St. Petersburg, Russia

PÉREZ HERNÁNDEZ, J.B., Laboratory of Tropical Crop Improvement, Catholic University of Leuven, Kardinaal Mercierlaan 92, 3001 Heverlee, Belgium

RAMBOUR, S., Physiologie et Génétique Moléculaire Végétales, Université des Sciences et Technologies de Lille, 59655 Villeneuve d'Ascq cedex, France

RECH, E.L., Embrapa Recursos Genéticos e Biotecnologia, Parque Estação Biológica, Final Av. W3 Norte, 70.770-900 Brasília, DF, Brazil

REMY, S., Laboratory of Tropical Crop Improvement, Catholic University of Leuven, Kardinaal Mercierlaan 92, 3001 Heverlee, Belgium

ROCKHOLD, D.R., United States Department of Agriculture, Agricultural Research Service, 800 Buchanan Street, Albany, California 94710, USA

SÁGI, L., Laboratory of Tropical Crop Improvement, Catholic University of Leuven, Kardinaal Mercierlaan 92, 3001 Heverlee, Belgium

SALTS, Y., Department of Plant Genetics, Institute of Field and Garden Crops, The Volcani Center, A.R.O., Bet-Dagan 50250, Israel

SCHÖPKE, C., Department of Botany and Plant Sciences, University of California, Riverside, California 92621, USA

SHABTAI, S., Department of Plant Genetics, Institute of Field and Garden Crops, The Volcani Center, A.R.O., Bet-Dagan 50250, Israel

SHIMADA, T., Research Institute of Agricultural Resources, Ishikawa Agricultural College, Nonoichi-machi, Ishikawa 921-8836, Japan

SIMPSON, J., Department of Plant Genetic Engineering, Cinvestav-Irapuato. Apdo, Postal 629, 36500 Irapuato, Gto. México

SPANÒ, L., Department of Basic and Applied Biology, State University of L'Aquila, via Vetoio 67010 Coppito AQ, Italy

SWENNEN, R., Laboratory of Tropical Crop Improvement, Catholic University of Leuven, Kardinaal Mercierlaan 92, 3001 Heverlee, Belgium

TAYLOR, N.J., ILTAB/Donald Danforth Plant Science Center, 8001 Natural Bridge Road, St. Louis, Missouri 63121, USA

TONE, S., Department of Chemical Science and Engineering, Osaka University, Toyonaka, Osaka 560-8531, Japan

YAZAWA, M., Faculty of Agriculture, Meiji University, 1-1-1, Higashimita, Tamaku, Kawasaki 214-8571, Japan

1 Transgenic Kiwi Fruit (*Actinidia deliciosa*)

M. Yazawa, T. Matsuyama, and T. Akihama

1 Introduction

Kiwi fruit (*Actinidia deliciosa*), originating from the Yangtze Valley in China, was initially cultured in New Zealand in the 1900s as a fruit tree. Its breeding history is short compared with many other plants. Presently, kiwi fruit is not only valued for its good taste, but also contains various vitamins and proteinases; therefore, the demand for kiwi fruit is increasing worldwide. In addition, further improvements are needed since there have been problems such as more diseases caused by bacteria, insect damage and decreased sweetness and softening during transportation (International Kiwifruit Organization Conference 1988).

Crossbreeding for genetic improvement in woody plants including kiwi fruit requires a long time because of genetic heterogeneity and long life cycles. Therefore, it is anticipated that advanced techniques such as genetic engineering will be powerful tools for solving such problems and such techniques are expected to increase efficiency by being able to induce single useful traits in kiwi fruit breeding. *Agrobacterium*-mediated transformation, which is one of the genetic transformation methods, has been mainly used for genetic engineering in dicotyledonous annual species, because long DNA fragments are stably and easily integrated into the plants, compared with other methods. We describe here the successful transformation of kiwi fruit with *A. rhizogenes* which induced a 'hairy root', and demonstrate the agricultural availability of the clones regenerated from them.

2 Genetic Transformation

Uematsu et al. (1991) and Rugini et al. (1991) have previously reported the *Agrobacterium*-mediated transformation in kiwi fruit employing *A. tumefaciens*. In the former report, the transformants were regenerated from hypocotyl segments and stem cuttings by infection with *A. tumefaciens* har-

Faculty of Agriculture, Meiji University, 1-1-1, Higashimita, Tamaku, Kawasaki, 214-0033, Japan

Table 1. Summary of various studies on the transformation of kiwi fruit

Reference (year)	Explant used	Vector used	Morphological observations on transformants
1. Uematsu et al. (1991)	Hypocotyl and stem segments	pLAN411 and pLAN421 contained the npt II gene and the GUS gene	No changes in relation to control plants
2. Rugini et al. (1991)	Leaf discs	pBin19 contained the npt II gene and *rol* A,B,C genes of Ri plasmid	Shorter internodes, dark green wrinkled leaves, and high rooting ability
3. Yazawa et al. (1995a)	Hypocotyl segments	Ri plasmid contained in *A. rhizogenes* NIAES 1724	Shorter internodes, dark green wrinkled leaves, and high rooting ability

boring a binary vector which contained the neomycin phosphotransferase 11 (npl II) gene and the β-glucuronidase (GUS) gene. In the latter report, transgenic kiwi fruit were obtained from the leaf disc callus with *A. tumefaciens* harboring a binary vector-contained T-DNA fragment encompassing the *rol* genes of *A. rhizogenes*. Physiological changes such as high root production in the resulting transformants have been reported (Table 1). Transformation with *A. rhizogenes* can be easily detected by the growth of hairy roots from wounded sites, and hairy root clones composed of only transformed cells may be obtained after a few cycles of root-tip subculturing on media without phytohormones. In some species, spontaneous shoot regeneration from hairy roots has been reported (David et al. 1984). Therefore, plants derived from hairy roots are considered to consist of transformed cells. In addition, the physiological characteristics of transformants that induce high rooting ability and dwarfing by *A. rhizogenes* are generally sought for the improvement of woody plants (Table 1).

3 Methodology

The authors have already reported the successful development of kiwi fruit transformants from the hypocotyls of seedlings by infection with a wild-type strain of *A. rhizogenes* (Yazawa et al. 1995a,b). We describe here the procedure for the acquisition of transfomants, analysis of transformation and morphological evaluation of acclimatized plants.

3.1 Plant Materials

Seeds were taken from mature samples of kiwi fruit [*Actinidia deliciosa* (A. Chev.) C.F. Liang et A.R. Ferguson var. deliciosa] cultivar "Hayward", soaked in 70% ethanol for 30 s, washed in 0.5% sodium hypochlorite solution with a few drops of neutral detergent for 10 min, and then rinsed with sterilized water. They were then plated on an MS medium elimination (Murashige and Skoog 1962) with half-strength inorganic salts containing 1.5% (w/v) sucrose and solidified with 0.2% Gelrite (1/2 MS medium), and were cultured under illuminated conditions (14-h photoperiod: 2000 lx) at 26 °C. After about 4 weeks, the hypocotyl segments were cut into 5-mm pieces and infected with *A. rhizogenes*.

3.2 Bacterial Strains

A wild-type strain of *A. rhizogenes* NIAES 1724, isolated from melon plants in Japan, was used. The bacteria were cultured overnight by shaking at 100 rpm on liquid YEB medium (0.5% Bact. beef extract, 0.5% Bact. yeast extract, 0.5% Bact. peptone, 2 mM $MgSO_4 \cdot 7\ H_2O$, 0.5% sucrose) supplemented with 20 mg/l acetosyringone (Sheikholeslam and Weeks 1987).

3.3 Infection and Induction of Hairy Roots

A 5-ml bacterial suspension was diluted with 20 ml liquid 1/2 MS medium and shaken with hypocotyl segments at 100 rpm for 20 min. After co-cultivation for 2 says on a 1/2 MS medium, the hypocotyls were plated on an MS medium containing 500 mg/l Claforan (Cefotaxime sodium, Hoechst Aktiengesellschaft) to remove any bacteria, and subcultured three times at 2-week intervals. They were then transferred to an MS medium for induction of adventitious roots and subcultured in an MS medium at 3-week intervals under dark conditions. Induced adventitious roots from the hypocotyl were cut into 5-cm root tips and subcultured on the same medium at 3-week intervals. Vigorously growing roots were selected, and an opine, mikimopine, was detected from these roots after proliferation. For the detection of mikimopine (Isogai et al. 1990), fresh root tissues were ground with a glass stick in a microtube, and a 3- to 5-μl crude extract obtained after centrifugation at 7500 g for 10 min was subjected to high-voltage paper electrophoresis according to Tanaka's method (Tanaka 1990).

3.4 Plant Regeneration from Hairy Roots

Adventitious shoots induced from the hairy roots were excised and cultured on a 1/2 MS medium under illuminated conditions. These plants were confirmed to be transformants by the detection of mikimopine from the leaf extracts. Shoot tips, including one to two leaves and/or internodes with lateral buds, were excised from these plants and repeatedly subcultured on the medium at 2-month intervals to produce a larger number of clones. To confirm the absence of bacteria in plant cultures, leaf cuttings were cultured on a YEB medium solidified with 1.2% agar at 26 °C under dark conditions. They were acclimatized for subsequent T-DNA detection. For acclimatization, the plants were transferred to pots containing mixed soil (vermiculite:sand, 1:1), cultured for 1 month under illuminated conditions, and transferred to a greenhouse. Seedlings cultured for 2 months after seeding on a 1/2 MS medium were propagated and acclimatized in the same way and used as control plants.

3.5 Detection of T-DNA from the Transformants

Total DNA was extracted from leaves of in vitro plants according to the procedure of Dellaporta et al. (1989). Samples were amplified through 25 cycles, each cycle consisting of exposure to 94 °C for 1 min, 55 °C for 2 min and 72 °C for 2 min in the reaction solution according to instructions (Perkin Elmer Cetus). Two pairs of primers, 1724A–1724B and 1724C–1724D, capable of amplifying the *rol* genes (Kiyokawa et al. 1992) were used for T-DNA detection from kiwi fruit transformants. The amplified products were electrophoresed on 1.0% agarose gel and hybridized with T-DNA clones pAr 24 containing a portion of *rol* genes A and B, and pAr 2 containing a portion of *rol* genes C and D derived from the Ri plasmid of *A. rhizogenes* NIAES 1724. Southern hybridization was performed with the ECL gene detection system (Amersham).

3.6 Rooting Test

To assess rooting ability, the third to fifth leaves of the transformants were cut into leaf cuttings about 2 cm^2 after the removal of leaf edges and petioles and plated on an MS medium of a 1/2 MS medium without phytohormones. The percentage of rooted leaf cuttings and the number of roots per cutting were then observed after about a 2-month culture period under dark conditions.

4 Results and Discussion

About 2 weeks after inoculation, adventitious shoots and roots emerged from the cut surface of the hypocotyl segments. There were two adventitious roots on the inoculated sections (2/140:number of adventitious roots/number of hypocotyl segments), and three on the non-inoculated sections (3/60). The two adventitious roots, which emerged from the inoculated sections, grew slowly during the first stage of culture, but began to grow rapidly with frequent lateral

Fig. 1. Shoot regeneration from hairy roots. *Arrow* indicates the site of shoot formation. (Yazawa et al. 1995a)

branching after the fifth subculture. At the same time, the growth of the three roots which emerged from the non-inoculated sections, tended to cease. Based on paper electrophoresis, mikimopine was detected only from the inoculated sections, which were considered to be the hairy roots induced by transformation with *A. rhizogenes*. These lines were designated as T_1 and T_2 lines, respectively. Shoot regeneration from the hair roots occurred during the 2-month period of subculture (Fig. 1). More than two shoots per hairy root emerged from both lines. Shoots were differentiated from the hairy roots, resulting in the growth of a small amount of calli and were transferred to a 1/2 MS medium in order to establish complete plants by inducing roots (under illumination, 26 °C). Mikimopine was detected from the extracts of the plants' leaves derived from the bairy roots, and they were considered to be transformants (Fig. 2a).

The in vitro characteristics of the transformants, compared with the control plants, included the presence of shorter internodes and dark green wrinkled leaves, active root formation, and reduction of root geotropism (Fig. 3). Aerial roots also appeared on the stems of some transformants. Leaf cuttings cultured on YEB medium for about 1 month did not exhibit any sign of bacterial growth. As bacteria were confirmed to be completely removed from the transformants, detection of T-DNA and acclimatization of the transformants were performed.

The roots of the transformants protruded on the soil surface in addition to obvious root growth in the soil. About 4 months after acclimatization, a decrease in apical dominance and dwarfing were observed in the transformants (Fig. 4, Table 2). In addition, a reduction in leaf size was observed in the transformants with 5% significant differences (Duncan's new multiple range test) in the AB – leaf area index, although the A/B – leaf shape index showed no significant difference (Table 3). These data show that the leaves of the transformants were miniaturized, with a similar shape.

When total DNAs extracted from the leaves were subjected to PCR analysis, DNA fragments from the transformants coincided with those from the amplified DNA derived from Ri plasmids, whereas no such amplified DNA

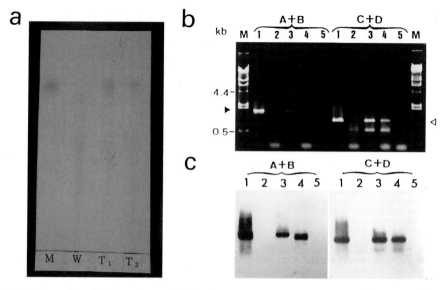

Fig. 2. Detection of mikimopine and T-DNA (Yazawa et al. 1995a). **a** Electrophoretic analysis of mikimopine *Lane M* mikimopine standard, *lane W* the control plant, *lane T₁, T₂* transformation derived from hairy root lines T₁ and T₂, respectively. **b** Electrophoretic patterns of amplified DNA products by 1724A–1724B (= *A* + *B*) and 14724C–1724D (= *C* + *D*) primers, *lane 1 A. rhizogenes*; *2* the control plant; *3,4* transformants derived from hairy root lines T₁, T₂, respectively; *5* PCR water; *M* λ/Hind III size marker. Each *arrow* indicates the same length within *A. rhizogenes* and transformants. ▲ DNA bands seemed to correspond with *rol* A + B genes. △ DNA bands seemed to correspond with *rol* C + D genes. **c** Southern bybridization patterns detected with the clone pAr 24 (*left*) and pAr 2 (*right*). Lane numbers are the same as those used in **b**

Table 2. Comparison of average number of lateral branches between potted transformants and wild type plants investigated after 4 months. (Yazawa et al. 1995b)

Plant material[a]	Number of lateral branches (±SE)
W	0.0 ± 0.0
T₁	3.2 ± 0.4
T₂	3.2 ± 0.7

[a] W is derived from four different wild type plants. T₁ and T₂ are nine transformants derived from each hairy root line T₁ and T₂.

fragments were detected among the control plants (Fig. 2b). The amplified DNA bands of the transformants also hybridized to the two kinds of probes for the *rol* genes clone (pAr2, 24) isolated from Ri plasmids (Fig. 2c). Therefore, it was demonstrated that the plants regenerated from hairy roots were transformants which harbored the *rol* genes of T-DNA.

In some herbs, it has been reported that hairy roots induced by infection with *A. rhizogenes* showed vigorous growth, and obvious branching on the medium without phytohormones, less geotropism and spontaneous plant

a

b

Fig. 3. Morphological comparison of the in vitro plants and their leaves (Yazawa et al. 1995b). **a** The in vitro plant, after about 2 months in culture on 1/2 MS medium. (*Left* the control plant, *right* transformant.) **b** The leaves from the in vitro plants (*left* the control plant, *right* transformant)

Fig. 4. Transformants after acclimatization for 4 months and roots in the pots (Yazawa et al. 1995a). **a** Transformant (*right*) with shorter internodes than the control plant (*left*). **b** The roots of the transformant (*right*) with greater growth than the control plant (*left*)

Table 3. Morphological comparison of average leaf length and width between potted wild-type plants and transformants after 4 months (±SE; Yazawa et al. 1995b)

Plant material	Leaf length (mm) A	Leaf width (mm) B	A/B	AB
W	106.5 ± 5.6	74.0 ± 4.9	1.47 ± 0.04[a]	8007.3 ± 901.4[a]
T_1	52.0 ± 9.0	41.0 ± 2.0	1.35 ± 0.36[a]	2348.7 ± 1101.5[b]
T_2	44.0 ± 1.4	33.0 ± 16.7	1.37 ± 0.03[a]	1612.3 ± 508.6[b]

Figures followed by the same letter are not significantly different at $P < 0.05$ Duncan's new multiple range test.

Table 4. Rooting test for leaf cutting of transformants. (Yazawa et al. 1995b)

Medium	Plant material[a]	Percentage of rooted leaf cuttings[b] (%)	Roots per cutting[b] (±SE)
1/2MS	W	0.0	0.0 ± 0.0
	T_1	77.0	2.8 ± 0.8
	T_2	42.3	1.3 ± 0.6
MS	W	6.5	0.1 ± 0.1
	T_1	87.9	4.7 ± 0.9
	T_2	74.1	2.9 ± 0.6

[a] W is two different wild-type plants. T_1 and T_2 are at least four transformants derived from each hairy root line T_1 and T_2. All plants were randomly chosen. At least 60 plants per experiment were used.
[b] The percentage of rooted leaf cuttings and the average root per cutting from two experiments were evaluated after 60 days culture.

regeneration on their own. The hairy root of the kiwi fruit obtained in the present study also exhibited a similar growth, and shoot regeneration was easily achieved. For the hairy root syndrome of the transformants obtained by infection with *A. rhizogenes*, it has become increasingly clear that the morphological changes are caused by a change in the endogenous hormonal balance due to the function of the *rol* genes group (Schmulling et al. 1993). Some reports have already focused on the enzymes produced by the *rol* genes (Estruch et al. 1991a,b) and on the investigation of the hormonal content of the transformants expressing single *rol* genes (Schmulling et al. 1993). Based on these reports, the morphological changes, including the presence of shorter internodes, dark green wrinkled leaves, active root formation and reduction in the geotropism of roots in the kiwi fruit transformants obtained in the present study also considered to represent the hairy root syndrome caused by changes in the endogenous hormonal balance due to the expression of the *rol* genes.

A 60-day observation of the cultured leaf segments cut from the plant cultures showed that the transformants were inclined to increase their rooting ability compared with the control plants (Fig. 5). The rate of root formation of the leaf cuttings and the number of roots per leaf cutting were higher in the transformants derived from the T_1 line of hairy roots than from the T_2 line (Table 4).

Fig. 5. Rooting test using leaf cuttings (Yazawa et al. 1995b). Compared with the control plant (*right*), the leaf cuttings of the transformant (*left*) have a higher rooting ability

This difference of rooting ability in the rooting test between transformants derived from two hairy root lines was estimated to be caused by differences in the site of T-DNA insertion and the number of copies. However, it is suggested that a trait such as high-rooting ability in the transformants may be useful for the development of root stock variety.

In addition, as the acclimatized transformants grew older, some of them reverted to the wild-type plants. This may suggest that the methylation of T-DNA occurred in the transformants. At present, efforts are in progress to determine the differences between the T_1 and T_2 lines and to detect the modification of T-DNA insertion using a molecular technique, as well as to observe the acclimatized transformants.

5 Summary and Conclusions

The hypocotyl of kiwi fruit seedlings inoculated with a domestic wild strain of *A. rhizogenes* NIAES 1724 produced two hairy root lines that grew vigorously on MS medium without phytohormones. Shoots were easily differentiated from the hairy roots and gave rise to complete plants. These plants developed shorter internodes, darker green wrinkled leaves and displayed active root formation under in vitro conditions. The number of branches increased and miniaturization of the leaves was promoted after acclimatization, compared with the control plants. They were found to be transformants because of the detection of mikimopine in the leaf extracts and the presence of *rol* genes from the Ri plasmid in the plant total DNA. In addition, a rooting test with leaf cuttings showed high rooting ability in the transformants. These results suggest that *A. rhizogenes* NIAES 1724 could be utilized as a vector for the genetic transformation of kiwi fruit. Furthermore, such characteristics as the tendency of dwarfing and high rooting ability in the transformants would be useful for the breeding of kiwi fruit, especially the root stock variety.

5.1 Present Status

After 3 years acclimatization, the development of the roots and lateral branches of the transformants tended to be greater than that of the control plants. The control plants showed normal differentiation in the flower bud, but the transformants did not show flower bud differentiation entirely. In addition, some of the transformants reverted to wild-type plants. Their vines emerged vigorously as well as those of the control plants. Further studies on these transformants and the wild-type plants are being performed by morphological and molecular biological analysis.

References

David C, Chilton MSD, Tempe J (1984) Conservation of T-DNA in plants regenerated from hairy root cultures. Bio/Technology 2:73–76

Dellaporta SL, Wood J, Hick JB (1989) A plant DNA minipreparation: version II. Plant Mol Biol Rep 1:19–21

Estruch JJ, Chriqui D, Grossmann K, Shell J, Spena A (1991a) The plant oncogene *rol* C is responsible for the release of cyiokinins from glucoside conjugates, EMBO J 10:2889–2895

Estruch JJ, Shell J, Spena A (1991b) The protein encoded by the *rol* B plant oncogene hydrolyses indole glucosides. EMBO J 10:3125–3128

International Kiwifruit Organization Conference (1988) Record of Proc of the 15th Int Kiwifruit Conf, Hong Kong

Isogai A, Fukuchi N, Kamada H, Harada H, Suzuki A (1990) Mikimopine, an opine in hairy roots of tobacco induced by *Agrobacterium rhizogenes*. Phytochemistry 29:3135–3139

Kiyokawa S, Kikuchi Y, Kamada H, Harada H (1992) Detection of *rol* genes of Ri plasmid by PCR method and its application to confirmation of transformation. Plant Tissue Cult Lett 9:94–98

Murashige T, Skoog F (1962) A revised medium for rapid growth and bioassays with tobacco tissue cultures. Physiol Plant 15:473–497

Rugini E, Pellegrineschi A, Mencuccini M, Mariotti D (1991) Increase of rooting ability in woody species kiwi (*Actinidia deliciosa* A. Chev.) by transformation with *Agrobacterium rhizogenes rol* genes. Plant Cell Rep 10:291–295

Schmulling T, Fladung M, Grossman K, Schell J (1993) Hormonal content and sensitivity of transgenic tobacco and potato plants expressing single *rol* genes of *Agrobacterium rhizogenes* T-DNA. Plant J 3:371–382

Sheikholeslam SN, Weeks DP (1987) Acetosyringone promotes high efficiency transformation of *Arabidopsis thaliana* explants by *Agrobacterium tumefaciens*. Plant Mol Biol 8:291–298

Tanaka N (1990) Detection of opines by paper electrophoresis. Plant Tissue Cult Lett 7:25–28
Uematsu C, Murase M, Ichikawa H, Imamura J (1991) *Agrobacterium*-mediated transformation
    and regeneration of kiwi fruit Plant Cell Rep 10:286–290
Yazawa M, Suginuma C, Ichikawa K, Kamada H, Akihama T (1995a) Regeneration of kiwi fruit
    (*Actinidia deliciosa*) induced by *Agrobacterium rhizogenes*. Breed Sci 45:241–245
Yazawa M, Matsuyama T, Akihama T (1995b) Characteristics of kiwifruit transformants induced
    by *Agrobacterium rhizogenes*. Bull Fac Agric Meiji Univ 106:17–25 (in Japanese)

2 Transgenic Amaranth (*Amaranthus hypochondriacus*)

A.E. JOFRE-GARFIAS, J.L. CABRERA-PONCE, R.M. ADAME-ALVAREZ, L. HERRERA-ESTRELLA, and JUNE SIMPSON

1 Introduction

Amaranths are dicotyledonous plants that belong to the Amaranthaceae family which comprises more than 50 genera, distributed in tropical and sub-tropical zones of the world (Kigel and Rubin 1985). Most of these genera are native to America, and they grow in disturbed areas and produce large seedheads with small seeds. Some of them, like *Amaranthus spinosus* L. and *A. powellii* S. Watts. are considered to be weeds, while others like *A. hypochondriacus* L., *A. cruentus* L., *A. retroflexus* L. and *A. caudatus* L. are consumed as a grain and *A. dubis* Mart and *A. dubis* L. are used as fresh vegetables when young. Additionally, some other species are used as ornamentals because of the bright colors of their leaves and inflorescences (Kiegel and Rubin 1985).

Amaranthus species are important in several parts of the world, including México, Central and South America, India and Africa, where they are consumed as vegetable or grain crops. The agronomic potential of amaranth, especially in developing countries, is due to its high nutritional value, resistance to drought, heat and pests, and adaptability to environments inhospitable to conventional cereal crops (National Research Council 1984). *Amaranthus cruentus*, *A. caudatus* and *A. hypochondriacus* produce an abundance of seeds, with a higher lysine and methionine content, than cereals or legumes (National Research Council 1984). In addition, the Amaranthaceae are of interest since photosynthesis in these species may be carried out by the C_3 pathway, the C_4 pathway or an intermediate C_3–C_4 pathway.

Amaranth was an ancient crop under cultivation 5000 to 7000 years ago (Kulakow and Hauptli 1994). This plant was one of the basic foods of pre–Columbian cultures. Thousands of hectares of Aztec, Inca and other farmland were planted, and it was nearly as important as corn and beans for them (National Research Council 1984).

"The story of ritual use of *Huauhtli* (amaranth) shows more clearly than any other evidence how deeply these plants were embedded in the memories and lives of various Mexican tribes" (Sauer 1950). Apparently, its use in pagan

Departamento de Ingeniería Genética de Plantas, Unidad Irapuato, CINVESTAV-IPN. Apdo. Postal 629, 36500 Irapuato, Gto. Mexico

Biotechnology in Agriculture and Forestry, Vol. 47
Transgenic Crops II (ed. by Y.P.S. Bajaj)
© Springer-Verlag Berlin Heidelberg 2001

rituals and human sacrifices was the main reason the conquistadors eliminated it from native crop cultivation. It became confined to very small areas and was consumed locally.

Amaranthus hypochondriacus is the most robust, highest yielding of the grain types, and was probably domesticated in central Mexico, in the Teotihuacan valley where it was found by archeologists in caves from about 1500 years ago. In this species the seeds have a high protein content (about 16%) and the good amino acid balance is closer than cereals and legumes to the optimum recommended for human consumption, especially in the case of lysine (Saunders and Becker 1983).

There are a few studies of in vitro plant regeneration of some *Amaranthus* species. Shoot formation has been reported from calli derived from hypocotyl segments of *A. hypochondriacus, A. cruentus, A. tricolor* and *A. paniculatus* has been reported (Flores et al. 1982; Flores and Teutonico 1982; Bagga et al. 1987; Bennici et al. 1992), from petiole and leaf segments of *A. caudatus, A. cruentus* and *A. hybridus* (Bennici et al. 1997), from hypocotyl and inflorescences of *A. paniculatus*, (Bagga et al. 1987; Arya et al. 1993), from shoot tips of *A. gangeticus, A. hypochondriacus, A. caudatus, A. viridis and A. retroflexus* (Tisserat and Galletta 1993) and from thin cell layers (van Le et al. 1998). *Agrobacterium*-mediated transformation is the most widespread and efficient method for DNA transfer to dicotyledonous species. In the case of *Amaranthus* species only a few data are available in the literature (Table 1). The development of an amaranth transformation system would widen the possibilities for the exploitation of this plant as an alternative crop. Another advantage would be the possibility of developing amaranth as a model system for studying photosynthetic pathways and the interactions between them. Reported here is a method for the regeneration of *A. hypochondriacus* and an *Agrobacterium*-mediated system for amaranth transformation. This chapter is based on our published work (Villegas-Sepúlveda 1995; Jofre-Garfias et al. 1997; Jofre-Garfias 1997).

2 Methodology

2.1 Tissue Culture

Two approaches were used: (1) shoot organogenesis and (2) somatic embryogenesis. For both, amaranth (*Amaranthus hypochondriacus*, cv. 'Azteca') seeds, obtained from a commercial source, were surface sterilized by dipping them in 70% ethanol for 1 min, and in 0.01% $HgCl_2$ for 15 min, (applying vacuum for the first 5 min), followed by extensive washing with sterile distilled water. The seeds were then germinated on half strength Murashige and Skoog (1962) (MS) medium supplemented with 1% sucrose and incubated at 25 ± 2 °C under a regime of 16 h light at an irradiance of $50 \mu mol.m^{-2}.s^{-1}$ from cool white fluorescent lamps, for isolation of cotyledon and hypocotyl segments for

Table 1. Summary of various studies conducted on transformation/molecular biology of *Amaranthus*

	Reference	Species (explant/ culture used)	Vector/method used	Observations/ remarks
1	Lopatin (1936); Höhn and Helfrich (1963)	*A. caudatus* (Stems of green house growing plants)	*Agrobacterium tumefaciens* strain B6/Direct inoculation	Tumor induction
2	De Cleene and Otten (1973)	*A. retroflexus* (greenhouse growing plants)	*Agrobacterium tumefaciens* strain B6/Direct inoculation	No tumors were found 106 days after inoculation
3	Gadgil and Roy (1961)	A. viridis (greenhouse growing plants)	*Agrobacterium tumefaciens* strain B23/Direct inoculation	No tumors were found
4	Villegas-Sepúlveda (1995)	*A. hypochondriacus*, (leaves from growth chamber-grown 3-week-old plants)	Biolistic/DNA carrying *rbc*S Ah1-GUS transciptional fusions under 35S promoter	Transient expression of constructs carrying different portions of the *rbc*S Ah1 gene
5	Jofre-Garfias et al. (1997)	*A. hypochondriacus* (mature embryos dissected from surface-sterilized seeds)	C58C1Rif (pGV2260) (pEsc4)/ Co-culture with *A. tumefaciens*	Regenerated stable transformants/ Mendelian inheritance of the transgenes
6	Jofre-Garfias (1997)	*A. hypochondriacus* (hypocotyl, cotyledon and mature embryo; in vitro germinated plantlets on MS basal medium)	pTiC58 and A281pGA471/ Co-culture with *A. tumefaciens*	Tumor induction with both strains/ Opines (nopaline and octopine) and NPT-II detection

the organogenic pathway, or soaked in water for 2 to 5 h before dissecting the embryos, under a stereomicroscope, for somatic embryogenesis.

For shoot regeneration, hypocotyl and cotyledon sections were cultured on MS medium supplemented with combinations of 6-bencil-adenine (BA) and 3-indole-acetic acid at concentrations ranging between 0–10 μM for IAA and 0–40 μM for BA.

For somatic embryogenesis, mature embryo tissue was used which was easily obtained by cutting the seed longitudinally through the embryo into two halves with a sharp blade and then removing embryo pieces. Embryo explants were cultured on MS basal medium supplemented with 10 μM 2,4-D (medium M008) or Dicamba (medium M009), 10% coconut liquid endosperm (CLE) and 3% sucrose. All explants were incubated under the same conditions of light and temperature described above. Embryogenic calli were transferred to

MS basal medium containing 2% sucrose but without growth regulators and incubated under the same conditions.

2.2 Genetic Transformation

2.2.1 Bacterial Strains

Two oncogenic *Agrobacterium tumefaciens* strains were used to test the susceptibility of Amaranth to *Agrobacterium* infection: C58pTiC58 (Van Larebeke et al. 1974) and A281(pGA471) (An et al. 1985). Tumors induced by oncogenic strains were grown on MS basal medium with 500 mg/l Claforan, or Claforan plus kanamycin in the case of tumors induced with A281 (pGA471). In order to produce transgenic amaranth plants the *Agrobacterium* strain C58C1Rif(pGV2260) (Deblaere et al. 1985) containing plasmid pEsc4 was used. The components of pEsc4 can be seen in Fig. 1.

2.2.2 Co-culture Conditions

Explants were soaked for 5 to 10 min in a bacterial suspension containing 10^7 cells/ml, blotted onto filter paper discs and then co-cultured on plates of MS without growth regulators for oncogenic strains and on M008 or M009 for C58C1Rif(pGV2260)(pEsc4), under 24 h darkness at $25 \pm 2\,°C$. Following co-cultivation, explants were transferred to selective media: MS basal medium with or without growth regulators, 300 mg/l kanamycin and/or

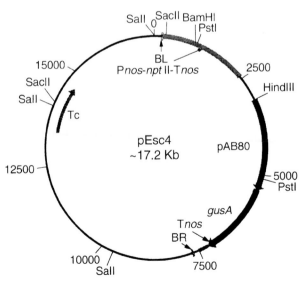

Fig. 1. pEsc4 plasmid map. pEsc4 was constructed by inserting a *Hind*III/*Nco*I fragment derived from the plasmid pPAR4 (Cashmore 1984), containing the AB80 cab promoter before the *gus* A coding region from pRAJ275 (Jefferson 1987). The new plasmid containing both fragments was digested with *Eco*RI and *Hind*III and subcloned to the *Hind*III/*Eco*RI fragment of pGA492 (An et al. 1985). The resulting plasmid was digested with *Eco*RI/*Sca* I, blunt ended, and ligated to remove unwanted sequences. (After Jofre-Garfias et al. 1997)

"500 mg/l Claforan, depending upon the strain used. Explants were incubated under light conditions as described above, and subcultured to adequate fresh selective medium at 1-month intervals. Calli obtained using C58C1Rif(pGV2260)(pEsc4), were transferred to MS medium supplemented with 2% sucrose for plant regeneration and incubated as described. Plantlets were obtained after 1.5 months. Regenerated plants were transferred to soil and grown under a 12-h light photoperiod at $25 \pm 2\,°C$ in a Conviron Model E7/2 growth chamber in order to stimulate the development of inflorescences and seed production in a short time.

2.2.3 Enzymatic Assays

Neomycin phosphotransferase type II (NPT-II) activity was detected according to Van den Broek et al. (1985). β-Glucuronidase activity from calli and regenerated shoots, grown in selective media, was determined by histochemical and fluorometric methods as described by Jefferson (1987) and McCabe et al. (1988). Following histochemical staining, 20 μm sections were prepared for microscopy as described by Fowke and Rennie (1995).

2.2.4 GUS Activity in Progeny from Transgenic Plants

A small section of a cotyledon was removed to determine which were transgenic by histological assay, then germinated plantlets were dissected into roots and cotyledons plus hypocotyls and frozen in liquid nitrogen. Roots or cotyledons plus hypocotyls were grouped, homogenized and analyzed by the β-glucuronidase fluorometric assay at $37\,°C$, according to Jefferson (1987). Assays were made on triplicate samples. Specific activities were recorded as the amount of 4-methyl-umbelliferone (pmol) produced by 1 mg of protein in 1 min. GUS reactions were stopped after 30 min of incubation at $37\,°C$. Protein was measured by the Bradford (1976) method.

2.2.5 DNA

Plant DNA was isolated using the method of Rogers and Bendich (1994). DNA samples were digested with either *Pst*I or *Hind*III restriction enzymes, and separated in 0.8% agarose gels, transferred to nylon membranes (Hybond N, Amersham), and hybridized with a [^{32}P]-labeled fragment of the AB80 promoter present in pEsc4 (see Fig. 1). Standard procedures were used throughout (Sambrook et al. 1989). Labeling was performed following the Life Technologies Random Primer DNA labeling system instructions. Membranes were exposed to Kodak X-ray film for 24 h.

3 Results and Discussion

3.1 Tissue Culture Experiments

Initial experiments to determine optimum culture conditions for the induction of shoots on cotyledon and hypocotyl segments were unsuccessful. Regeneration was achieved only when the apical meristem was present in the original explant (Fig. 2A,B). However, when used in genetic transformation experiments, this was the only explant which was not transformed using *Agrobacterium tumefaciens*. To resolve this problem, other explants, culture media and incubation conditions were investigated, including different growth regulators, basal media formulations, and environmental conditions to induce somatic embryogenesis. The best results were obtained using embryos from mature seeds cultured on MS basal medium supplemented with 10% (v/v) CLE, 3% (w/v) sucrose, 10 µM 2,4-D (M008) or Dicamba (M009), and incubated as described in materials and methods. After 4 to 5 weeks nearly 100% of the cultured embryos produced calli (Fig. 2C,D). The calli were fully organized and somatic embryos were clearly distinguishable.

After three to four propagation steps, calli containing embryos were transferred to hormone-free MS basal medium with 2% sucrose. After 3–4 weeks of growth on this medium, shoots started to develop from individual somatic embryos and from non-embryogenic callus tissue (Fig. 2E,F). Plantlets derived from callus tissue were transferred to soil in a growth chamber after 4 weeks, when they were 1.5 to 3 cm high, to produce mature plants (see Fig. 2G,H).

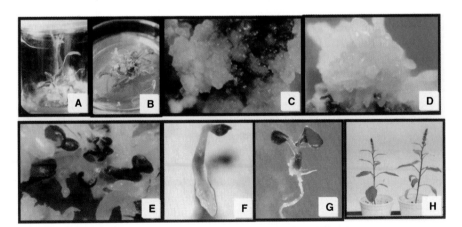

Fig. 2.A–F Development of an amaranth plant regeneration system from shoot apical meristem. **A** and **B** Shoot formation from explants containing the apical meristem. **C** Callus with globular structures that develop into somatic embryos or shoots. **D–E** Calli with globular and embryo-like structures and **F** A germinating somatic embryo. **H** Mature regenerated plants in soil. **G** Seedling with developed root system. (Jofre-Garfias 1997; Jofre-Garfias et al. 1997)

Plant regeneration was apparently achieved by both shoot-root organogenesis and somatic embryogenesis. The presence of CLE seems to be an essential component for induction of plant regeneration in both cases since no regeneration was obtained when CLE was omitted from the media (data not shown). This confirms earlier observations of Flores et al. (1982), who suggested that CLE was important in the production of embryogenic calli, although they did not report plant regeneration.

3.2 Transformation Experiments

3.2.1 Amaranth Is Susceptible to A. tumefaciens Infection

Previous reports suggested that amaranth species were not susceptible to infection by *Agrobacterium* (De Cleene and De Ley 1976), therefore in order to determine the susceptibility of *A. hypochondriacus* to infection by *Agrobacterium tumefaciens*, we used two oncogenic strains, C58C1pTiC58 and A281(pGA471), to infect hypocotyl segments and cotyledons from 5-day-old seedlings germinated on MS medium. Both bacterial strains produced tumors on explants cultured in media without growth regulators. It was also observed that tumors induced by the strain A281(pGA471) were able to grow on kanamycin-containing media, as expected for a strain harboring the *nos-npt*(II)-*nos* kanamycin resistance gene.

All C58C1pTiC58-induced tumors produced nopaline (Fig. 3A,B), whereas all A281(pGA471)-induced tumors (Fig. 3C) showed NPT-II activity (Fig. 3D). These results demonstrate that the amaranth genotype (*A. hypochondriacus* cv. 'Azteca') used in this study is susceptible to genetic transformation by *Agrobacterium tumefaciens*.

3.2.2 Regeneration of Transgenic Plants

In order to regenerate transgenic amaranth plants, we co-cultivated mature embryos with the non-oncogenic *A. tumefaciens* strain pTiB6S3(pGV2260) (pEsc4), and then transferred them to selective media as described in Section 2. After 2–3 weeks of culture in selective media, kanamycin-resistant calli developed from 10–15% of the co-cultivated embryos. Initial development on kanamycin-containing medium was slow, but later resistant calli showed rapid proliferation.

To test whether the kanamycin-resistant calli were indeed transgenic, they were assayed for NPT-II and GUS activities. All kanamycin-resistant calli tested showed both NPT-II and β-glucuronidase activity (Fig. 3E,F). Kanamycin-resistant embryogenic calli were transferred to selective regeneration media and shoots and small plants were obtained after 2–3 weeks of growth. Plants were transferred to soil and further grown in a growth chamber. All plants were fertile and produced seeds.

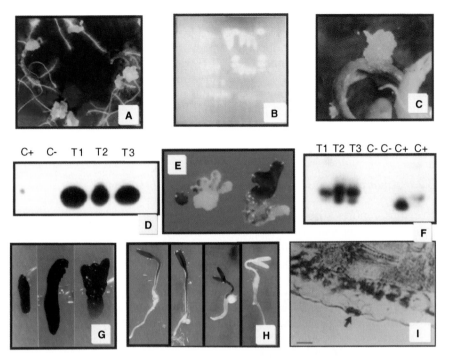

Fig. 3. Amaranth transformation by *Agrobacterium tumefaciens* strains. **A** Tumors induced by the strain C58pTiC58 growing on MS medium supplemented with sucrose 30 g/l and Claforan 500 mg/l. **B** Electrophoretogram of nopaline from tumors induced in **A**. **C** Tumor induced from a cotyledon by the strain A281pGA471 growing on MS medium supplemented with sucrose 30 g/l, kanamycin 300 mg/l and Claforan 500 mg/l. **D** NPT-II assay of three different tumors, compared with a callus grown on MS medium supplemented with 5 µM IAA and 5 µM BA and antibiotics as in **C**. **F** NPT-II assay of three different transformation events with strain C58C1Rif(pGV2260)(pEsc4) (T1, T2 and T3), compared with Tobacco transformed plants (C⁺) and two non-transgenic regenerated plants (C⁻). **E**, **G** and **H** β-Glucuronidase assay in transgenic calli, developing somatic embryos at different stages and segregating T1 transgenic seedlings respectively. **I** Transversal sections of 20 µm from transgenic leaves stained for β-glucuronidase. *Bar* represents 0.2 mm and *arrow* points to positive signals in stomatal cells. (Jofre-Garfias 1997; Jofre-Garfias et al. 1997)

To demonstrate that the plants regenerated from kanamycin and GUS-positive calli were indeed transgenic, genomic DNAs were isolated from four plants derived from independent transformation events and Southern blot hybridizations were performed. In Fig. 4A, DNA digested with *Hind*III from callus (lanes 3, 5, 9 and 11) and regenerated plants (4, 6, 10 and 11) from transformed (lanes 3, 4, 9 and 10) and non transformed (lanes 5, 6, 11 and 12) tissues were run and hybridized against the 2.8 kb *Bam*HI-*Hind*III fragment of the AB80 promoter and the actin 2 gene from *Arabidopsis thaliana*, to demonstrate the presence of foreign genes incorporated into the amaranth genome. Similar experiments were done using DNA from four plants, digested separately with *Pst*I and *Hind*III, and run in parallel with an undigested sample,

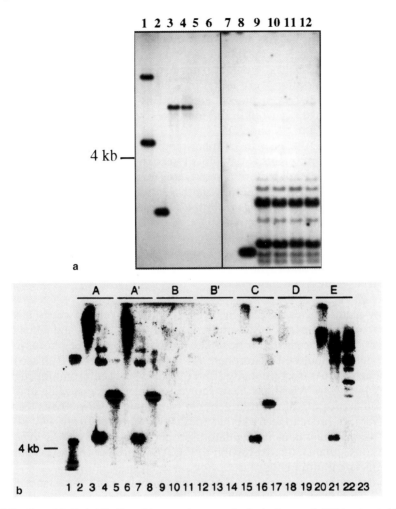

Fig. 4 Southern blot hybridization of transgenic amaranth plants. Amaranth DNA extracted from transgenic and non-transformed plants were digested with *Pst*I or *Hind*III and compared with an undigested sample of the same DNA. In **A**, *lanes 1* and *7* pEsc4 plasmid digested with *Pst*I; *lanes 2* and *8*, 1 kb ladder; *lanes 3,* and *9* transgenic callus; *4* and *10*, regenerated plant; *5* and *11* callus *6* and *12* regenerated non-transformed control. Hybridizations were carried out against the *Bam*HI-*Hind*III fragment (2.8 kb) of the AB80 promoter from pPAR4 which was labeled by random priming using [32]P (*lanes 1–6*) or the *Arabidopsis thaliana* Actin2 gene labeled in the same manner (lanes 7–12). **B** *Lane 1 Pst*I-digested pEsc4 plasmid. Samples from four independent transformation events *A and A'* parental and T1 plants from transformation event 1, *B and B'*, parental and R1 non-trangenic plants, *C, D, E* plants from transformation events 2, 3 and 4 were run grouped in contiguous wells, labeled as undigested, and *Pst*I or *Hind*III digested. *B* is a non-transformed negative control. *A' and B'* are plants derived from seeds from *A and B*. The probe used was the same as in **A** (Bam HI-Hind III fragment of the AB80 promoter). (Jofre-Garfias 1997; Jofre-Garfias et al. 1997)

using the 2.8 kb *Bam*HI-*Hin*dIII fragment of the AB80 promoter as a probe. The expected 4 kb *Pst*-I fragment was present in the genome of all analyzed plants (Fig. 4B, lanes 3, 6, 15, 18 and 21). When DNA samples were digested with *Hin*d-III, which has a unique site in the T-DNA of pEsc4, different patterns of hybridization were found in each assayed transformation event, confirming that they were indeed independent transformation events and that integration had occurred in different sites within the amaranth genome (Fig. 4C, lanes 4, 7, 16, 19 and 22).

To determine the pattern of inheritance of the integrated T-DNAs in transgenic amaranth plants, 7-day-old T1 seedlings were examined for GUS activity. The segregation pattern in most cases was 3:1 as expected for a single functional insertion within the amaranth genome. This is illustrated in Fig. 2D, where 9 out of 12 seedlings from the T1 progeny of one of the amaranth transgenic lines show GUS activity.

3.3 Tissue Specific and Light Regulation of the AB80 Promoter in Transgenic Amaranth Plants

In many plant species, regulation of photosynthesis related genes has been shown to occur primarily at the transcriptional level (Moses and Chua 1988). However, in amaranth it has been shown that at least one photosynthesis related gene, a ribulose bisphosphate carboxylase oxygenase (Rubisco) gene, is regulated post-transcriptionally (Berry et al. 1985). The promoter region of the pea lhcp gene AB80 has been shown to direct tissue-specific and light-regulated expression of reporter genes in transgenic tobacco plants (Simpson et al. 1985). In order to test whether the AB80 promoter (from a C_3 plant) is functional and maintains its regulatory properties in *Amaranthus hypochondriacus*, (a C_4 plant), the expression of the AB80 promoter was examined in transgenic amaranth plants.

To determine whether the AB80 promoter maintained tissue-specific regulation in amaranth, β-glucuronidase activity in seedlings and fully developed leaf tissue of transgenic plants was examined by histochemical GUS assays. In seedlings, GUS activity was found to be present only in the cotyledons and the green part of the hypocotyl but not in the lower hypocotyl or roots (Fig. 2E). Examination of 20-μm sections of mature leaves showed that the AB80 promoter directs GUS expression exclusively in chloroplast-containing cells, including the bundle sheath, companion cells and stomata, but not in cells lacking active chloroplasts such as the epidermal layer (Fig. 2E,F). These results demonstrate that normal tissue-specific regulation of the AB80 promoter is maintained in amaranth. To determine whether light regulation by the AB80 promoter is also maintained in amaranth, T1 seeds were germinated under light or darkness, and levels of GUS activity in each seedling type was determined by fluorometric assay. As can be seen in Table 2, a 15- to 20-fold higher GUS activity was detected in light-grown seedlings compared with those germinated and maintained in darkness, demonstrating that the AB80 promoter also retains the capacity for light regulation in amaranth.

Table 2. β-Glucuronidase activity of amaranth progeny. (After Jofre-Garfias et al. 1997)

T1 plants	Cotyledons (light)	Cotyledons (darkness)	Roots (light)	Roots (darkness)
Glucuronidase (+)	135.69 ± 1.81	8.44 ± 0.47	1.64 ± 1.36	1.25 ± 0.11
Glucuronidase (+)	91.51 ± 3.32	5.96 ± 2.33	1.22 ± 0.47	0.64 ± 0.11
Glucuronidase (−)	0.14 ± 0.12	0.08 ± 0.074	0.20 ± 0.197	0.07 ± 0.086

These results demonstrate that the AB80 promoter maintains its functional properties in amaranth, and implies that the signal transduction and transcription factors that regulate the transcriptional activity of photosynthesis-related genes are evolutionarily conserved in amaranth and pea although each species utilizes different photosynthetic pathways (C_3 or C_4).

3.4 Conclusions and Perspectives

Amaranth (*Amaranthus hypochondricus* L.) is susceptible to transformation by *Agrobacterium tumefaciens* strains. This was demonstrated by tumor production in hypocotyl and cotyledon segments co-cultured with the oncogenic strains C58C1(pTiC58) and A281(pGA471). Those tumors produced nopaline and NPT-II respectivelly.

Regenerated transgenic plants were obtained from calli induced from mature embryos transformed with the strain pTiB6S3(pGV2260)(pEsc4). Those plants were fertile and produced viable seeds from which plants were germinated. The activity of the reporter gene, β-glucuronidase, was measured in those seedlings by histological and fluorometric methods, and they showed a segregation ratio of 3:1.

Since Amaranth is a plant that photosynthesizes by the C_4 pathway, by using the procedure of transformation reported here, it can be used as a model system for further studies on photosynthesis.

Acknowledgments. We thank Martha Martínez, Mireya Sánchez and Azucena Mendoza for their invaluable advice for the DNA work and to Antonio Cisneros for the photographic work. AJ is indebted to CONACYT for a doctoral fellowship. We would also like to thank Grupo Hoechst Marion Roussel S.A. de C.V. for its kind donation of Claforan.

References

An G, Watson DB, Stachel S, Gordon MP, Nester EW (1985) New cloning vehicles for transformation of higher plants. EMBO J 4:277–288

Arya ID, Chakravarty TN, Sopory SK (1993) Development of secondary inflorescences and in vitro plantlets from inflorescence cultures of *Amaranthus paniculatus*. Plant Cell Rep 12:286–288

Bagga S, Venkateswari K, Sopory SK (1987) In vitro regeneration of plants from hypocotyl segments of *Amaranthus paniculatus*. Plant Cell Rep 6:183–184

Bennici A, Schiff S, Bovelli R (1992) In vitro culture of species and varieties of four Amaranths species. Euphytica 42:181–184

Bennici A, Grifoni T, Schiff S, Bovelli R (1997) Studies on growth and morphogenesis in several species and lines of *Amaranth*. Plant Cell Tissue Organ Cult 49:29–33

Berry JO, Nikolau BJ, Carr JP, Klessig DF (1985) Transcriptional and post-transcriptional regulation of ribulose 1,5-bisphosphate carboxylase gene expression in light- and dark-grown amaranth cotyledons. Mol Cell Biol 5:2238–2246

Bradford M (1976) A rapid and sensitive method for the quantitation of microgram quantities of protein utilizing the principle of protein-dye binding. Anal Biochem 72:248–254

Cashmore AR (1984) Structure and expression of a pea nuclear gene encoding a chlorphyll a/b binding polypeptide. Proc Natl Acad Sci USA 81:2960–2964

Deblaere R, Bytebier B, De Greve H, Deboeck F, Schell J, Van Montagu M, Leemans J (1985) Efficient octopine Ti plasmid-derived vectors for *Agrobacterium*-mediated gene transfer to plants. Nucleic Acids Res 13:4777–4788

De Cleene M, De Ley J (1976) The host range of crown gall. Bot Rev 42:389–466

De Cleene M, Otten K (1973) Het voorkomen van *Agrobaterium-kanker* (crown gall) bij economisch belangrijke plantengeslachten. De Belgische Tuinbouw 54:196–197

Flores HE, Teutonico RA (1982) Amaranth (*Amaranthus* spp.): potential grain and vegetable crops. In: Bajaj YPS (ed) Biotechnology in agriculture and forestry, vol 2. Crops. Springer, Berlin Heidelberg New York, pp 568–578

Flores HE, Thier A, Galston AW (1982) In vitro culture of grain and vegetable amaranths (*Amaranthus* spp.). Am J Bot 69:1049–1054

Fowke LC, Rennie PJ (1995) Botanical microtechniques for plant cultures. In: Gamborg OL, Phillips GC (eds) Plant cell, tissue and organ culture. Fundamental methods. Springer, Berlin Heidelberg New York, pp 271–228

Gadgil VN, Roy SK (1961) Studies on crown gall tumour. I. Host susceptibility of the causal organism, *Agrobacterium tumefaciens*, strain B-23. Trans Bose Res Inst 24:141–146

Höhn K, Helfrich O (1963) Beiträge zum Krebsproblem der höheren Pflanzen. I. Untersuchungen über die Korrelation zwischen Krebsdisposition und Blühinduktion. Beitr Biol Pflanz 38:83–98 (cited in De Cleene M, De Ley J (1976) The host range of Crown Gall. Bot Rev 42:389–466)

Jefferson RA (1987) Assaying chimeric genes in plants: The gus gene fusion system. Plant Mol Biol Rep 5:387–405

Jofre-Garfias A (1997) Transformación genética de *Amaranthus hypochondriacus* L. con cepas de *Agrobacterium tumefaciens*. PhD Thesis. Centro de Investigación y de Estudios Avanzados del IPN Unidad Irapuato, México 111 pp

Jofre-Garfias A, Villegas-Sepúlveda, N, Cabrera-Ponce JL, Adame-Alvarez RM, Herrera-Estrella L, Simpson J (1997) *Agrobacterium*-mediated transformation of *Amaranthus hyponcondriacus*: light and tissue specific expression of a pea chlorophyll a/b binding protein promoter. Plant Cell Rep 16:847–852

Kigel J, Rubin B (1985) *Amaranthus*. In: Abraham H, Halevy H (eds) Handbook of flowering plants, vol 1. CRC Press, Boca Raton pp 427–433

Kulakow PA, Hauptli H (1994) Genetic characterization of grain amaranth. In: Paredes-López O (ed) Amaranth biology, chemistry and technology. CRC Press, Boca Raton pp 9–19

Lopatin MI (1936) The susceptibility of plants to *Bact. Tumefaciens*, the causative agent of the root-cancer of plants. Mikrobiologia (Moskwa) 5:57–75

McCabe DE, Martinell BJ, Christou P (1988) Stable transformation of soybean (*Glycine max*) by particle acceleration. Bio/Technology 6:923–926

Moses PB, Chua NH (1988) Light switches for plant genes. Sci Am 258:64–69

Murashige T, Skoog F (1962) A revised medium for rapid growth and bioassays with tobacco tissue cultures. Physiol Plant 15:473–497

National Research Council (1984) Amaranth: modern prospects for an ancient crop. National Academy Press, Washington, DC, pp 3, 5, 14

Rogers SO, Bendich AJ (1994) Extraction of total cellular DNA from plants, algae and fungi. In: Gelvin SB, Schilperoort RA, Verma DPS (eds) Plant molecular biology manual. Kluwer, Dordrecht, pp D1:1–8

Sambrook J, Fritsch EF, Maniatis T (1989) Molecular cloning. A laboratory Manual. Cold Spring Harbor Laboratory Press, Plainview

Sauer JD (1950) The grain amaranths: a survey of their history and classification. Ann Mo Bot Gard 37:561–632

Saunders RM, Becker R (1983) *Amaranthus*: a potential food and feed resource. In: Pomeranz Y (ed) Advances in cereal science and technology, vol VI. American Association of Cereal Chemists, St Paul, pp 357–396

Simpson J, Timko MP, Cashmore AR, Schell J, Van Montagu M, Herrera-Estrella L (1985) Light-inducible and tissue-specific expression of a chimaeric gene under control of the 5′ flanking sequences of a pea chlorophyll a/b-binding protein gene. EMBO J 11:2723–2729

Tisserat B, Galletta DP (1993) In vitro flowering in *Amaranthus*. Hortic Sci 23:210–212

Van den Broek G, Timko MP, Kausch AP, Cashmore AR, van Montagu M, Herrera-Estrella L (1985) Targeting of a foreign protein to chloroplasts by fusion of the transit peptide from the small subunit of ribulose 1, 5-bisphosphate carboxylase. Nature 313:258–263

Van Larebeke NG, Engler M, Holsters S, Van den Elsacker I, Zaenen RA, Schilperoort RA, Schell J (1974) Large plasmid in *Agrobacterium tumefaciens* essential for crown gall-inducing ability. Nature (Lond) 252:169–170

van Le B, Domy NT, Jeanneau M, Sadik S, Tu S, Vidal J, Tran Thanh Van K (1998) Rapid plant regeneration of a C4 dicot species: *Amaranthus edulis*. Plant Sci 132:45–54

Villegas-Sepúlveda N (1995) Aislamiento y caracterización de clonas *cab* y de la subunidad pequeña de Rubisco de *Amaranthus hypochondriacus*. Ph D Thesis. Centro de Investigación y de Estudios Avanzados del IPN Unidad Irapuato, México 84 pp

3 Transgenic Horseradish (*Armoracia rusticana*)

Y. Mano

1 Introduction

Horseradish (*Armoracia rusticana*) plants produce large amounts of peroxidase (EC 1.11.1.7, donor: hydrogen peroxide oxidoreductase), which is a glycoprotein containing protoheme IX. Horseradish roots are an important source of peroxidase, which is widely used as a reagent for clinical diagnosis and microanalytical immunoassays. Horseradish roots are also used as a relish, traditionally accompanying roast beef and other meats. Since horseradish plants scarcely bear seeds, they are vegetatively propagated by various means including tissue culture techniques and planting fresh root segments in soil. Therefore, it would be desirable to produce transgenic plants expressing useful properties such as rapid root growth and high peroxidase content. The development of effective propagation systems for horseradish plants would also be desirable. Such studies are reviewed here and summarized in Table 1.

2 Genetic Transformation of Horseradish

2.1 *Agrobacterium*-Mediated Transformation

The genetic modification using the Ri plasmid constitutes a feasible approach for the improvement of plant properties because transformed cellular clones with new traits have been produced (Fig. 1).

Horseradish hairy roots were induced by inoculation with *Agrobacterium rhizogenes* strain 15834 and established as described previously (Noda et al. 1987; Mano and Matsuhashi 1995). After culturing hairy roots for seven passages in phytohormone-free medium, hairy root clones were established and subcultured on fresh medium every 3 weeks. These processes are important in establishing stable transformants (Mano et al. 1986, 1989; Mano 1993). The horseradish hairy root clones grew rapidly with extensive lateral branching, and their properties were stably maintained for at least 4 years.

Department of Biological Science and Technology, Tokai University, 317 Nishino, Numazu, Shizuoka 410-0321, Japan

Biotechnology in Agriculture and Forestry, Vol. 47
Transgenic Crops II (ed. by Y.P.S. Bajaj)
© Springer-Verlag Berlin Heidelberg 2001

Table 1. Summary of *Agrobacterium*-mediated transformation studies on horseradish

Subject	Transformants/bacteria used[a]	Observations and remarks	Reference
Shoot development and plant regeneration	HR/Ar A4	Light requirement for shoot formation from HR. Variation in morphology of regenerants.	Noda et al. (1987)
	HR/Ar 15834	Variation in morphology of regenerants. Peroxidase isozyme patterns in regenerants.	Saitou et al. (1991)
	HR/Ar 15834	Light requirement (wavelength, duration, intensity) for shoot formation from HR. Inhibition of shoot formation by far-red light.	Saitou et al. (1992, 1993)
	HR/Ar 15834	Novel life cycle arising from leaf segments in regenerants. Variations in morphology and peroxidase activity of regenerants.	Mano and Matsuhashi (1995)
	CG/At B6S3	Shoot morphogenesis from teratoma tissues. Peroxidase isozyme patterns in CG tumours.	Peskan et al. (1996)
Peroxidase production	HR/Ar	Two step culture system of HR and HR cells.	Taya et al. (1989)
	HR/Ar 15834	Peroxidase activity in HR and culture medium.	Saitou et al. (1991)
	HR/Ar LBA9402	Effects of abiotic and biotic elicitors on peroxidase production by HR culture. Fungal extracts of *Verticillium* sp. Peroxidase isozyme patterns in elicited samples.	Flocco et al. (1998)
Artificial seeds	HR/Ar	Encapsulated HR in alginate beads. Plantlet development from HR with an apical meristem in the beads.	Uozumi et al. (1992, 1994)
	HR/Ar A4	Encapsulation of plantlet derived from HR fragment. Plantlet development.	Nakashimada et al. (1995, 1996)
Cryopreservation	HR/Ar A4	Cryopreservation of encapsulated shoot primordia derived from HR. Plant regeneration from encapsulated shoot primordia.	Phunchindawan et al. (1997)

[a] Ar, *Agrobacterium rhizogenes*; At, *Agrobacterium tumefaciens*; HR, hairy root; CG, crown gall.

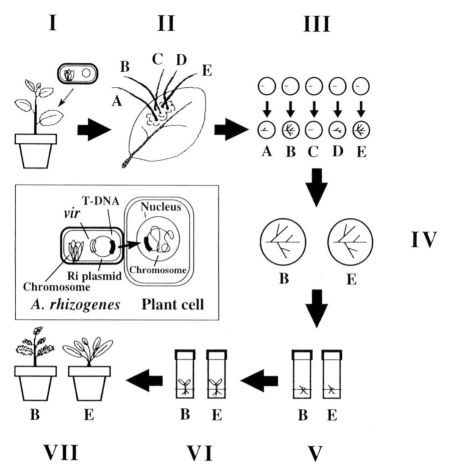

Fig. 1. Genetic modification of plants with the Ri plasmid in *Agrobacterium rhizogenes*. **I** Infection. **II** Hairy root induction. **III** Isolation, selection and cloning of rapidly growing roots in phytohormone-free medium. **IV** Establishment of hairy root clones. **V** Plant regeneration. **VI** Examination of useful properties. **VII** Establishment and evaluation of transgenic plants

Characterization of the horseradish hairy root clones showed that there was considerable variation in growth rate from clone to clone in a similar manner to that described in *Scopolia japonica* (Mano et al. 1986) and *Duboisia leichhardtii* (Mano et al. 1989) hairy root clones. The hairy root clones were cultured in a modified Gamborg's (G) medium (Mano and Matsuhashi 1995) and a modified Heller's (HF) medium (Mano et al. 1989) at 25 °C in the dark. They were classified into three categories according to growth rate (Table 2). Hairy root clones AR182, AR450 and AR455, which were the representatives in each group, produced both agropine and mannopine (Fig. 2), indicating the integration of TR-DNA of pRi15834 (De Paolis et al. 1985).

Table 2. Classification of horseradish hairy root clones by growth rates. (Mano and Matsuhashi 1995)

Group[a]	Hairy root clone
A	AR450, AR449, AR78
B	AR455, AR454, AR445, AR444, AR441, AR419, AR307, AR284, AR232, AR66, AR34
C	AR436, AR182, AR47

[a] A, Very rapid growth in both G and HF media; B, rapid growth in G medium but slow growth in HF medium; C, rapid growth in G medium but very slow growth in HF medium.

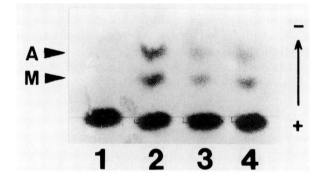

Fig. 2. Opine assay. Paper electrophoretic analysis of extracts from transgenic plants AR182P, AR450P and AR455P derived from horseradish hairy root clones. *1* Non-transformed plants, *2* AR182P, *3* AR450P, *4* AR455P, *A* agropine, *M* mannopine. (Mano and Matsuhashi 1995)

2.2 Characteristics of Horseradish Hairy Roots

Since plant hairy roots have many advantageous properties such as high growth rate, clonal origin and genetic stability, and the complete differentiation of root tissue, they have the potential for in vitro production of root-specific biochemicals like primary and secondary metabolites (Bajaj 1993; Charlwood and Rhodes 1996). Horseradish hairy roots have been reported to have peroxidase activity comparable to that of the original plant root tubers. By a two-step culture of the hairy roots, peroxidase productivity has been enhanced several times over the conventional culture (Taya et al. 1989). Peroxidase activity was also detected in the culture medium and the activity was 1/4 of that in the hairy roots (Saitou et al. 1991).

The combined addition of auxin and cytokinin have caused a significant change in morphology from hairy roots to cell aggregates, and effective regeneration of hairy roots from the cell aggregates has been obtained in a culture with the addition of auxin only. The properties of growth rate and peroxidase productivity of hairy roots have been stably maintained, even after a series of the cell aggregate formation and hairy root formation processes have been conducted (Repunte et al. 1993). A reversible process of morphology change between hairy roots and cell aggregates would be a means of extending the application of hairy root cultures to the production of useful compounds on an industrial scale.

3 High Potency of Regeneration in Transgenic Horseradish Plants

3.1 Plant Regeneration from Hairy Roots

Plant regeneration from hairy roots was first shown in tobacco, carrot and morning glory. Adventitious shoot formation could be induced by light in hairy roots of these plants (Tepfer 1984). Plant regeneration from root tissues has also been observed in horseradish hairy roots (Noda et al. 1987; Saitou et al. 1991, 1992, 1993; Mano and Matsuhashi 1995). In horseradish hairy roots, adventitious shoot formation from root tissues is regulated by both the intensity and the duration of light, with the duration being more important (Saitou et al. 1992). The promotive effect of red light and the inhibitory effect of irradiation with far-red light in the induction of adventitious shoots indicate that phytochrome is probably involved in light-induced formation of adventitious shoots in horseradish hairy roots (Saitou et al. 1992, 1993).

3.2 Hairy-Root Based Artificial Seeds

The advantageous properties described above have stimulated interest in developing regeneration and delivery systems for horseradish hairy roots. Fragmented hairy root segments (Uozumi et al. 1992) and adventitious shoot primordia of hairy roots (Uozumi et al. 1994) of horseradish have been studied for their application in the production of artificial seeds encapsulated in alginate beads. In order to develop a hairy-root-based artificial seed system for commercial applications, hairy roots fragmented mechanically by blender were treated with growth regulators and dehydrated before the encapsulation, resulting in the improvement of plantlet productivity (Nakashimada et al. 1995, 1996). Encapsulated shoot primordia induced in hairy roots can be successfully cryopreserved (Phunchindawan et al. 1997).

3.3 Characteristics of Transgenic Plants Regenerated from Horseradish Hairy Roots

The transgenic plants AR182P, AR450P and AR455P regenerated from hairy root clones displayed normal leaf morphologies and normal stem length (Fig. 3). On the other hand, the growth of roots and stems in transgenic plants was more active than in non-transformed plants. The transgenic plants had thicker roots and stems with extensive lateral branches compared with non-transformed plants (Fig. 3B, C). The transgenic plant AR450P grew most rapidly, followed in order by AR455P, AR182P, and non-transformed plants (Table 3). The growth index of AR450P was 2.6 times that of non-transformed plants.

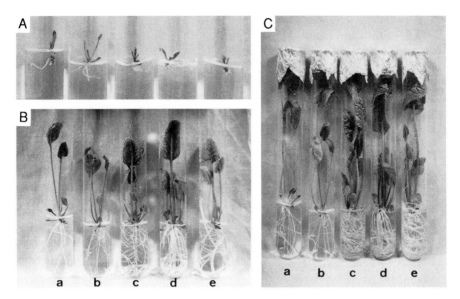

Fig. 3. Growth and typical morphologies of transgenic horseradish plants. Plantlets of AR182P, AR450P, AR455P, and non-transformed horseradish were cultured in G medium containing 0.15% (w/v) Gelrite at 25 °C in light. Culture at initial time (**A**), and after 18 days (**B**) and 32 days (**C**) of cultivation. *a* and *b* Non-transformed plant, *c* AR182P, *d* AR450P, *e* AR455P. (Mano and Matsuhashi 1995)

Table 3. Growth index and peroxidase activity of transgenic horseradish plants (Mano and Matsuhashi 1995). Plantlets of transgenic plants AR182P, AR450P, AR455P, and non-transformed plants (WT) were cultured in G liquid medium at 25 °C in light at 50 rpm for 35 days. Growth index (harvest fresh weight at 35 days per unit inoculum fresh weight), peroxidase (POD) activity and protein contents were determined as described previously (Mano and Matsuhashi 1995). Experiments were repeated twice with three replications

Plant	Growth index	Protein contents (μg/mg FW[a] tissue)		POD contents (10^{-3} Unit/mg FW[a] tissue)	POD activity (unit/mg protein)
WT	201	Root	0.883	3.12	3.53
		Leaf	14.6	3.12	0.214
AR182P	222	Root	1.82	10.8	5.93
		Leaf	12.3	3.60	0.293
AR450P	530	Root	1.30	3.52	2.71
		Leaf	10.6	6.00	0.568
AR455P	350	Root	1.20	5.40	4.50
		Leaf	11.7	3.12	0.266

[a] Fresh weight.

Peroxidase activities in the transgenic plants were determined after 35 days of culture in G liquid medium (Table 3). Total peroxidase contents of transgenic plants were 1.4 to 2.3 times those of non-transformed plants. Peroxidase activities in roots were 4.8 to 20 times those in shoots. Thus, it is possible to produce transgenic plants expressing a high growth rate and with a high content of peroxidase (Mano and Matsuhashi 1995).

The transgenic plants regenerated from horseradish hairy roots can be classified into three categories, i.e., normal, wrinkle and rooty according to their respective morphologies (Saitou et al. 1991). In all three types, peroxidase activities in roots were two to ten times those in shoots. Peroxidase activities in roots and shoots of normal and wrinkle types were comparable to those of non-transformed plants. But peroxidase activities of the rooty type were several times those of non-transformed plants.

3.4 Plant Regeneration from Leaves of Transgenic Plants

Capabilities for plant regeneration from leaves were examined in transgenic plant AR450P, the most rapidly growing clone, when leaves were cut with a scalpel into small squares approximately 0.15×0.3 cm in size. Adventitious roots were generated at the leaf sections of AR450P within 14 days after culturing on G agar medium at $25\,°C$ in light (Fig. 4A). The roots grew rapidly with extensive lateral branching, and adventitious shoots were generated mainly at the cut ends of the leaf sections after 25 days of culture. Adventitious shoots grew and developed into complete plants within 38 days (Fig. 4B). Approximately 20 plants were regenerated per square centimeter of leaf. Leaf sections of the regenerated plants also retained the abilities to form adventitious roots and shoots, and to produce the complete plants (Mano and Matsuhashi 1995).

These properties have been maintained after successive transfers. No adventitious shoots were generated in the non-transformed plants even after 38 days of culture. These results indicate that the technique of culturing leaf sections of transgenic plants can be applied to the micropropagation of horseradish.

3.5 Novel Life Cycle Arising from Leaf Segments in Transgenic Plants

The transgenic plant AR450P was easily transferable to soils from sterile conditions such as culturing in phytohormone-free media. AR450P also grew rapidly in soils, with normal morphology, and formed shoots from the root systems. Leaves of AR450P were excised, immersed in a liquid fertilizer, and cultured under non-sterile conditions on a bench in the laboratory at $25\,°C$.

After 16 days of culture, adventitious roots were generated at the cut ends of the leaves (Fig. 5A, B). Adventitious shoots were generated at the bound-

Fig. 4. Plant regeneration from leaf sections of transgenic plant AR450P in phytohormone-free medium. Leaf sections (ca. 0.15 × 0.3 cm in size) of AR450P were cultured on G agar medium at 25 °C in light. Regeneration after 14 days (**A**) and 38 days (**B**) of cultivation. (Mano and Matsuhashi 1995)

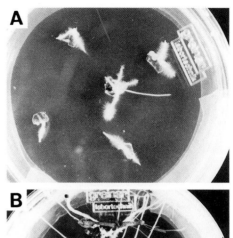

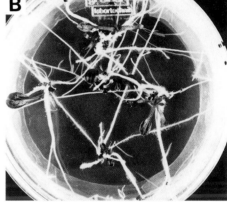

1 cm

ary between the leaf and the adventitious roots after 35 days of culture (Fig. 5C, D). Adventitious shoots grew (Fig. 5E) and increased in number, then developed into complete plants after 7 to 9 weeks (Fig. 5F, G, H) in a manner similar to that observed under sterile conditions. Plants were regenerated from leaf segments at 100% frequency, but not regenerated from small pieces of leaf sections under non-sterile conditions. Similar results were obtained with culture in water. The transgenic plants AR34P, AR47P and AR441P also showed similar properties, but no adventitious roots and shoots were generated in the other transgenic plants, including AR182P and AR455P, and non-transformed plants. Sexual transmission experiments could not be performed in either transgenic plants or non-transformed plants because they did not bear seeds.

Thus, a novel life cycle in which plants were autonomously regenerated from leaf segments was induced reproducibly (Mano and Matsuhashi 1995). This is a unique property of transgenic plants derived from the hairy root clones AR450, AR34, AR47 and AR441. The above results show that plants

Fig. 5. Novel life cycle arising from leaves of transgenic plant AR450P under non-sterile conditions. Leaves of AR450P were immersed in a liquid fertilizer and cultured under non-sterile conditions on a bench in the laboratory at 25 °C. Culture after 16 days (**A** and **B**), 35 days (**C** and **D**), 39 days (**E**), 48 days (**F** and **G**) and 63 days (**H**) of cultivation. *Arrowheads* indicate adventitious shoots. (Mano and Matsuhashi 1995)

transformed with the Ri plasmid possibly acquire new traits that allow plants to survive after serious injuries.

3.6 Evaluation of Greenhouse-Grown Transgenic Plants

The transgenic plants regenerated from hairy root clones were grown in pots in the greenhouse to examine morphologies (Fig. 6). The leaf morphologies in the transgenic plants were similar to those in non-transformed plants. The leaf wrinkling described in other plants was not observed. On the other hand, the growth of roots and stems in the transgenic plants was more active than in non-transformed plants. The transgenic plants had the increased leaf and branch production. As a result, the number of leaves in transgenic plant T3 was five times that of non-transformed plants. Transformation with the Ri plasmid thus produces altered phenotypes that display rapid growth and extensive branching.

Fig. 6. Typical morphologies of transgenic horseradish plants grown in pots in the greenhouse. Non-transformed plant (**A**); transgenic plants T1 (**B**), T2 (**C**) and T3 (**D**)

4 Protocols

4.1 Induction of Hairy Roots

Leaves of *Armoracia rusticana* were sterilized as described elsewhere (Mano et al. 1989) and cut with a sterile scalpel into squares approximately 1.5×1.5 cm in size. Leaf sections were immersed in G medium containing a diluted (10^5–10^6 cells/ml) overnight culture of *Agrobacterium rhizogenes* 15834. After culturing overnight at $25\,°C$ in the dark, the leaf sections were rinsed three times with sterile distilled water and placed on sterile filter paper to remove excess bacterial cells. The inoculated leaf sections were placed on G medium containing 0.1 mg/ml of sodium cefotaxim and 1% agar (G Cef agar medium) and cultured at $25\,°C$ in the dark for 1–2 weeks.

4.2 Establishment of Hairy Root Clones

Adventitious roots that appeared at the cut ends of the inoculated leaf sections were excised and each cultured separately on G Cef agar medium at 25 °C in the dark for 1–2 weeks. Tips (0.5 cm long) of rapidly growing roots were excised and each one cultured three times on G Cef agar medium to eliminate bacteria, then four times on G agar medium without cefotaxim. Culture was performed at 25 °C in the dark for 1–2 weeks.

4.3 Plant Regeneration from Hairy Root Clones

Hairy root clones cultured in G medium at 25 °C in the dark were transferred to the light, with a photoperiodic cycle of 16 h light (at 2500 lx from white fluorescent lamps) followed by 8 h darkness, to produce the transgenic plants.

4.4 Cultures of Transgenic Plants

Transgenic plants were cultured in G medium at 25 °C in the light. Leaf sections of transgenic plants were cultured on G agar medium to regenerate plants. The transgenic plants were transferred to soil [a mixture of Vermiculite and ordinary soil (1:1)] in pots. Leaf segments of transgenic plants were immersed in water or in a liquid fertilizer [0.5 mg/ml solution of Hyponex (Hyponex Corporation, USA)] to regenerate plants via adventitious roots and shoots.

5 Summary and Conclusions

Transgenic horseradish plants have been regenerated from hairy roots which are transformed with the Ri plasmid in *Agrobacterium rhizogenes* 15834. The transgenic plants have thicker roots with extensive lateral branches and thicker stems, and grow faster compared with non-transformed plants. They have peroxidase activity comparable to or more than that of the original plant root tubers. The selected transgenic plants display unique properties in that complete plants are autonomously regenerated from leaf segments in vivo as well as in in vitro cultures. Such a novel life cycle arising from leaf segments has been induced reproducibly. These results show that genetic modification using the Ri plasmid possibly contributes to the improvement of plant properties (Mano and Matsuhashi 1995). The high potency of regeneration in the selected transgenic plants probably opens the way for the clonal propagation of horseradish plants.

The properties of facilitated formation of adventitious shoots in horseradish hairy roots (Noda et al. 1987; Saitou et al. 1991, 1992, 1993; Mano and

Matsuhashi 1995) have been applied to the production of artificial seeds (Uozumi et al. 1992, 1994; Nakashimada et al. 1995, 1996). A hairy-root based artificial seed system may contribute to elite transgenic plant propagation in agronomic and industrial fields.

The addition of phytohormones to the hairy root culture causes cell aggregate formation and hairy root regeneration for the reversible morphology change (Repunte et al. 1993). Peroxidase activity of repeatedly regenerated hairy roots has been stably maintained, indicating that the reversible process of morphology change can be applied to the hairy root culture system to enhance the peroxidase productivity.

Transgenic plants regenerated from hairy roots grow actively in pots in the greenhouse. The roots, stems and leaves in the transgenic plants are more numerous than those in non-transformed plants. Transformation with the Ri plasmid thus produces improved phenotypes that display desired traits.

References

Bajaj YPS (ed) (1993) Biotechnology in agriculture and forestry, vol 22. Plant protoplasts and genetic engineering III. Springer, Berlin Heidelberg New York

Charlwood BV, Rhodes MJC (eds) (1996) Secondary products from plant tissue culture. Oxford Science Publications, Oxford

De Paolis A, Mauro ML, Pomponi M, Cardarelli M, Spano L, Constantino P (1985) Localization of agropine-synthesizing functions in the TR region of the root-inducing plasmid of *Agrobacterium rhizogenes*. Plasmid 13:1–7

Flocco CG, Alvarez MA, Giulietti AM (1998) Peroxidase production in vitro by *Armoracia lapathifolia* (horseradish) – transformed root cultures: effect of elicitation on level and profile of isoenzymes. Biotechnol Appl Biochem 28:33–38

Mano Y (1993) Transformation in *Duboisia* spp. In: Bajaj YPS (ed) Biotechnology in agriculture and forestry, vol 22. Plant protoplasts and genetic engineering III. Springer, Berlin Heidelberg New York, pp 190–201

Mano Y, Matsuhashi M (1995) A novel life cycle arising from leaf segments in plants regenerated from horseradish hairy roots. Plant Cell Rep 14:370–374

Mano Y, Nabeshima S, Matsui C, Ohkawa H (1986) Production of tropane alkaloids by hairy root cultures of *Scopolia japonica*. Agric Biol Chem 50:2715–2722

Mano Y, Ohkawa H, Yamada Y (1989) Production of tropane alkaloids by hairy root cultures of *Duboisia leichhardtii* transformed by *Agrobacterium rhizogenes*. Plant Sci 59:191–201

Nakashimada Y, Uozumi N, Kobayashi Y (1995) Production of plantlets for use as artificial seeds from horseradish hairy roots fragmented in a blender. J Ferment Bioeng 79:458–464

Nakashimada Y, Uozumi N, Kobayashi Y (1996) Efficient culture method for production of plantlets from mechanically cut horseradish hairy roots. J Ferment Bioeng 81:87–89

Noda T, Tanaka N, Mano Y, Nabeshima S, Ohkawa H, Matsui C (1987) Regeneration of horseradish hairy roots incited by *Agrobacterium rhizogenes* infection. Plant Cell Rep 6:283–286

Peskan T, Perica MC, Krsnik-Rasol (1996) Biochemical characterization of horseradish tumour and teratoma tissues. Plant Physiol Biochem 34:385–391

Phunchindawan M, Hirata K, Sakai A, Miyamoto K (1997) Cryopreservation of encapsulated shoot primordia induced in horseradish (*Armoracia rusticana*) hairy root cultures. Plant Cell Rep 16:469–473

Repunte VP, Kino-Oka M, Taya M, Tone S (1993) Reversible morphology change of horseradish hairy roots cultivated in phytohormone-containing media. J Ferment Bioeng 75:271–275

Saitou T, Kamada H, Harada H (1991) Isoperoxidase in hairy roots and regenerated plants of horseradish (*Armoracia lapathifolia*). Plant Sci 75:195–201

Saitou T, Kamada H, Harada H (1992) Light requirement for shoot regeneration in horseradish hairy roots. Plant Physiol 99:1336–1341

Saitou T, Tachikawa Y, Kamada H, Watanabe M, Harada H (1993) Action spectrum for light-induced formation of adventitious shoots in hairy roots of horseradish. Planta 189:590–592

Taya M, Yoyama A, Nomura R, Kondo O, Matsui C, Kobayashi T (1989) Production of peroxidase with horseradish hairy root cells in a two step culture system. J Ferment Bioeng 67:31–34

Tepfer DA (1984) Transformation of several species of higher plants by *Agrobacterium rhizogenes*: sexual transmission of the transformed genotype and phenotype. Cell 37:959–967

Uozumi N, Nakashimada Y, Kato Y, Kobayashi T (1992) Production of artificial seed from horseradish hairy root. J Ferment Bioeng 74:21–26

Uozumi N, Asano Y, Kobayashi T (1994) Micropropagation of horseradish hairy root by means of adventitious shoot primordia. Plant Cell Tissue Organ Cult 36:183–190

4 Transgenic Asparagus (*Asparagus officinalis* L.)

B. Delbreil[1], A. Limanton Grevet[2,3], and M. Jullien[3]

1 Introduction

1.1 The Plant

Asparagus officinalis L. is a perennial monocotyledon, a member of the Liliaceae family. It grows in temperate climates and under subtropical conditions. Growth of spears normally takes place in sandy soils. There are two culture types: the white asparagus, in which the spears are harvested from earthed up plants, where spears are cut at the crown level just as they emerge through the soil, and the green asparagus, in which the spears are harvested at or about ground level when they reach a fixed height above the soil. The main producers are the USA, Spain, France, and Taiwan. White asparagus production predominates in Europe (France and Spain) and Taiwan, while green asparagus is essential in North America and Australasia. In European countries, white asparagus is produced mainly for the fresh market, whereas canned white asparagus is produced in Taiwan.

Asparagus is a dioecious species (2n = 20) with a sex ratio of 1 : 1; however, plants are andromonoecious to hermaphroditic. Male flowers bear six stamens and a rudimentary pistil; female flowers have collapsed anthers surrounding an ovule. Genetic experiments have demonstrated that asparagus female flowers are homogametic (XX), and male and hermaphroditic flowers heterogametic (XY) (Rick and Hanna 1943). The dioecity means that the old common varieties are populations exhibiting a high degree of natural heterozygosity with very irregular yields.

Improvement in asparagus breeding was obtained in two steps with the help of in vitro cell culture techniques. The first step was the production of hybrid varieties: double hybrids (Corriols-Thévenin 1979) and clonal hybrids (Doré 1975; Javouhey 1990). Micropropagation was developed (reviewed by Desjardins 1992) to provide the large number of clones required

[1] Laboratoire de Physiologie de la Différenciation Végétale UPRES 2702. Université des Sciences et Technologies de Lille, 59655 Villeneuve D'Ascq CEDEX, France
[2] Laboratoire "in vitro", J. Marionnet GFA, Rout de Courmenin, 41230 Soings-en-Sologne, France
[3] UMR Biologie des Semences, INRA-INA-PG, Route de Saint-Cyr, 78026 Versailles CEDEX, France

Biotechnology in Agriculture and Forestry, Vol. 47
Transgenic Crops II (ed. by Y.P.S. Bajaj)
© Springer-Verlag Berlin Heidelberg 2001

for the parental production of hybrids. The second step was the use of haploids to create F_1 hybrids (Thévenin 1968). Haploids are obtained by polyembryony (Thévenin 1968) or in vitro anther culture (reviewed by Doré 1990).

1.2 Need for Transformation

Because of its perennial character, asparagus demands heavy financial input at the plantation stage; plants are productive for 10 years or more, and the benefits are obtained only after 5 or 6 years. Each factor limiting the production and the longevity of the crop is highly damaging.

A disease syndrome known as asparagus decline or crown and root rot, caused by *Fusarium oxysporum* Schlecht f. sp. *asparagi* Cohen and *F. moniliforme* Sheldon, decreases yields and necessitates the removal of fields worldwide (Cohen and Heald 1941; Endo and Burkholder 1971). This disease is the major limiting factor in asparagus production (Grove 1976). The French harvest decreased to 20% in 1989–1990 and 2500 ha were removed in 1989–1991. These fungi associated with the Asparagus viruses I, II, or tobacco streak virus also significantly reduce the productivity of asparagus in Mexico.

These diseases have initiated important breeding programs to tackle this problem. Unfortunately, no highly resistant cultivars were available in *Asparagus officinalis*. Crosses with the resistant species *A. sprengeri* (Lewis and Shoemaker 1964) have been unsuccessful, probably due to incompatibility barriers (Elmer et al. 1989). Plant chemical treatments have not been successful, and soil fumigations offer no long-term effectiveness (Lacy 1979). Alternative strategies are needed.

One strategy is to produce transgenic plants which express cloned resistance genes for improving resistance to some pathogens. The first report of success with this approach was transgenic tobacco and rape containing a chitinase with a constitutive promoter. These plants have been shown to exhibit higher basal levels of chitinase and concomitant increased resistance to pathogenic soil fungi when compared with control plants (Broglie et al. 1991). One other gene, encoding a ribosome-inactivating protein in transgenic plants, confers partial protection against fungal attacks (Logemann et al. 1992).

Since higher yield and precocity are correlated with staminate plants (Ellison et al. 1960), there have been efforts in asparagus breeding to produce all male varieties. Regeneration of dihaploids after anther culture is the only way to obtain supermale (YY) plants, homozygous for all characters. Such homozygous plants, crossed with female (XX), produced all male F_1 plants (Doré 1974). This research is very complicated, and an alternative could be to tag and select the male plants with a marked gene after transformation of existing varieties.

2 Transformation

2.1 Review of *Asparagus* Transformation (Table 1)

Asparagus officinalis L. is sensitive to *Agrobacterium tumefaciens*. Hernalsteens et al. (1984) obtained a tumor tissue that grew on hormone-free medium and produced two opines; nopaline and agrocinopine, after inoculation with *A. tumefaciens*. They used stem pieces cultured in vitro and infected on the upper end with a wild-type *A. tumefaciens* strain C58. Molecular proofs of the *A. tumefaciens* T-DNA integration in asparagus tumor tissues were given by the works of Bytebier et al. (1987) and Prinsen et al. (1990).

Asparagus was the first monocotyledon to be transformed and regenerated (Bytebier et al. 1987). These authors obtained calli after inoculation of stem slices by *A. tumefaciens* strain C58C1 pGV3850::1103neo(dim) which exhibited kanamycin resistance. The callus showed multiple T-DNA insertions. Plants regenerated from resistant callus integrated *A. tumefaciens* T-DNA and showed no major T-DNA rearrangements in comparison with the transformed callus.

Those two reports demonstrate that *Agrobacterium*-mediated transformation was possible in asparagus. Although it was long considered that *Agrobacterium*-mediated transformation was limited to dicotyledonous plant species, it is now clear that *Agrobacterium* can also transfer DNA to cells of monocotyledonous plants (Hernalsteens et al. 1984; Schafër et al. 1987; Raineri et al. 1990). The limitation in producing transgenic plants from monocotyledonous species with *Agrobacterium* still remains in the difficulty of finding sensitive tissue, and in the regeneration of plants from infected explants. The sensitivity of monocotyledonous zygotic embryo tissues to *Agrobacterium* (Raineri et al. 1990; Mooney et al. 1991), and the fact that somatic embryogenesis seems to be the best pathway for regenerating monocots and especially asparagus (Reuther 1984), favor the use of somatic embryos for transformation experiments in asparagus. Transgenic asparagus can be obtained by three other methods: electroporation, DNA uptake by PEG treatments, and particle gun. The first two methods are delicate. First, they need the establishment of an efficient system of plant regeneration from protoplasts. Plant regeneration from asparagus protoplasts was first reported from cladophylls by Bui Dang Ha and Mackenzie (1973). Other authors have described protoplast isolation and regeneration from calli (Bui Dang Ha et al. 1975; Kong and Chin 1988; Elmer et al. 1989, Dan and Stephens 1991) or somatic embryos (Kunitake and Mii 1990). But in all these cases the plating efficiency was very low. Mukhopadhyay and Desjardins (1994) described the use of protoplast electroporation to obtain transformed asparagus. They selected resistant calli with GUS expression, but no plant regeneration occurred. The lack of regeneration was certainly due to the choice of a non-embryogenic callus as source of protoplasts.

The use of a particle gun combined with somatic embryogenesis is easier to handle and is probably the only alternative method available today to the

Table 1. Summary of work done on transformation in asparagus

Varieties	Explant used	Method used	Strain or plasmid	Result	Reference
Roem van Brunswijk	Stem cuts	*A. tumefaciens*	Wild C58	Tumor lines	Hernalsteens et al. (1984)
Not done	Stem cuts	*A. tumefaciens*	C58C1 pTiB6S3 C58C1pGV3850	Tumor lines Plant regeneration	Bytebier et al. (1987)
Limbras 26	Stem cuts	*A. tumefaciens*	Wild C58 C58C1pTiB6S3	Tumor lines Tumor lines	Prinsen et al. (1990)
Andreas, three Ma clones[a]	Somatic embryos	*A. tumefaciens*	C58 p35SGUSINT	Plant regeneration	Delbreil and Jullien (1993)
UC72, UC157	Somatic embryos	Particle bombardment	PWRG1515 pGPTV-Bar	Plant regeneration	Cabrera-Ponce et al. (1997)
G171, G447, G457, G203[b]	Cell suspensions	Particle bombardment	pKGUS	Transient gus expression Plant regeneration	Li and Wolyn (1997)
G203, G171[b]	Protoplasts	Electroporation	pBI121	Callus	Mukhopadhyay and Desjardins (1994)

[a] The Ma clones were provided by Jacques Marionet GFA.
[b] The G plants were provided by D.J. Wolyn, University of Guelph.

co-culture described here. Cabrera-Ponce et al. (1997) regenerated more plants using this method than by *A. tumefaciens* inoculation after bombardment of embryogenic calli. Unfortunately, molecular analysis showed multiple independent and long concatemeric plasmid chain insertions. Li and Wolyn (1997) bombarded a cell suspension. They reported four genotypes of transient gus expression in cell suspensions, but only plants regenerated from one transformation event could be regenerated after somatic embryogenesis induction. In this case, inheritance of gus expression was detected.

In *Asparagus officinalis*, somatic embryogenesis can be induced in vitro from several explant sources: hypocotyls, stems, excised buds, in vitro or young seedling crowns, cladophylls, and cladophyll cell cultures (reviewed by Reuther 1984; Levi and Sink 1995). In some cases, asparagus tissue cultures can produce long-term embryogenic callus (Jullien 1974). These embryogenic calli can be maintained on a hormone-free medium and grow by unicellular secondary embryogenesis (Delbreil et al. 1994). They have been used to develop efficient transformation in asparagus (Delbreil and Jullien 1993).

2.2 Material and Methods

2.2.1 Somatic Embryo Cultures

In embryogenesis induction experiments from different explants of *Asparagus officinalis* L., such as cladophylls, cells, and apices cultivated first on an inducing medium and then on hormone-free medium, several long-term embryogenic calli were isolated (Fig. 1). These embryogenic calli were habituated, and continuously produced somatic embryos by secondary embryogenesis from the embryo epidermis. These embryos appeared to be of single-cell origin (Delbreil et al. 1994). These embryogenic lines were maintained on B medium (Bourgin et al. 1979) in a growth chamber with 16 h/day fluorescent light providing 40 to 70 E/m^2/s, at 25 °C and 70% relative humidity. Embryogenic lines had a yellow, friable, and nodular appearance. Four embryogenic lines (L_1, L_2, L_3, and L_4) were used. They were isolated from three male genotypes. The lines

Fig. 1. Morphology of L_1 embryogenic line after 3 weeks on subculture medium. *Bar* 1 mm

L_1 and L_2 were isolated from clone Ma8 and line L_3 from clone Ma5. The Ma clones were provided by Jacques Marionnet GFA. The line L_4 was isolated from one INRA F_1 hybrid (Andreas). Plants were recovered after embryo germination on B medium and their transplantation into soil.

2.2.2 Agrobacterium Strain and Inoculation

An *A. tumefaciens* C58 suspension (p35SGUSINT; Vancanneyt et al. 1990) was used. Between the border sequences are the following genes: a chimeric gene encoding the neomycin phosphotransferase II containing the nopaline synthase promoter and terminator, and a chimeric *gus* gene harboring an intron and containing the 35S terminator and promoter from cauliflower mosaic virus. This construct was chosen for the robustness of the GUS reporter gene system and the efficiency of the *nos* promoter in asparagus (Bytebier et al. 1987).

The bacteria were grown overnight at 28 °C in LB medium and then washed with B medium. Prior to co-cultivation experiments, bacteria were diluted in B liquid medium to an OD_{600} of 0.6–0.8. Before inoculation, somatic embryos were sieved on various meshes: 0.2, 0.4, 0.8, and 1.6 mm, to separate globular (0.2–0.4 mm) and cylindrical (0.8–1.6 mm) embryos. Somatic embryos, 1 g fresh weight, were immersed in 5 ml of the diluted bacteria suspension for 15 min. Embryos were then blotted dry, plated (0.5 g/dish) on B liquid or solid medium (co-culture medium) and cultured for 48 h in the dark. Embryos were then transferred to a B selection medium containing cefotaxime (400 mg/l) and supplemented with kanamycin (100 mg/l). The resistant lines were maintained by subculturing monthly on B medium supplemented with kanamycin (100 mg/l) and cefotaxime (200 mg/l).

2.2.3 Histochemical GUS Assay

The histochemical localization of the GUS activities in kanamycin-resistant embryogenic lines or regenerated shoots was performed as described by Jefferson et al. (1987) with some modifications. Samples were incubated overnight at 37 °C in a 0.5 g/l solution of X-Gluc in 50 mM sodium phosphate (pH 7).

2.2.4 DNA Isolation and Analysis

Genomic DNA was isolated from embryogenic lines or cladophylls according to Dellaporta et al. (1983). Ten micrograms of DNA was digested with the restriction endonuclease *Hind*III (Bethesda Research Laboratories, UK), separated by electrophoresis (0.8% agarose gel), blotted, and UV-cross-linked to Hybond-N filters (Amersham, UK). Prehybridization and hybridization were performed according to standard procedures at 65 °C (Sambrook et al. 1989).

The [32]Plabeled probes were synthesized using random oligonucleotide primers and [32]PdCTP as described by Feinberg and Vogelstein (1983). The *npt*II gene was detected by using the 1.2-kb *Eco*RV fragment of pABD 1 as a probe (Paszkowski et al. 1984) and the *gus* gene with the 2-kb *PstI/Eco*RI fragment of pBI 121 plasmid (Jefferson et al. 1987). The filters were washed 10 min in 6 × SSC (Sambrook et al. 1989) at room temperature, 10 min in 2 × SSC, 0.1% sodium dodecyl sulfate at 65 °C, 5 min in 0.1 × SSC, 0.1% sodium dodecyl sulfate at 65 °C. Filters were exposed to Kodak X-ARS films for 1 week with one Quanta III intensifying screen at −80 °C.

2.3 Results and Discussion

2.3.1 Effect of Cefotaxime and Kanamycin on the Growth of Embryogenic Lines

Preliminary studies were carried out to determine the experimental concentrations of selecting and decontaminating agents. Cefotaxime is used as a means of eliminating *Agrobaterium tumefaciens* from asparagus callus in co-cultivation experiments. Some authors claim that cefotaxime stimulates callus growth in bread wheat (Mathias and Boyd 1986), barley (Mathias and Mukasa 1987), and regeneration in bread wheat (Mathias and Boyd 1986), durum wheat (Borrelli et al. 1992), barley (Manthias and Mukasa 1987), and finger millet embryogenic calli (Eapen and George 1990). In our case, for the concentrations employed, there was no influence of cefotaxime on the embryogenic callus growth (Fig. 2A) and morphology (Fig. 3A,B). The aminoglycoside kanamycin was used to select transformed cells with the neomycin phosphotransferase selection gene. The asparagus somatic embryos turned brown and died after 3 weeks' exposure at 25 mg/l of kanamycin. This death was visualized by the arrest of embryogenic line growth (Fig. 2B) and embryo necrosis (Fig. 3C). Asparagus showed no high natural kanamycin resistance, as observed in some monocotyledons (Hauptmann et al. 1988). We chose to use

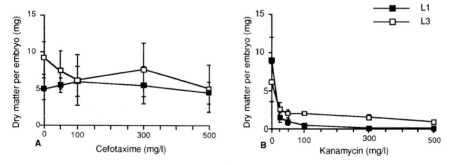

Fig. 2A,B. Influence of cefotaxime (**A**) and kanamycin (**B**) on embryogenic growth of different lines after 1 month on culture medium

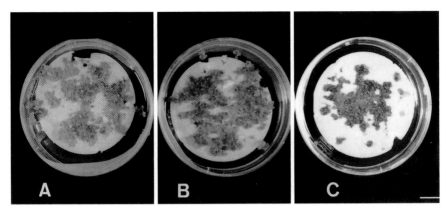

Fig. 3A–C. Morphology of embryogenic line L_1 after 1 month on B medium (**A**), supplemented with cefotaxime (**B**) or kanamycin (**C**). *Bar* 1 cm

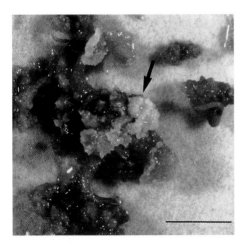

Fig. 4. Isolation of kanamycin-resistant embryogenic line (*arrow*) 1 month after inoculation of L_1 somatic embryos. *Bar* 1 cm. (Delbreil and Jullien 1993)

kanamycin at 100 mg/l because under the experimental conditions clusters of somatic embryos were used.

2.3.2 Selection of Kanamycin-Resistant Embryogenic lines

Resistant embryogenic lines appeared 2 or 3 months after the beginning of the selection period. They were characterized by the emergence of white secondary embryos on necrosed primary embryos (Fig. 4). Kanamycin (100 mg/l) inhibited but did not completely block new embryo formation when somatic embryos were cultured en masse (0.5 g/dish). Therefore, identification of transformants required repetitive selection on Kanamycin for several months. These results were similar to those of McGranahan et al. (1988), who transformed walnut by co-culture of somatic embryos and *Agrobacterium*.

Twenty-five kanamycin-resistant lines from four embryogenic lines were isolated (Table 2). For the L_1 line (24 g inoculated in four manipulations), we obtained 14 kanamycin-resistant lines. Five of these lines did not regenerate, six regenerated only shoots, and three others regenerated whole plants (Table 2). The L_2 line (1 g inoculated) produced two kanamycin-resistant lines which regenerated shoots (Table 2). The L_3 line (1 g inoculated) produced three kanamycin-resistant lines. Two lines that produced only calli without visualized embryos died after five subcultures, and one regenerated shoots (Table 2). The L_4 line (2 g inoculated) produced eight kanamycin-resistant lines, three of which regenerated shoots (Table 2). This last line seemed to be more sensitive to the agrobacteria than the other three (Table 2). Some resistant lines ($L_{1.1}$ and $L_{1.3}$) show increased growth when cultured on media supplied respectively with 50 and 100 mg/l kanamycin (Fig. 5). Other lines were not affected by these kanamycin concentrations. For all the resistant lines, kanamycin at 500 mg/l induced a decrease in growth (Fig. 5).

Table 2. Regeneration capacity of kanamycin-resistant lines selected after cocultivation of somatic embryos isolated from four embryogenic lines (L_1, L_2, L_3, L_4) with *Agrobacterium tumefaciens* C58 (p35SGUSINT)

Embryogenic line	No. of kanamycin-resistant lines selected	No. of resistant lines regenerating shoots	No. of resistant lines regenerating plants	Transformation rate[a]
L_1	14	6	3	0.6
L_2	2	2	0	2
L_3	1	1	0	1
L_4	8	3	0	4

[a] The transformation rate is the number of kanamycin resistant embryogenic lines produced per g (fresh weight) of inoculated somatic embryos.

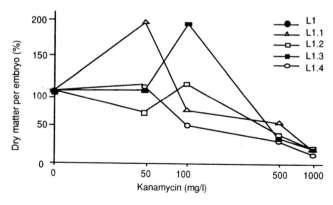

Fig. 5. Influence of kanamycin on growth of resistant embryogenic lines obtained after inoculation with *Agrobacterium tumefaciens* p35SGUSINT. Growth was measured after 1 month on culture medium

Embryos (Fig. 6A) and shoots from uninoculated lines did not express GUS activity. All the kanamycin-resistant embryogenic lines and all the plantlets or shoots regenerated from them, exhibited a positive blue coloration when tested for β-glucuronidase activity. The 35S promoter was expressed in a constitutive manner in tissues exhibiting a highly transcriptional activity: embryogenic cells (Fig. 6B), meristematic tissues, and growing stem cells (Fig. 6C). The GUS coloration did not show any evident chimeric patterns (transformed and nontransformed cells) in the kanamycin-resistant embryos and derived plants. This phenomenon could be due to the single-cell origin of asparagus secondary somatic embryos (Delbreil et al. 1994). The regenerated plants have the same phenotype as plants regenerated from untransformed somatic embryos.

In order to optimize the transformation process, we tested the co-culture medium, selection timing, embryo wounding by a superficial blade cut, and embryo developmental stage. It appeared that co-culture in liquid medium was less effective than in solid medium for obtaining kanamycin-resistant embryogenic lines (Table 3). Early kanamycin exposure (0–7 days after co-culture) enhanced emergence of resistant embryos (Table 3). The co-culture of globular embryos with *Agrobacterium* was ineffective in obtaining resistant embryogenic lines compared with the co-culture of cylindrical embryos (Table 3).

A similar response to transformation of globular and cylindrical embryos was reported by Cabrera-Ponce et al. (1997). This could be due to a greater capacity for secondary embryogenesis shown by cylindrical embryos, or to the deleterious effect of the bacterial lysis during the co-culture phase on the very young embryos. The wounded embryos died rapidly on kanamycin. They were unable to regenerate kanamycin-resistant lines. The wound could have a deleterious effect.

The transformation rates obtained were low. This poor efficiency did not represent a limiting factor in our transformation process due to the

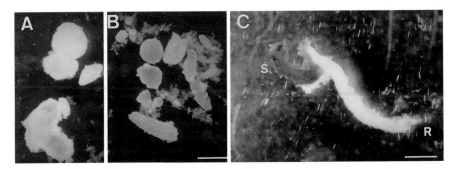

Fig. 6A–C. Gus expression in different transgenic asparagus tissues. **A** Uninoculated somatic embryos from L_1 line. **B** Embryos from a kanamycin-resistant line (L_1) derived from L_1 inoculated with *Agrobacterium tumefaciens* strain C58 (p35SGUSINT). *Bar* 1 mm. **C** Plantlet derived from a kanamycin-resistant line ($L_{1.4}$), shoot (*S*), and root meristem (*R*). *Bar* 1 mm. (Delbreil and Jullien 1993)

Table 3. Influence of the coculture medium, the stage of embryo development, and the timing of selection on the transformation rate of embryogenic line L_1. (Delbreil and Jullien 1993)

Embryo stage	Coculture medium	Timing of selection[b] (days) Transformation rate[a]			
		0	7	15	30
Cylindrical	Liquid	0	0.5 (1)[c]	0	0
	Solid	1 (4)	2 (8)	0	0.5 (1)
Globular	Liquid	0	0	0	0
	Solid	0	0	0	0

[a] Transformation rate is the number of kanamycin resistant embryogenic lines produced per g (fresh weight) of inoculated somatic embryos.
[b] Timing of selection is the time between the coculture period and the beginning of the selection.
[c] The number of kanamycin-resistant lines isolated is indicated in brackets.

great number of embryos used in co-cultivation experiments: the embryogenic lines produced about 300 cylindrical embryos per gram of fresh weight.

Southern hybridizations were performed to confirm the presence of the T-DNA from the C58 p35SGUSINT in kanamycin-resistant embryogenic lines showing GUS activity and in a plant regenerated from line $L_{1.4}$. Embryos and shoots from an uninoculated embryogenic line did not contain either *npt*II or *gus* sequences (Fig. 7). The enzyme *Hind*III cut the end of the *nos* terminator of the *npt*II gene and in the plant DNA. Consequently, each variant fragment visualized after the use of an NPT II probe corresponded to a single integration event, i.e., one T-DNA copy. In the hybridization pattern (Fig. 7A), the lines $L_{1.1}$, $L_{1.2}$, $L_{1.3}$ showed two integration events of the *npt*II gene of respectively 13 and 10 kb in length, 5.1 and 3 kb, 4.2 and 3.4 kb. Line $L_{1.4}$ showed only one fragment of 12 kb. A plant regenerated from the $L_{1.4}$ line presented the same integration pattern (Fig. 7B).

The asparagus genomic DNA was also tested with a GUS probe. Based on the restriction map of the T-DNA region of the p35 GUSINT, it could be predicted that the digestion by HindIII should give a 2.8-kb internal fragment of the T-DNA containing the *gus* gene sequence with the 35S promoter and terminator. All the lines tested with the GUS probe presented a hybridization signal corresponding to a fragment of 2.8 kb in length (Fig. 7C). In conclusion, it appeared that the kanamycin-resistant lines $L_{1.1}$, $L_{1.2}$, $L_{1.3}$, and $L_{1.4}$ have integrated the whole sequence of the T-DNA (Fig. 7A,C) with the *gus* and *npt*II genes.

The unique DNA hybridization pattern with *Hind*III and GUS probe (Fig. 7A) for each transgenic line confirmed independent insertion events. The hybridization patterns indicated that one to two copies of T-DNA have been incorporated in the asparagus DNA. This is in agreement with the results of Bytebier et al. (1987) on asparagus and in dicotyledonous plants (Weising et al. 1988). The number of insertions will be confirmed by the progeny analysis.

50 B. Delbreil et al.

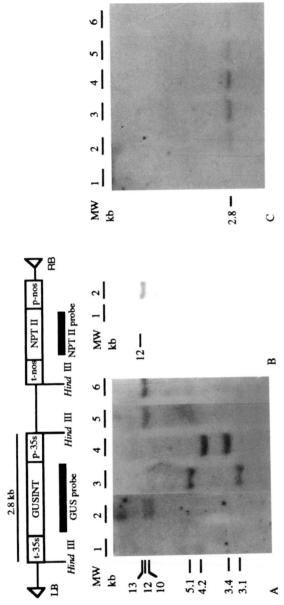

Fig. 7A–C. Southern blots of kanamycin-resistant embryogenic *Asparagus* lines and derived plants. (Delbreil and Jullien 1993) **A** DNA (10 mg) from kanamycin-resistant lines was digested by the restriction enzyme *Hind*III and hybridized with an NPTII probe. *Lane 1* Uninoculated control line (L1); *lanes 2 to 5* different kanamycin-resistant embryogenic lines isolated from L1, respectively lines $L_{1.1}$, $L_{1.2}$, $L_{1.3}$, $L_{1.4}$; *lane 6* digestion from DNA corresponding to one copy of the p35GUSINT plasmid. **B** DNA (10 mg) from plants was digested by the restriction enzyme *Hind*III and hybridized with an NPTII probe. *Lane 1* Uninoculated plant (C1) control; *lane 2* plant regenerated from kanamycin-resistant embryogenic line $L_{1.4}$ isolated from L1 **C** DNA (10 mg) from kanamycin-resistant lines was digested by the restriction enzyme *Hind*III and hybridized with a *gus* probe. *Lane 1* Uninoculated control line (L1); *lanes 2 to 5* different kanamycin-resistant embryogenic lines isolated from L1, respectively, lines $L_{1.1}$, $L_{1.2}$, $L_{1.3}$; *lane 6* digestion from DNA corresponding to one copy of p35GUSINT plasmid; *LB* left T-DNA border; *RB* right T-DNA border. *t-35S* 35S terminator from cauliflower mosaic virus; *p-35S* 35S promoter from cauliflower mosaic virus; *t-nos* terminator from nopaline synthase gene; *p-nos* promoter from nopaline synthase gene

3 Summary and Conclusion

Twenty-five independent kanamycin-resistant lines were obtained after co-cultivation of long-term embryogenic cultures of three *Asparagus officinalis* L. genotypes with an *Agrobacterium tumefaciens* strain harboring β-glucuronidase and neomycin phosphotransferase II genes. All the lines showed β-glucuronidase activity by histological staining. DNA analysis by Southern blots of the kanamycin-resistant embryogenic lines and of a plant regenerated from one of them confirmed the integration of the T-DNA.

Only a few transformed embryos germinated, but low rates of germination have often been reported for somatic embryos of asparagus (Delbreil et al. 1994) and various other species (Gray and Purohit 1991); dehydration methods could improve the germination of asparagus somatic embryos (Saito et al. 1991).

Application of this method to different genotypes will require the isolation of embryogenic somatic lines for each of them or the transfer of the foreign genes by sexual crossing. The creation of embryogenic lines, even if rare, is not difficult to obtain (Delbreil et al. 1994), and their use in co-culture appeared to be an efficient method of transformation. We also have used another *A. tumefaciens* strain AGLI gin (Lazo et al. 1991) on three new genotypes. Numerous plants have been regenerated and sexual transmission of kanamycin and gus expression has been confirmed (A. Limanton-Grevet, pers. comm.). The transformation and regeneration of asparagus offers the opportunity to transfer foreign genetic material which could be useful for plant improvement. The transfer of genes encoding chitinase and β 1,3-glucanase now seems a realistic strategy to fight *Fusarium*, the walls of which contain high levels of chitin (Sivan and Chet 1989). The expression of the gene could be improved and monitored by the use of asparagus wound-responsive promoter involved in the control of pathogenesis-related genes (Warner et al. 1993).

Three other transformation methods could be used: electroporation, DNA uptake by PEG treatments, and particle gun. The first two methods are delicate. Preliminarily, they need the establishment of an efficient system of plant regeneration from protoplasts. Plant regeneration from asparagus protoplasts was first reported from cladophylls by Bui Dang Ha and Mackenzie (1973). Other authors described protoplast isolation and regeneration from calli (Bui Dang Ha et al. 1975; Kong and Chin 1988; Elmer et al. 1989; Dan and Stephens 1991, 1994), or somatic embryos (Kunitake and Mii 1990); but in all these cases the plating efficiency was very low. The use of a particle gun on embryogenic calli is easy to handle, and is probably today the only alternative method to the coculture described here.

Acknowledgments. This work was supported by Jacques Marionnet GFA. The authors thank C. Lefrançois for reading the English.

References

Borrelli GM, Di Fonzo N, Luputto E (1992) Effect of cefotaxime on callus culture and plant regeneration in durum wheat. J Plant Physiol 140:372–374

Bourgin JP, Chupeau Y, Missonier C (1979) Plant regeneration from mesophyll protoplasts of several *Nicotiana* species. Physiol Plant 45:288–292

Broglie K, Chet I, Holliday M, Cressman R, Biddle P, Knowlton S, Mauvais CJ, Broglie R (1991) Transgenic plants with enhanced resistance to the fungal pathogen *Rhizoctonia solani*. Science 254:1194–1197

Bui Dang Ha D, Mackenzie I (1973) The division of protoplasts from *Asparagus officinalis* L. and their growth and differentiation. Protoplasma 78:215–221

Bui Dang Ha D, Norrel B, Masset A (1975) Regeneration of *Asparagus officinalis* L. through callus derived from protoplasts. J Exp Bot 26:263–270

Bytebier B, Deboeck F, De Greve H, Van Montagu M, Hernalsteens JP (1987) T-DNA organization in tumor cultures and transgenic plants of the monocotyledon *Asparagus officinalis*. Proc Natl Acad Sci USA 84:5345–5349

Cabrera-Ponce J-L, Lopez L, Assad-Garcia N, Medina-Arevalo C, Bailey A, Herrera-Estrella L (1997) An efficient particle bombardment system for the genetic transformation of asparagus (*Asparagus officinalis* L.). Plant Cell Rep 16:255–260

Cohen SI, Heald FD (1941) A wilt and root rot of asparagus caused by *Fusarium oxysporum* (Schlecht.). Plant Dis Rep 25:503–509

Corriols-Thévenin L (1979) Different methods in asparagus breeding. In: Reuther G (ed) 5th Int Asparagus Symp. Eucarpia section vegetables, Geisenheim. Forschungsanstalt, Germany, pp 8–20

Dan Y, Stephens CT (1991) Studies of protoplast culture types and plant regeneration from callusderived protoplasts of *Asparagus officinalis* L. cv. Lucullus 234. Plant Cell Tissue Organ Cult 27:321–331

Dan Ying-hui, Stephens C (1994) Regeneration of plants from protoplasts of *Asparagus officinalis* L. In: Bajaj YPS (ed) Biotechnology in agriculture and forestry, vol 29. Plant protoplasts and genetic engineering V. Springer, Berlin Heidelberg New York, pp 3–15

Delbreil B, Jullien M (1993) *Agrobacterium*-mediated transformation of *Asparagus officinalis* L. long-term embryogenic callus and regeneration of transgenic plants. Plant Cell Rep 12:129–132

Delbreil B, Jullien M (1994) Evidence for in vitro induced mutation which improves somatic embryogenesis in *Asparagus officinalis* L. Plant Cell Rep 13:372–376

Delbreil B, Goebel-Tourand I, Lefrancois C, Jullien M (1994) Isolation and characterization of long-term embryogenic lines in *Asparagus officinalis* L. J Plant Physiol 144:194–200

Dellaporta SL, Wood J, Hicks JB (1983) A plant DNA minipreparation: version II. Plant Mol Biol Rep 1:19–21

Desjardins Y (1992) Micropropagation of Asparagus (*Asparagus officinalis* L.). In: Bajaj YPS (ed) Biotechnology in agriculture and forestry, vol 19. High-tech and micropropagation III. Springer, Berlin Heidelberg New York, pp 26–41

Doré C (1974) Production de plantes homozygotes mâles et femelles à partir d'anthères d'asperge, cultivées in vitro (*Asparagus officinalis* L.). CR Acad Sci 278:2135–2138

Doré C (1975) La multiplication clonale de l'Asperge (*Asparagus officinalis* L.) par culture in vitro: son utilistion en sélection. Ann Amélior Plantes 25:201–244

Doré C (1990) Asparagus anther culture and field trials of dihaploids and F₁ hybrids. In: Bajaj YPS (ed) Biotechnology in agriculture and forestry, vol 12. Haploids in crop improvement. I. Springer, Berlin Heidelberg New York, pp 322–345

Eapen S, George L (1990) Influence of phytohormones, carbohydrates, amino acids, growth supplements and antibiotics on somatic embryogenesis and plant regeneration in finger millet. Plant Cell Tissue Organ Cult 22:87–93

Ellison JH, Scheer DF, Wagner JJ (1960) Asparagus yield as related to plant vigor, earliness and sex. Proc Am Soc Hortic Sci 75:411–415

Elmer WH, Ball T, Volokita M, Stephens CT, Sink KC (1989) Plant regeneration from callus-derived protoplasts of asparagus. J Am Soc Hortic Sci 1147:1019–1024

Endo RM, Burkholder EC (1971) The association of *Fusarium moniliforme* with the crown rot complex of asparagus. Phytopathology 61:891

Feinberg AP, Volgelstein B (1983) A technique for radiolabeling DNA restriction endonuclease fragments to high specific activity. Anal Biochem 132:6–13

Gray DJ, Purohit A (1991) Somatic embryogenesis and development of synthetic seed technology. Crit Rev Plant Sci 10:33–61

Grove MD (1976) Fusarial disease of asparagus. Proc Am Phytopathol Soc 3:313

Hauptmann RM, Vasil V, Ozias-Akins P, Tabaeizadeh Z, Rogers SG, Fraley RT, Horsch RB, Vasil IK (1988) Evaluation of selectable markers for obtaining transformation in the Gramineae. Plant Physiol 86:602–606

Hernalsteens JP, Thia-Toong L, Schell J, Van Montagu M (1984) An *Agrobacterium*-transformed cell culture from the monocot *Asparagus officinalis*. EMBO J 3:3039–3041

Javouhey M (1990) Fifty years of asparagus breeding valorized through twelve years of vitroculture. I – Production and agronomical performances of the new French clonal hybrids. In: Falavigna A, Schiavi M (eds) 7th Int Asparagus Symp. Acta Hortic 271:55–62

Jefferson RA, Kavanagh TA, Bevan MW (1987) GUS fusions: β-glucuronidase as a sensitive and versatile gene fusion marker in higher plants. EMBO J 6:3901–3907

Jullien M (1974) La culture in vitro de cellules du tissu foliaire d'*Asparagus officinalis* L.: obtention de souches à embryogenèse permanente et régénération de plantes entières. CR Acad Sci Paris Sér D 279:747–750

Kong Y, Chin CK (1988) Culture of asparagus protoplasts on porous polypropylene membrane. Plant Cell Rep 7:67–69

Kunitake H, Mii M (1990) Somatic embryogenesis and plant regeneration from protoplasts of asparagus (*Asparagus officinalis* L.). Plant Cell Rep 8:706–710

Lacy ML (1979) Effects of chemicals on stand establishment and yields of asparagus. Plant Dis Rep 63:612–616

Lazo GR, Stein PA, Ludwig RA (1991) A DNA transformation-competent *Arabidopsis* genomic library in *Agrobacterium*. Biotechnology 9:963–967

Levi A, Sink KC (1995) Somatic embryogenesis in asparagus. In: Bajaj YPS (ed) Biotechnology in agriculture and forestry, vol 31. Somatic embryogenesis and synthetic seed II. Springer, Berlin Heidelberg New York, pp 117–124

Lewis GD, Shoemaker PB (1964) Resistance of asparagus species to *Fusarium oxysporum* f. *asparagi*. Plant Dis Rep 48:364–365

Li B, Wolyn DJ (1997) Recovery of transgenic asparagus plants by particle gun bombardment of somatic cells. Plant Sci 126(1):59–68

Logemann J, Jach G, Tommerup H, Mundy J, Schell J (1992) Expression of a barley ribosome-inactivating protein leads to increased fungal protection in transgenic tobacco plants. Bio/Technology 10:305–308

Mathias RJ, Boyd LA (1986) Cefotaxime stimulates callus growth, embryogenesis and regeneration in hexaploid bread wheat (*Triticum aestivum* L. em. Thel.). Plant Sci 46:217–223

Mathias RJ, Mukasa C (1987) The effect of cefotaxime on the growth and regeneration of callus from four varieties of barley (*Hordeum vulgare* L.). Plant Cell Rep 6:454–457

McGranahan GH, Leslie CA, Uratsu SL, Martin LA, Dandekar AM (1988) *Agrobacterium*-mediated transformation of walnut somatic embryos and regeneration of transgenic plants. Bio/Technology 6:800–804

Mooney PA, Goodwin PB, Dennis ES, Llewellyn DJ (1991) *Agrobacterium tumefaciens*-gene transfer into wheat tissues. Plant Cell Tissue Organ Cult 25:209–218

Mukhopadhyay S, Desjardins Y (1994) Direct gene transfer to protoplasts of two genotypes of *Asparagus officinalis* L. by electroporation. Plant Cell Rep 13:421–424

Nichols AM (1990) Asparagus – the world scene. In: Falavigna A, Schiavi M (eds) 7th Intl Asparagus Symp. Acta Hortic 271:25–33

Paszkowski J, Shillito RD, Saul M, Mandak V, Hohn T Hohn B, Potrykus I (1984) Direct gene transfer to plants. EMBO J 3:2717–2722

Prinsen E, Bytebier B, Hernalsteens J, De Greef J, Van Onckelen H (1990) Functional expression of *Agrobacterium tumefaciens* T-DNA *onc*-genes in *Asparagus* crown gall tissues. Plant Cell Physiol 31:69–75

Raineri DM, Bottino P, Gordon MP, Nester EW (1990) *Agrobacterium*-mediated transformation of rice (*Oryza sativa* L.). Bio/Technology 8:33–38

Reuther G (1984) Asparagus. In: Evans DA, Sharp WR, Ammirato PV, Yamada Y (eds) Handbook of plant cell culture vol 2. Macmillan, New York, pp 211–242

Rick CM, Hanna GC (1943) Determination of sex in *Asparagus officinalis* L. Am J Bot 30: 711–714

Saito T, Nishizawa S, Nishimura S (1991) Improved culture conditions for somatic embryogenesis from *Asparagus officinalis* L. using an aseptic ventilative filter. Plant Cell Rep 10:230–234

Sambrook J, Fritsh EF, Maniatis T (eds) (1989) Molecular cloning: a laboratory manual. Laboratory Press, Cold Spring Harbor

Schafër W, Görz A, Kahl G (1987) T-DNA integration and expression in a monocot crop plant after induction of *Agrobacterium*. Nature 327:529–532

Sivan A, Chet I (1989) Cell wall composition of *Fusarium oxysporum*. Soil Biol Biochem 21:869–871

Thévenin L (1968) Les Problèmes d'amélioration chez *Asparagus officinalis* L. II/Haploïdie et amélioration. Ann Amélior Plantes 18:327–366

Vancanneyt G, Schmidt R, O'Connor-Sanchez A, Willmitzer L, Rocha-Sosa M (1990) Construction of an intron-containing marker gene: splicing of the intron in transgenic plants and its use in monitoring early events in *Agrobacterium*-mediated plant transformation. Mol Gen Genet 220:245–250

Warner SAJ, Scott R, Draper J (1993) Isolation of an asparagus intracellular PR gene (AoPR1) wound-responsive promoter by the inverse polymerase chain reaction and its characterization in transgenic tobacco. Plant J 3:191–201

Weising K, Schell J, Kahl G (1988) Foreign genes in plants: transfer, structure, expression and applications. Annu Rev Genet 22:421–477

5 Transgenic Beetroot (*Beta vulgaris*)

M. KINO-OKA and S. TONE

1 Introduction

Beta vulgaris L. (red beet) is a vegetable grown in temperate and subfrigid zones. The tuberous root of this plant is well known as the material for food-stuff in Europe. The root has received attention because of its high content of water-soluble pigments, which belong to the betalain family. They consist of red-violet betacyanin (mainly betanin) and yellow betaxathin (mainly vulgax-anthin-I) (Fig. 1). Since the root powder of *B. vulgaris* is permitted as a food colorant under the 1960 Color Additive Amendment, it has been used in the food industries as a natural color additive for some food processes (von Elbe et al. 1974b).

B. *vulgaris* is distributed in North America, Russia, Europe, Israel, Sudan, Kenya, Morocco and South America and is harvested twice a year in some regions. The root of *B. vulgaris* L. var. *rubra* in particular is used as a food colorant (von Elbe and Maing 1973) in the manufacture of sausage (von Elbe et al. 1974a), yogurt, ice cream, and sherbet (Pasch et al. 1975). The demand for the *B. vulgaris* pigment in the world is huge (for example in Japan; 130–230 × 10^6 tonnes of *B. vulgaris* powder per year).

Plant hairy roots transformed by *Agrobacterium rhizogenes*, which involves the integration of a root-inducing plasmid (Ri) into the plant cell genome have become of great interest as material for the production of plant-derived metabolites (Hamill et al. 1987). Compared with callus cells (Table 1), the hairy roots generally exhibit active propagation in a phytohormone-free medium and have a metabolite content comparable to the original plant roots (Bhadra et al. 1993). The metabolite production is stable after repeated sub-cultures because of the high genetic stability. In terms of bioreactor design for high cell density culture, hairy roots can be regarded as immobilized cells with a highly branched morphology.

The hairy roots of *B. vulgaris* were used for several purposes (Table 2). The kinetics of growth and pigment formation (Hamill et al. 1986) and the mechanism of pigment release (Taya et al. 1992; DiIorio et al. 1993) were inves-tigated, advancing both biological interests and engineering developments such as metabolic pathway (Hempel and Böhm 1997) and bioreactor design (DiIorio et al. 1992a), respectively.

Department of Chemical Science and Engineering, Osaka University, Osaka 560–8531, Japan

Biotechnology in Agriculture and Forestry, Vol. 47
Transgenic Crops II (ed. by Y.P.S. Bajaj)
© Springer-Verlag Berlin Heidelberg 2001

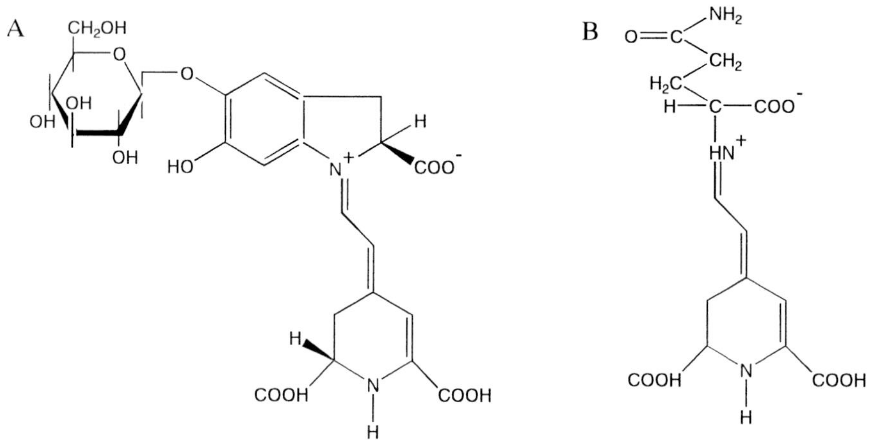

Fig. 1. Structures of pigments in tuberous root in original plant of *B. vulgaris*. **A** Betanin, **B** Vulgaxanthin-I

Table 1. Comparison of some characteristics between plant hairy roots and callus cells

Biochemical or biotechnological characteristics	Hairy roots	Callus cells
Genetic property	Genetically homogenous	Genetically heterogenous
Product formation	Mostly comparable to original plant	Often lower than original plant
Medium	Simple, without phytohormone	Complex, with phytohormone
Cell morphology	Structured, branched, fibrous	In particles or aggregates
Specific growth rate	0.1 to 0.3 day^{-1}	0.05 to 1 day^{-1}
High cell density culture	Relatively easy to maintain	Difficult to meet O_2 demand
Biomass handling	Not easily pumped	Pumpable

2 Genetic Transformation

2.1 Protocol

Hairy roots in *B. vulgaris* can be induced by different procedures. One method is direct infection, in which a colony of *A. rhizogenes* is cultured on slanting agar medium and then is pasted onto the plant specimens (especially stems and roots). The specimens are then incubated for the induction of hairy root. Another method is the co-culture method in which the plant seed is incubated in suspension with *A. rhizogenes* until roots are germinated (for several days). The most common procedure is the leaf disk method (Tanaka et al. 1985) in which the plant specimens (especially leaves) and *A. rhizogenes* cells are sus-

Table 2. Summary of various studies on hairy roots of *B. vulgaris*

Reference	Explant	*A. rhizogenes* strain	Infection method	Interest
Hamill et al. (1986)	Shoot	LBA9402	Direct infection	Kinetics of growth and pigment formation
Paul et al. (1990)	Seeds	LBA9402	Co-culture	Resistance to nematode
Taya et al. (1992)	Leaf	A4	Leaf disk	Pigment release
DiIorio et al. (1992a)	–	15834	–	Bioreactor design
Ramakrishnan and Curtis (1994)	–	–	–	Root morphology
Xing et al. (1996)	Seeds	*A. tumefaciens* C58C1 harboring rhizogenic plasmid (pRiA4b)	Direct infection	Gene expression of phytohormone
Hempel and Böhm (1997)	–	–	–	Metabolic pathway
Hammouri et al. (1998)	Leaf	–	Direct infection	Sugar consumption

pended together in a medium for a certain period (less than 1 h) and the specimens are then incubated on agar medium with antibiotics to remove *A. rhizogenes*.

In our laboratory (Taya et al. 1992), the leaf disk method was selected for the induction of hairy roots. Seeds of *B. vulgaris* L. cv. Detroit dark red were purchased from a local market. After sowing and growth outdoors for about 2 months, suitable specimens of the plant (leaves, petioles and roots) were subjected to hairy root induction. Hairy roots were induced by inoculating the plant specimens with *A. rhizogenes*. An original plant harvested from outdoors was sterilized by immersing it in 75% EtOH and 5% hypochlorite for 20 s and 10 min, respectively. After the sterilization with hypochlorite, further manipulations were done under aseptic conditions. After rinsing three times with sterilized water, the specimens were prepared by cutting the plant with a cork borer and a scalpel. The specimens were suspended for 10 min with *A. rhizogenes* A4 (10^8 cells/ml). After removal of remaining droplets on specimens, they were incubated at 25 °C for 3–5 days on agar plates without any nutrients (10 g/l agar). The specimens were transferred to an agar plate with MS medium (Murashige and Skoog 1962) containing 30 g/l sucrose, 10 g/l agar and antibiotics (500 mg/l carbenicillin and 200 mg/l vancomycin) for 7–20 days. After the emergence of the adventitious roots from the specimens, the roots were cut and put on new agar plates with MS medium containing sucrose, agar and antibiotics for 7 days. After the removal of *A. rhizogenes*, the adventitious roots were transferred to MS agar medium without antibiotics.

Axenic hairy roots were maintained on MS medium containing 30 g/l sucrose and 10 g/l agar by subculturing every month. Preculture of hairy roots was conducted at 25 °C in the dark using a 200 ml Erlenmeyer flask with 80 ml of liquid MS medium containing 20 g/l sucrose. The flask was shaken at 100 rpm on a rotary shaker.

2.2 Results and Discussion

2.2.1 Characterization of Hairy Roots and Pigments Produced

By infecting the *B. vulgaris* specimens (leaves, petioles, and roots) with *A. rhizogenes*, many adventitious roots appeared, mainly at the vein cut-ends of leaf sections (Taya et al. 1992). After repeated transfers and elongation of the root tips on MS agar medium, some axenic root clones were established. Figure 2 shows one of the clones cultured in liquid MS medium. The root clone of no. 5 exhibited more active proliferation, with frequent lateral branching, than the other clones. The difference in growth rates among clones may be attributable to the positional effects and/or copy number of the T-DNA region integrated into the plant genomes. In the shake culture with liquid MS medium, the hairy root clone of no. 5 showed active growth although the pigment contents in the cells were almost the same among the clones. Therefore, the root clone of no. 5 was selected for subsequent work. This clone contained agropine and mannopine, which are synthesized as unique opines (amino acid derivatives) in transformants by enzymes encoded on the T-DNA of the Ri plasmid (Tanaka 1990). Thus, this clone was concluded to be a transformant induced by the plasmid. The hairy roots exhibited active growth in liquid MS medium without phytohormones, even after successive subcultures.

To analyze the kinetics of growth and pigment formation, the cultures of the hairy roots in MS medium containing 20 kg/m³ sucrose started with the inoculum of 0.2 g dry cell weight (DW)/l at 25 °C. Figure 3 shows the changes in the concentration of hairy roots and content of betanin pigment during the

Fig. 2. *B. vulgaris* hairy roots (clone no. 5)

1 cm

Fig. 3. Profiles of growth and betanin content in culture of *B. vulgaris* hairy roots. The culture was in MS medium with 20 g/l sucrose

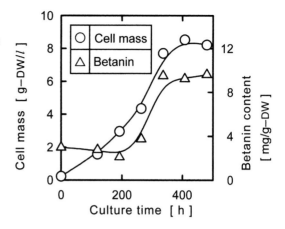

Table 3. Comparison of pigment contents between hairy roots and original plant

Plant material	Pigment content [mg/g DW]	
	Betanin	Vulgaxanthin-I
Hairy root		
Clone no. 5 (this work)	2.1–9.6	3.9–15.1
Clone reported by Hamill et al.	7.6[a]	14.1[a]
Original plant		
Leaf[b]	4.4	3.8
Petiole[b]	6.3	3.5
Root[b]	5.5	2.9

[a] Maximum values calculated from reference (Hamill et al. 1986).
[b] Data from reference (Taya et al. 1992).

culture. At the beginning of the culture, the hairy roots were grown without formation of betanin. The betanin formation occurred during the active growth after 200 h of culture time. As shown in Table 3, the pigment contents of betanin and vulgaxanthin-I were compared between *B. vulgaris* hairy roots and the original plant. The contents in the hairy root cells cultured for 20–30 days were approximately comparable to those in the original plant, although they varied somewhat depending on culture time. Hamill et al. (1986) induced the hairy roots of *B. vulgaris* by means of a leaf disk method and reported that the pigment contents in *B. vulgaris* hairy roots were comparable to those of original plant. The pigment contents of the clone no. 5 induced in the present study were similar to those of the clone induced by Hamill et al.

The pigments produced by the hairy root culture were characterized. Figure 4 shows HPLC profiles of the pigments in the hairy roots and those in the original *B. vulgaris* roots. The chromatographic pattern of the pigments in hairy roots was substantially identical to those extracted from the original roots. Here, the main peak of no. 1 shown in Fig. 4 was identified as betanin, although the other peaks were unknown. Table 4 shows the Hunter color properties of pigments using a color difference meter. From this analysis, distinc-

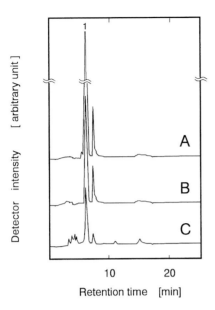

Fig. 4. HPLC profiles of pigments from hairy roots and original plant of *B. vulgaris*. *A* Extract from original plant (root). *B* Extract from hairy roots. *C* Culture broth with released pigments. *Peak no. 1* betanin. Column; Nucleosil 5C18 (4.6φ × 250 mm, Macherey and Nagel Co., FRG), eluent; 18% (v/v) methanol in 1 mM KH$_2$PO$_4$ (pH = 2.5) and detection; absorbance at 530 nm

Table 4. Hunter color properties of pigments from the original plant and hairy root of *B. vulgaris* (Taya et al. 1992). L_H lightness (0 = black and 100 = white), a_H redness, b_H yellowness, and $tan^{-1}(b_H/a_H)$ color tone

Pigment source	L_H	a_H	b_H	$tan^{-1}(b_H/a_H)$
Extract from original plant (root)	68.7	36.1	12.0	18.4
Extract from hairy root cells	65.1	32.7	11.9	20.0

tive differences were not found between the pigments in the extracts from hairy roots and original roots in terms of the values of lightness (L_H), redness (a_H), yellowness (b_H), and color tone ($tan^{-1}(b_H/a_H)$). Therefore, the pigments in extracts from the hairy roots were regarded to be substantially the same as the original ones.

2.2.2 Effect of Medium Components on Growth and Pigment Formation in Hairy Root Cultures

Hairy roots were cultivated using MS medium containing various sugars for the choice of carbon source suitable for growth. As shown in Table 5, the values of cell mass were relatively high in the cultures using sucrose and fructose as carbon sources. Therefore, fructose was used as a carbon source for the following cultures of hairy roots.

Table 5. Effects of carbon sources on cell growth and pigment formation in the culture of *B. vulgaris* hairy roots. The cultures were carried out for 290h in Erlenmeyer flasks containing the MS medium with 20g/l sugar

Carbon source	Cell mass [g DW/l]	Betanin content [mg/g DW]
Glucose	3.2	3.5
Fructose	9.1	7.5
Xylose	0.4	–
Galactose	0.4	–
Maltose	1.8	2.3
Lactose	1.3	3.8
Sucrose	6.4	9.9
No sugar	0.4	–

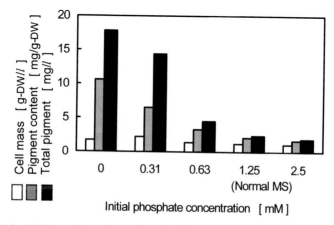

Fig. 5. Growth and pigment formation of *B. vulgaris* hairy roots cultivated in media with various initial concentrations of phosphate (Taya et al. 1994). The hairy roots were cultured in modified MS medium containing 20 g/l fructose and 0, 0.31, 0.63, 1.25 and 2.5 mM phosphate

When hairy roots were cultivated in a diluted MS medium containing all the MS medium constituents at one-eighth strength, the pigment content in the hairy roots was found to increase significantly in spite of the suppressed growth due to the possible deficiency of some nutrients. This finding suggested that the concentration(s) of certain constituent(s) in MS medium may have a negative effect on pigment formation by the hairy roots. Through the investigation of the effects of individual constituents in MS medium on the hairy root culture, it was found that phosphate was a key nutrient responsible for decreasing pigment accumulation in *B. vulgaris* hairy roots (Taya et al. 1994). The effect of the initial phosphate level in the medium on the hairy root culture was examined in detail. Figure 5 presents the results after 240h in the hairy root cultures using the media containing 0–2.5 mM phosphate. The concentrations of the other nutrients were the same as those of normal MS medium. Higher values of content and total amount of betanin were obtained at lower phosphate concentrations of 0–0.25 mM, and the highest values were achieved in the phosphate-free medium. It was also observed that root growth did not decline in the phosphate-free medium. The amount of cell mass was 1.7 g DW/l in the phosphate-free medium, which was larger than the value of 1.2 g DW/l in the normal MS medium with 1.25 mM phosphate. The exclusion of phosphate from the medium resulted in 5- and 7.4-fold increases in the values of content and total amount of betanin, respectively, compared with the control culture at the 1.25 mM phosphate level.

In higher plant cells, it is generally recognized that the cells store phosphorus in vacuoles or other locations in the form of polyphosphate, phytic acid etc., and that the availability of stored phosphorus compounds supports the growth of plant cells even under phosphate-deficient culture conditions (Bieleski 1973). Based on the assay of phosphorus in the hairy roots used as inocula in the present study, the phosphorus content was about 0.52 mmol of phosphorus equivalent per g DW. This value was larger than the phosphorus contents of cultured plant cells in stationary phases (ca. 0.1 mmol of phosphorus equivalent per g DW) as reported by Curtis et al. (1991). It is presumed that *B. vulgaris* hairy roots grown in phosphate-free medium utilize phosphorus compounds stored in the roots.

Although the metabolic role of phosphate (or phosphorus compounds) in plant cells is still unclear, there have been several studies on the enhanced production of secondary metabolites by callus tissues cultivated in phosphate-limited media; e.g., the formation of alkaloids and phenolics by *Catharanthus roseus* cells, L-3,4-dihydroxyphenylalanine by *Stizolobium hassjoo* cells, harmalol and harmine by *Peganum harmala* cells, and anthocyanins by *Vitis* cells (Dougall 1980, Rokem and Goldberg 1985). In hairy root culture, Dunlop and Curtis (1991) reported that in *H. muticus* there was a 28-fold higher content of sequiterpene in cultures with a phosphate-free medium, compared with control cultures using media with phosphate levels of 1.1 mM.

From these results, the use of the phosphate-free Ms medium was proposed for the production of pigment in the culture of *B. vulgaris* hairy roots in which phosphorus was accumulated in advance in the preculture using the normal MS medium. The kinetic model of growth and pigment production for hairy roots has been described in our previous work (Kino-oka et al. 1995b).

2.2.3 Pigment Production in Hairy Roots Associated
with Repeated Processes of Cell Growth and Pigment Release

Kilby and Hunter (1990, 1991) reported that sonic exposure resulted in pigment release from *B. vulgaris* callus. DiIorio et al. (1993) tried to release the pigment by the combined treatment of *B. vulgaris* hairy roots with thermal and ionic stresses. In these methods, however, it seems likely that the products are contaminated with the additives and that the locational heterogeneity against the cell suspension in the bioreactor retards the effective release of the products.

It was reported that pigment release from roots of field-grown *B. vulgaris* occurred under anaerobic conditions due to loss of integrity of the tonoplast and the cell membrane, accompanying the irreversible decay of cells (Zhang et al. 1992). In a previous paper (Taya et al. 1992), it was found that a significant amount of pigment was released into the medium by intermittent oxygen supply to the hairy root cells in a shake culture. As shown in Fig. 4, the pigments released from the hairy roots were recognized to be the same as those in the extract from the original plant.

To recover the released pigments from the medium, a bioreactor was constructed as depicted in Fig. 6 (Kino-oka et al. 1992). The bioreactor consisted of two units for cell culture and pigment recovery. As a cell culture unit, a turbine-blade fermentor (Model TBR-2, Sakura Seiki Co., Tokyo) was used. The culture process for *B. vulgaris* hairy roots comprised of periods of growth and pigment release. The culture started with an inoculum of about 1 g DW/l at 25 °C. During the growth period, the dissolved oxygen (DO) concentration in the fermentor was kept at 5 ppm and the hairy roots were cultured with pigment formation in the cells. To prevent a lack of nutrients, sucrose solution (300 g/l) was occasionally added to the culture, and the culture broth was exchanged with fresh medium when necessary. The concentration of hairy roots was estimated by measuring the electric conductivity in the medium. During the pigment release period, including oxygen starvation, the operation was carried out as follows. The air supply was first stopped and then pure N_2 gas was introduced into the fermentor for an hour to decrease the DO level substantially to zero. After pigment release for 16h, the hairy root growth was resumed by supplying air into the fermentor. To recover the released pigments, the culture broth was circulated through an adsorbent column packed with

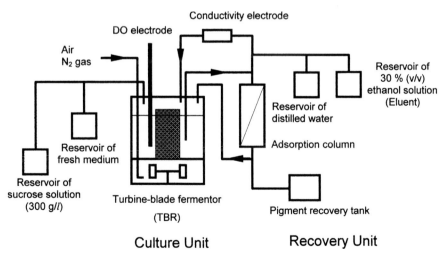

Fig. 6. Schematic diagram of bioreactor for culture of *B. vulgaris* hairy roots with pigment release. (Kino-oka et al. 1992)

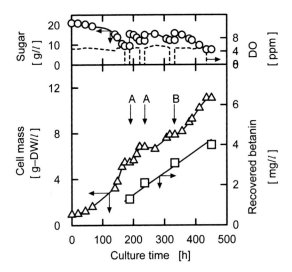

Fig. 7. Culture of *B. vulgaris* hairy roots accompanied by repeated operations of pigment release (Kino-oka et al. 1992). Culture was in MS medium with 20 g/l sucrose. *Arrows A* and *B* show the culture times of sucrose addition and medium exchange, respectively

styrene-divinyl benzene co-polymer resin. Two columns were connected to the fermentor and each one was used alternately for the in situ adsorption and desorption of the pigments. For the elution of pigments from the resin, 30% ethanol aqueous solution was introduced into the column, followed by washing with distilled water.

A typical time course of the hairy root culture is shown in Fig. 7. The hairy roots were grown initially for 172 h, and afterwards the operation of pigment recovery was repeated four times with pigment released by oxygen starvation for 16 h. The amount of betanin produced in each operation was 0.8–1.3 mg/l. In another experiment (Kino-oka and Tone 1996), this procedure was applied to the culture of hairy roots using the phosphate-free medium as described in Section 2.2.2 and enhanced production of the pigment was achieved.

2.2.4 Bioreactor for the Culture of Hairy Roots

Several workers reported that hairy roots could be successfully cultivated using fermentors in which hairy roots were anchored to supports and nutrients were supplied to them by means of medium flowing across them (Kondo et al. 1989; Hilton and Rhodes 1990; Rodriguez-Mendiola et al. 1992; Curtis 1993; McKelvey et al. 1993; Ramakrishnan and Curtis 1994). For *B. vulgaris* hairy roots, cultures were examined from a bioreactor where the nutrients were fed by means of supplying a mist of the medium (DiIorio et al. 1992b), and from a column-type reactor where medium was circulated through a packed bed of hairy root cells (Kino-oka et al. 1995a, 1996). In both types of bioreactors, it was found that the oxygen supply was important to support the active growth of the hairy roots.

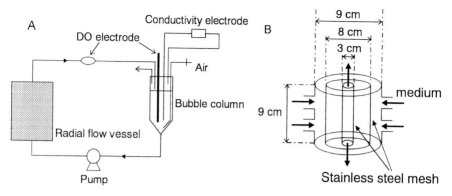

Fig. 8A–B. Schematic diagram of radial flow reactor (RFR) for hairy root culture (Tone et al. 1997). **A** Total RFR system, **B** radial flow vessel in RFR

As shown in Fig. 8A, a radial flow reactor (RFR) was constructed for the cultivation of hairy roots (Tone et al. 1997). The RFR consisted of a radial flow vessel for hairy root growth and a bubble column for aeration. Between these units, the medium was circulated using a tube pump. The total volume of medium in this culture system was 1.5 l. In the radial flow vessel (Fig. 8B), the volume of medium was 0.6 l and the hairy roots were inoculated in the compartment between two cylindrical mesh partitions. To minimize the variation in fluid flow rate through the compartment, the medium was made to flow from four ports on the side of the outer cylinder to its center through the compartment. During hairy root growth, DO concentration in the bubble column was controlled at 7.5 ppm by changing the air flow rate in the range of 0.03–0.4 l/min.

Figure 9 shows the time course of the culture of *B. vulgaris* hairy roots with MS medium containing 20 g/l fructose in the RFR. For reference, results from culture in an Erlenmeyer flask (FL) are also shown in Fig. 12. When the concentration of cell mass reached 5 or 10 g DW/l, the culture broth was exchanged with fresh medium in these cultures. In the culture with the RFR, active growth occurred in a manner of linear growth after about 100 h and the growth rate did not decline even at a high density of root mass over 10 g DW/l. At 280 h, a cell mass of 13.1 g DW/l was obtained in the culture with the RFR, the value of which was 2.2 times larger than that in the culture with the FL. Thus it was verified that successful culture of *B. vulgaris* hairy roots could be obtained using the RFR.

3 Summary and Conclusions

The hairy roots were induced by inoculating B. *vulgaris* (leaves, petioles and roots) with *A. rhizogenes* A4 by means of the leaf disk method. Several hairy

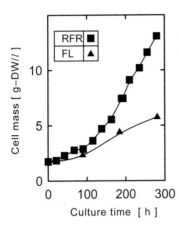

Fig. 9. Growth profiles of *B. vulgaris* hairy roots in cultures in radial flow reactor (RFR) and Erlenmyer flask (FL) (Tone et al. 1997). Cultures were in MS medium with 20 g/l fructose

root clones, which included unique opines in transformants, were obtained at vein cut-ends of leaf sections. The hairy roots exhibited active growth in conventional liquid medium without phytohormones. The pigment contents of hairy roots were comparable to those found in various parts (roots, stems and leaves) of the original plants. From analyses using HPLC and a Hunter color meter, the pigments in the hairy roots were confirmed to be substantially identical to those extracted from the original plant. Through the improvement of the culture medium, it was demonstrated that the phosphate level was important for enhancement of pigment productivity and that high levels of pigment were achieved in the phosphate-free medium.

The culture of hairy roots using repeated operations of growth and pigment release was performed. In the culture with oxygen starvation for a 16-h period, the extracellular production rate of pigment was 1.1×10^{-5} g/(l·h), on average. Taking into account oxygen transfer between the medium and hairy roots, the radial flow reactor (RFR) with high efficiency of oxygen supply was developed. The RFR facilitated high density culture of hairy roots without an appreciable reduction in the growth rate.

The studies described in this chapter, including induction and selection of transformants, improvement of the culture medium, development of a culture operation for metabolite release, and construction of a bioreactor for high density culture can offer a common strategy for the production of plant-derived metabolites using plant hairy roots.

References

Bhadra R, Vani SS, Shanks JV (1993) Production of indole alkaloids by selected hairy root lines of *Catharanthus roseus*. Biotechnol Bioeng 41:581–592

Bieleski RL (1973) Phosphate pools, phosphate transport, and phosphate availability. Annu Rev Plant Physiol 24:225–252

Brodelius P (1988) Permeabilization of plant cells for release of intracellularly stored products: viability studies. Appl Microbiol Biotechnol 27:561–566

Curtis WR (1993) Cultivation of roots in bioreactors. Curr Opin Biotechnol 4:205–210

Curtis WR, Hasegawa PM, Emery AH (1991) Modeling linear and variable growth in phosphate limited suspension cultures of opium poppy. Biotechnol Bioeng 38:371–379

DiIorio AA, Cheetham RD, Weathers PJ (1992a) Carbon dioxide improves the growth of hairy roots cultured on solid medium and in nutrient mists. Appl Microbiol Biotechnol 37:463–467

DiIorio AA, Cheetham RD, Weathers PJ (1992b) Growth of transformed roots in a nutrient mist bioreactor: reactor performance and evaluation. Appl Microbiol Biotechnol 37:457–462

DiIorio AA, Weathers PJ, Cheetham RD (1993) Non-lethal secondary product release from transformed root cultures of *Beta vulgaris*. Appl Microbiol Biotechnol 39:174–180

Dougall DK (1980) Nutrition and metabolism. In: Staba EJ (ed) Plant tissue culture as a source of biochemicals. CRC Press, Boca Raton, pp 21–58

Dunlop DS, Curtis WR (1991) Synergistic response of plant hairy-root cultures to phosphate limitation and fungal elicitation. Biotechnol Prog 7:434–438

Hamill JD, Parr AJ, Robins RJ, Rhodes MJC (1986) Secondary product formation by cultures of *Beta vulgaris* and *Nicotiana rustica* transformed with *Agrobacterium rhizogenes*. Plant Cell Rep 5:111–114

Hamill JD, Parr AJ, Rhodes MJC, Robins RJ, Walton NJ (1987) New routes to plant secondary products. Bio/Technology 5:800–804

Hammouri MK, Ereifej KI, Shibli RA (1998) Quantitive analysis of fructose fate in a plant fermentation system. J Agric Food Chem 46:1428–1432

Hempel J, Böhm H (1997) Betaxanthin pattern of hairy roots from *Beta vulgaris* var. *lutea* and its alteration by feeding of amino acid. Phytochemistry 44:847–852

Hilton MG, Rhodes MJC (1990) Growth and hyoscyamine production of "hairy root" cultures of *Datura stramonium* in a modified stirred tank reactor. Appl Microbiol Biotechnol 33:132–138

Hunter CS, Kilby NJ (1990) Betanin production and release in vitro from suspension culture of *Beta vulgaris*. Methods Mol Biol 6:545–554

Kilby NJ, Hunter CS (1990) Repeated harvest of vacuole-located secondary product from in vitro grown plant cells using 1.02 MHz ultrasound. Appl Microbiol Biotechnol 33:448–451

Kilby NJ, Hunter CS (1991) Towards optimization of the use of 1.02 MHz ultrasound to harvest vacuole-located secondary product from in vitro grown plant cells. Appl Microbiol Biotechnol 34:478–480

Kino-oka M, Tone S (1996) Extracellular production of pigment from red beet hairy roots accompanied by oxygen starvation. J Chem Eng Jpn 29:488–493

Kino-oka M, Hongo Y, Taya M, Tone S (1992) Culture of red beet hairy root in bioreactor and recovery of pigment released from the cells by repeated treatment of oxygen starvation. J Chem Eng Jpn 25:490–495

Kino-oka M, Taya M, Tone S (1995a) Culture of red beet hairy roots in a column-type reactor associated with pigment release. Plant Tissue Cult Lett 12:201–204

Kino-oka M, Taya M, Tone S (1995b) Kinetic expression for pigment production in culture of red beet hairy roots. J Chem Eng Jpn 28:772–778

Kino-oka M, Tsutsumi S, Tone S (1996) Oxygen transfer in bioreactor for culture of plant hairy roots. J Chem Eng Jpn 29:531–534

Kondo O, Honda H, Taya M, Kobayashi T (1989) Comparison of growth properties of carrot hairy root in various bioreactors. Appl Microbiol Bioeng 32:291–294

Mano Y (1989) Variation among hairy root clones and its application. Plant Tissue Cult Lett 6:1–9

McKelvey SA, Gehrig JA, Hollar KA, Curtis WR (1993) Growth of plant root cultures in liquid- and gas-dispersed reactor environments. Biotechnol Prog 9:317–322

Murashige T, Skoog F (1962) A revised medium for rapid growth and bioassays with tobacco tissue cultures. Physiol Plant 15:473–497

Nakajima H, Sonomoto K, Morikawa H, Sato F, Ichimura K, Yamada Y, Tanaka A (1986) Entrapment of *Lavandula vera* cells with synthetic resin prepolymers and its application to pigment production. Appl Microbiol Biotechnol 24:266–270

Pasch JH, von Elbe JH, Sell RJ (1975) Betalains as colorants in daily products. J Milk Food Technol 38:25–28

Paul H, van Deelen JEM, Henken B (1990) Expression in vitro of resistance of *Heterodera schachtii* in hairy roots of an alien monotelosomic addition plant of *Beta vulgaris*, transformed by *Agrobacterium rhizogenes*. Euphytica 48:153–157

Ramakrishnan D, Curtis WR (1994) Fluid dynamics studies in plant root cultures for application to bioreactor design. In: Ryu DD, Furusaki S (eds) Advances in plant biotechnology. Elsevier, Amsterdam, pp 281–305

Ramakrishnan D, Curtis WR (1995) Evaluated meristematic respiration in plant root cultures: implications to reactor design. J Chem Eng Jpn 28:491–493

Rodriguez-Mendiola MA, Stafford A, Cresswell R, Arias-Castro C (1992) Bioreactors for growth of plant roots. Enzyme Microb Technol 13:697–702

Rokem JS, Goldberg I (1985) Secondary metabolites from plant cell suspension cultures. In: Mizrahi A, Wezel AL (eds) Methods for yield improvement. Advances in biotechnological processes 4. Alan R Liss, New York, pp 241–274

Tabata M, Fujita Y (1985) Production of shikonin by plant cell cultures. In: Zaitln M, Day P, Hollaender A (eds) Biotechnology in plant science. Academic Press, New York, pp 207–218

Tanaka N (1990) Detection of opines by paper electrophoresis. Plant Tissue Cult Lett 7:45–47

Tanaka N, Hayakawa M, Mano Y, Ohkawa H, Matsui C (1985) Infection of turnip and radish storage roots with *Agrobacterium rhizogenes*. Plant Cell Rep 4:74–77

Taya M, Mine K, Kino-oka M, Tone S, Ichi T (1992) Production and release of pigment by culture of transformed hairy root of red beet. J Ferment Bioeng 73:31–36

Taya M, Yakura K, Kino-oka M, Tone S (1994) Influence of medium constituents on enhancement of red beet hairy roots. J Ferment Bioeng 77:215–217

Tone S, Kino-oka M, Hitaka Y, Taya M (1997) High-density culture of hairy roots with a radial flow reactor. In: Nienow AW (ed) Bioreactor and bioprocess fluid dynamics. Mechanical Engineering Publications, London, pp 115–125

Uozumi N, Kato Y, Nakashimada Y, Kobayashi T (1992) Excretion of peroxidase from horseradish hairy root in combination with ion supplementation. Appl Microbiol Biotechnol 37:560–565

von Elbe JH, Maing IY (1973) Betalains as possible food colorants of meat substitutes. Cereal Sci Today 18:263–265

von Elbe JH, Klement JT, Amundson CH, Cassens RG, Lindsav RC (1974a) Evaluation of betalain pigments as sausage colorants. J Food Sci 39:128–132

von Elbe JH, Maing IY, Amundson CH (1974b) Color stability of betanin. J Food Sci 39:334–337

Xing T, Zhang D-Y, Hall JF, Blumwald E (1996) Auxin level and auxin binding protein availability in rol B transformed *Beta vulgaris* cells. Biol Plant 38:351–362

Zhang Q, Lauchli A, Greenway H (1992) Effect of anoxia on solute loss from beetroot storage tissue. J Exp Bot 43:897–905

6 Transgenic Broccoli (*Brassica oleracea* Var. *italica*) and *Cabbage* (Var. *capitata*)

T.D. METZ

1 Introduction

1.1 Distribution and Importance

Broccoli (*Brassica oleracea* var. *italica*) and cabbage (*Brassica oleracea* var. *capitata*) are two of the many economically important vegetable crops of the *B. oleracea* species. Broccoli is grown for its large, edible, very young inflorescences and has two different distinct forms. One variation is the single-headed type, known as "calabrese" in Great Britain. It differs from cauliflower (*B. oleracea* var. *botrytis*), consisting of clusters of fully differentiated green to purple flower buds less densely arranged in one large main head on a short central stalk. The other form of broccoli, "sprouting broccoli," is multibranched, bearing many relatively small flowerheads with green, purple or white flower buds. Cabbage forms include white, red and Savoy heading varieties. The edible part of the plant consists of a mass of thick overlapping leaves that tend to form a head around a central growing point. White and red varieties are known as hard-heading cabbage with smooth, flattened leaves forming a compact head. Savoy type cabbages have loose heads of curled and puckered leaves.

Broccoli has become popular both as a fresh and as a frozen vegetable, particularly in the United States and Europe. Worldwide production of broccoli is estimated at nearly 45000 ha (Siemonsma and Piluek 1993). The United States is the largest producer of sprouting broccoli, with production of 26000 ha at a value of $278 million (USDA 1994). Cabbage is grown in large quantities in North America, Europe and Asia. It is considered to be one of the most important vegetable crops of northern Europe. Worldwide production of cabbage exceeds 800000 ha (Siemonsma and Piluek 1993). Broccoli is highly nutritious, containing high amount of carotene and calcium, as well as vitamins C, A and B_2. White cabbage also has good nutritional value, but less so than leafy green vegetables. Both broccoli and cabbage, as well as other *Brassica* vegetables have been increasing in popularity with the health-conscious, as they contain compounds with anti-cancer properties (Beecher 1994).

Biology Department, Campbell University, P.O. Box 308 Buies Creek, North Carolina 27506, USA

Biotechnology in Agriculture and Forestry, Vol. 47
Transgenic Crops II (ed. by Y.P.S. Bajaj)
© Springer-Verlag Berlin Heidelberg 2001

1.2 Applications of Genetic Transformation

Broccoli and cabbage are both susceptible to several insect pests and diseases. Insect pests of agricultural importance include the cutworm (*Spodoptera littoralis*), cabbage looper (*Trichoplusia ni*), cabbage aphid (*Brevicoryne brassicae*), cabbage fly (*Delia brassicae*), imported cabbage worm (*Pieris rapae*), cabbage butterfly (*Pieris canidia*), cabbage moth (*Crocidolomia binotalis*) and the diamondback moth (*Plutella xylostella*). The diamondback moth is considered to be the major pest of the crucifers worldwide, and has become resistant to all major categories of insecticides (Tabashnik et al. 1991).

Important diseases include bacterial black rot (*Xanthomonas campestris* pv. *campestris*) and soft rot (*Erwinia carotovora*), downy mildew (*Peronospora parasitica*), clubroot (*Plasmodiophora brassicae*), *Alternaria* blight (A. *brassicae*), cabbage yellows (*Fusarium oxysporum*) and cauliflower and turnip mosaic virus. While some cultivars are available with resistance to some of the diseases listed above, the majority of present day cultivars are still susceptible (Siemonsma and Piluek 1993).

Genetic engineering could be used to develop insect and disease resistant varieties utilizing genes that cannot be transferred through traditional plant breeding techniques. The cloning of genes expressing the insecticidal proteins of *Bacillus thuringiensis* (Höfte and Whiteley 1989) has allowed development of transgene strategies for incorporation of resistance to a variety of lepidopteran and coleopteran insect pests in many crop plants (Fischhoff et al. 1987; Perlak et al. 1990; Bottrell et al. 1992; Fujimoto et al. 1993; Fromm et al. 1994). Resistance to viral infection has also been demonstrated in transgenic plants containing viral genes (Lomonossof 1995), including coat protein genes (Beachy et al. 1990) and more recently, viral movement genes (Cooper et al. 1995; Tacke et al. 1996).

Other traits of interest include delay of post-harvest senescence and improved quality and yield. Post-harvest senescence is quite rapid in broccoli and leads to a deterioration of crop quality. Should ethylene play a role in the post-harvest senescence of broccoli, as has been suggested (Pogson et al. 1995a), antisense strategies designed to block the production of ethylene might be used to delay the onset of senescence, in a similar manner to that demonstrated in tomato (Hamilton et al. 1990; Oeller et al. 1991) and cantaloupe (Ayub et al. 1996). The recent identification and cloning of flower development genes within *B. oleracea* (Pogson et al. 1995b) may lead to the development of higher yielding and higher quality varieties.

2 Genetic Transformation

2.1 Literature Review

An excellent review of the status of genetic transformation research in *Brassica oleracea* has recently been published (Puddephat et al. 1996). Table 1

Table 1. Summary of various studies on genetic transformation of broccoli and cabbage

Reference	Explant/culture type used	Vector/method used	Observations/remarks
Broccoli			
Hosoki et al. (1991)	Leaf/hairy root	Ri plasmid/ *A. rhizogenes*	Plants expressing mannopine recovered from two root clones
Toriyama et al. (1991)	Peduncles	Binary/ *A. tumefaciens*	7 transgenic plants recovered; express hygromycin phosphotransferase (*hpt*) gene and S-locus gene
Lee and Keller (1993)	Cotyledonary petioles	Binary/ *A. tumefaciens*	9% of explants produced plants expressing β-glucuronidase (*gus*), *hpt*, and phosphinothricin acetyltransferase (*pat*)
Metz et al. (1995)	Hypocotyls, cot. petioles, peduncles	Binary/ *A. tumefaciens*	2–10% of explants regenerated transformed plants; 144 plants express a *Bacillus thuringiensis* cryIA(c) gene
Christey et al. (1997)	Leaf, cot. petioles/ hairy root	Binary/ *A. rhizogenes*	1–8% transformation efficiency; transfer of *B. thuringiensis* cry1a(c), *gus*, and neomycin phosphotransferase (*npt*) II
Cao et al. (1999)	Hypocotyls, petioles	Binary/ *A. tumefaciens*	5% of explants regenerated transformed plants; 21 plants express a *Bacillus thuringiensis* Cry1C gene
Henzi et al. (1999)	Leaf, cot. petioles/ hairy root	Binary/ *A. rhizogenes*	0–2% transformation efficiency with cot. petioles; 17–35% efficiency with leaf explants; transfer of tomato antisense 1-aminocyclopropane-1-carboxylic acid oxidase and *nptII*
Cabbage			
Birot et al. (1987)	Hairy root	Ri plasmid/ *A. rhizogenes*	Oncogenic *A. rhizogenes* strain used; mannopine synthase gene transferred; numbers of plants obtained not reported
Jianping and Qiquan (1988)	Hypocotyls	Ti plasmid/ *A. tumefaciens*	No plants regenerated; nopaline detected in calli
Berthomieu and Jouanin (1992)	Leaf petioles/hairy root	Ri plasmid/ *A. rhizogenes*	Rapid cycling cabbage; 6 root clones regenerated transformed plants expressing either *hpt* or *nptII*
Berthomieu et al. (1994)		*A. tumefaciens*	Rapid cycling cabbage, most plants chimeric but expressed *gus* gene; progeny also expressed *gus*
Metz et al. (1995)	Hypocotyls, petioles	*A. tumefaciens*	10% of explants regenerated transformed plants; 92 plants express a *Bacillus thuringiensis* cryIA(c) gene
Christey et al. (1997)	Leaf, cot. petioles/ hairy root	Binary/ *A. rhizogenes*	18–33% transformation efficiency; transfer of *gus*, *npt* II, *B. thuringiensis* cry1a(c), antisense tomato ethylene (*efe*) genes

summarizes the transformation results achieved in broccoli and cabbage. While there have been reports of successful transformation of many of the *B. oleracea* vegetables, there have been few reports of successful recovery of phenotypically normal, genetically transformed broccoli and cabbage. Reliable, reproducible methods of transformation for most *Brassica oleracea* vegetables are not yet available. Lack of genetic uniformity within varieties and within the species, and difficulties in regeneration of plants under transforming conditions are among the problems that have been encountered (Puddephat et al. 1996).

2.1.1 Broccoli Transformation

Genetic transformation of broccoli has been achieved using both *Agrobacterium tumefaciens* and *A. rhizogenes*. Hosoki et al. (1991) utilized a mannopine strain of *A. rhizogenes* to induce hairy root cultures from inoculated leaf explants. Plants regenerated from two hairy root clones demonstrated the presence of mannopine. Toriyama et al. (1991) demonstrated genetic transformation of broccoli with *A. tumefaciens*. A modification of a method used to transform *B. napus* (Fry et al. 1987) involving the inoculation of flowering stalk (peduncle) explants from mature plants was used to transfer the *hpt* gene and an *S*-locus gene cloned from *B. campestris*. Seven diploid transgenic plants were recovered using Green Comet hybrid broccoli as the source material. The primary focus of this paper was the transfer and analysis of the *S*-locus gene. The transformation procedure or resulting efficiencies were not discussed. An abstract presented at a recent Crucifer Genetics Workshop (Lee and Keller 1993) reported broccoli transformation using *A. tumefaciens* to infect cotyledonary petioles. Approximately 9% of inoculated explants regenerated transgenic shoots. Genes transferred included GUS (β-glucuronidase), *neo* and *bar*. Metz et al. (1995) describe the use of *A. tumefaciens* to transfer a *Bacillus thuringiensis* insecticidal protein gene to broccoli, with the recovery of 144 insect-resistant plants. The details of this procedure will be discussed shortly. A binary vector approach using *A. rhizogenes* has been used to successfully transfer genes into broccoli varieties while reducing the undesirable morphological characteristics and fertility problems associated with transfer of the *rol* genes from the Ri plasmid (Christey et al. 1997). This binary *A. rhizogenes* approach has recently been improved and used to incorporate a tomato antisense 1-aminocyclopropane-1-carboxylic acid oxidase gene into two broccoli varieties with the goal of down-regulating ethylene biosynthesis to delay post-harvest senescence (Henzi et al. 1999).

2.1.2 Cabbage Transformation

As with broccoli, both *A. tumefaciens* and *A. rhizogenes* have been used in experiments to genetically transform cabbage. Much of the work has been conducted using rapid cycling varieties of cabbage. Birot et al. (1987) used oncogenic *A. rhizogenes* strains to transform cabbage. Transgenic plants were regenerated from hairy root cultures, but many displayed abnormal phenotypic characteristics typical of *A. rhizogenes* infection. The hairy root phenotype includes reduced apical dominance and fertility, shortened internodes, wrinkled leaves, late flowering, and plagiotropic roots (Tepfer 1989). No genes were transferred in these experiments other than the endogenous opine genes from the *Agrobacterium*. A rapid cycling cabbage line was transformed with both *hpt* and *neo* genes using a wild-type *A. rhizogenes* strain (Berthomieu and Jouanin 1992). Regenerated plants displayed a range of phenotypes including reduced or no apical dominance, wrinkled leaves, and varying degrees of male sterility. Christey et al. (1997) used a binary vector approach to transfer genes to several cabbage varieties using *A. rhizogenes*. Their approach was very successful, resulting in 18–33% of inoculated explants producing antibiotic resistant hairy roots that regenerated transgenic plants. Plants exhibited varying degrees of Ri-induced morphological changes.

Jianping and Qiquan (1988) produced transgenic cabbage callus cultures using an oncogenic strain of *A. tumefaciens* and seedling hypocotyl segments. Genes transferred were the endogenous opine production gene and the *neo* gene. The callus was maintained on medium containing kanamycin, but no transgenic plants were recovered. *A. tumefaciens*-mediated transformation of a rapid cycling cabbage has been achieved using a combination of disarmed and wild-type *Agrobacterium* strains (Berthomieu et al. 1994). Recovered transformants contained the transgene from the binary plasmid but not from the oncogenic T-DNA. Due to lack of selection imposed during regeneration, most recovered transformants proved to be chimeric, however, inheritance of the transgene sequences was observed in their progeny. *A. tumefaciens* has also been used to successfully transfer a *B. thuringiensis* insecticidal protein to a variety of head cabbage, with the recovery of 92 insect-resistant plants (Metz et al. 1995). Details of this work are given in this chapter.

2.2 Transformation Protocol

2.2.1 Plant Material and Vectors

Three field-grown cytoplasmic male-sterile broccoli lines, designated CMS nos. 1, 2, 3 and one male fertile broccoli line, Fertile no. 1, provided by Dr. Michael Dickson (New York State Agriculture Experiment Station, Geneva, NY) were used in the peduncle (flowering stalk) transformation, as was the commercial F_1 hybrid, Green Comet broccoli (W. Atlee Burpee and Co., USA). Green Comet broccoli and King Cole cabbage (Ferry Morse Seed Co., USA) were used in the seedling explant transformation. *Agrobacterium tumefaciens* strain ABI harboring the binary vector pMON10517-1 or pMON10837-1 containing the neomycin phosphotransferase gene (*neo*) under control of the CaMV 35S promoter and a *Bacillus thuringiensis* cryIA(c) gene was obtained from the Monsanto Company, St. Louis, MO, USA.

2.2.2 Plant Transformation

The peduncle transformation method (Fig. 1) was basically that of Toriyama (1992) with several modifications (Metz et al. 1995). Peduncle explants (5–7 mm) were excised from the top 5–10cm of surface-sterilized (5min 70% EtOH, 10min 1.58% NaOCl with 0.5% Tween20 (Sigma), three washes in sterile distilled water) flowering stalks which had buds and pedicels removed prior to sterilization. Explants were inoculated for 5–10min in a flask containing *Agrobacterium* grown 20–24h (OD_{600} = 1.6–1.8) in YEP medium and diluted 1:10 in liquid hormone-free Linsmaier-Skoog medium. Inoculated explants were placed horizontally on a sterile 7-cm Whatman no. 1 filter paper circle over 2ml of a 6- to 8-day-old tobacco suspension culture plated on

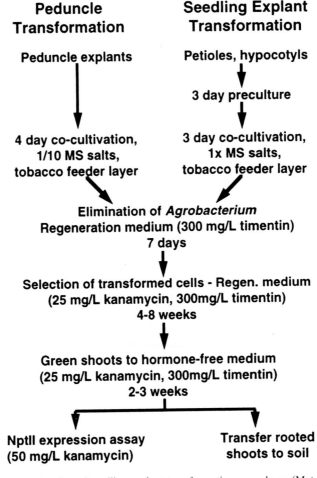

Fig. 1. Outline of peduncle and seedling explant transformation procedures. (Metz et al. 1995)

regeneration medium containing 1/10 concentration standard MS salts, B5 vitamins, 3% sucrose, 0.8% Phytagar (Gibco Laboratories), 2 mg/l benzyladenine, pH 5.7 for a 4 day co-cultivation in the dark at 25 °C. Petri dishes at all stages in the peduncle transformation were sealed with Micropore surgical tape (3M Corp.). Explants were transferred to regeneration medium as above except for full strength MS salts and 300 mg/l timentin [SmithKline Beecham Pharmaceuticals, 30:1 (w/w) ticarcillin: clavulanic acid formula] and cultured for 7 days to kill the *Agrobacterium,* followed by a transfer to the same medium with 25 mg/l kanamycin to select for transformed cells, both at 25 °C, 16 h photoperiod, $45–70 \mu E m^{-2} s^{-1}$. Explants were maintained on this medium for 4–6 weeks with transfer to new medium every 2 weeks. Regenerating green shoots were transferred to Magenta GA7 boxes (same medium) sealed with Micropore surgical tape and allowed to enlarge for 2 weeks before transfer to the same medium without BA for rooting. Rooted green shoots were transferred to soil in 2 in. plastic pots and covered with a plastic bag to maintain a high level of humidity. Holes were cut in the bag gradually over 7–10 days before removal.

For the seedling explant transformation (Fig. 1) seeds were surface-sterilized as described above and germinated on hormone-free Linsmaier-Skoog medium solidified with 0.8% Bacto-agar (Difco Laboratories) and cultured at the light and temperature conditions indicated above. Hypocotyl segments and petioles from the cotyledons (5–7 mm long) were excised 14–17 days after germination. Explants were precultured for 3 days on the regeneration medium as described above, containing standard MS salts (no feeder layer), under the same light and temperature conditions. During this preculture period, the plates were sealed with Parafilm. All subsequent plates were sealed with Micropore surgical tape Explants were removed from the plates and inoculated with *Agrobacterium* as described for the peduncle transformation. After inoculation, explants were treated in a similar manner to the peduncle explants except that the co-cultivation time with *Agrobacterium* was reduced from 4 to 3 days.

2.2.3 Characterization of Transformed Plants

NptII Expression Assay. Immediately before transfer of rooted plantlets from medium to soil, a small (5 mm^2) piece of leaf was removed, placed on expression assay medium (Fry et al. 1987) containing standard MS salts, B5 vitamins, 3% sucrose, 0.5 mM arginine, 5 mg/l BA, 0.5 mg/l NAA, 200 mg/l timentin, 50 mg/l kanamycin, and cultured at 25 °C, under a 16-h photoperiod, $45–70 \mu E m^{-2} s^{-1}$. The explants were scored for resistance to kanamycin after 2 weeks of culture, as evidenced by the ability to enlarge, produce callus and form roots.

Molecular Analysis. DNA for Southern blot analysis was isolated from leaves (Shure et al. 1983). Five micrograms of DNA was cut with BglII, electrophoresed through a 0.8% agarose gel and transferred to Hybond N$^+$ nylon membrane (Amersham protocol). Purified pMON10517–1 plasmid DNA or a

1351-bp fragment of the *Bt* gene was ^{32}P-labeled using the Boehringer Mannheim random priming kit and used as the probe. Nuclear DNA content of the transformants was determined using the flow cytometric method of Arumuganathan and Earle (1991b).

Progeny Analysis. Transformants derived from the cytoplasmic male-sterile broccoli lines were cross-pollinated using the commercial hybrid Green Comet broccoli. Transformants derived from Green Comet broccoli were selfed by bud pollination using 0.25 M NaCl to overcome self-incompatibility as described in Monteiro et al. (1988). Progeny plants were grown in styrofoam flats (72 cell) and were tested for segregation of kanamycin resistance using a spray containing 250 mg/l kanamycin and 0.05% TritonX100 (Weide et al. 1989). Plants were sprayed for 4 consecutive days at a rate of 50 ml/flat/day using a misting spray bottle. Plants were sprayed when two true leaves had developed. The assay was scored 2 weeks after spraying. Kanamycin-resistant plants remained green, whereas sensitive plants showed chlorotic regions on their leaves.

Insect Bioassays. Kanamycin-resistant transformants and their progeny were assayed for insect resistance using six to ten first instar diamondback moth (*Plutella xylostella*) larvae of *Bt*-susceptible strain Geneva88 (provided by Dr. Anthony Shelton, NYS Agriculture Experiment Station, Geneva, NY). Insect mortality was scored after 3–5 days. Plants were scored as insect-resistant if they provided 100% mortality of the larvae.

2.3 Results and Discussion

2.3.1 Characterization of Transgenic Plants

All five broccoli lines and King Cole cabbage were successfully transformed. The results from both procedures are presented in Table 2. A total of 181 kanamycin-resistant transformants were recovered from the flowering stalk transformation; 112 of 162 plants tested were also resistant to the diamondback moth (Fig. 2). The overall transformation efficiency for the flowering stalk transformation was 6.4%. A total of 166 kanamycin-resistant broccoli and cabbage transformants were recovered from the seedling explant transformation; 124 of 164 plants tested were also resistant to the diamondback moth. The overall transformation efficiency for the seedling explant transformation was 6.6%. Cabbage hypocotyls and petioles and broccoli hypocotyls resulted in transformation efficiencies of approximately 10%. Broccoli petioles produced transformants at only a 1.8% efficiency.

Transgenic shoots were obtained via callus and directly from the explants. Small green calli, from which transgenic shoots regenerated, developed from the cut ends of some explants approximately 2 weeks after initial kanamycin selection, and continued to develop for an additional 6 weeks. Many transgenic plants can be generated from such calli by removing the shoots and

Table 2. Transformation results. (Metz et al. 1995)

Plant material	Transformation efficiency[a]	Kanamycin-resistant plants	Insect-resistant plants[b]
CMS no. 1			
Peduncles	8.3% (10/120)	16	5/16 (31%)
CMS no. 2			
Peduncles	Not determined	60	46/51 (90%)
CMS no. 3			
Peduncles	9.2% (11/120)	29	16/28 (57%)
Fertile no. 1			
Peduncles	10.0% (12/120)	25	16/25 (64%)
King Cole cabbage			
Hypocotyls	9.4% (37/394)	68	54/66 (81%)
Petioles	10.3% (41/400)	52	38/52 (73%)
Green Comet broccoli			
Peduncles	5.1% (40/787)	51	29/42 (69%)
Hypocotyls	9.8% (19/194)	33	22/33 (67%)
Petioles	1.8% (12/668)	13	10/13 (77%)
Totals	6.5% (182/2803)	347	236/326 (72%)

[a] Transformation efficiency is the number of explants yielding kanamycin-resistant transformants per total number of explants inoculated.
[b] Transformed plants were scored resistant if 100% mortality of insects was observed after 3, 4, or 5 days.

Fig. 2. Transgenic broccoli plant (*right*) sustains very little damage compared with non-transformed control plant (*left*) when screened with five to ten first instar susceptible DBM larvae

returning the calli to selective regeneration medium for further shoot development. Not all shoots obtained from a single kanamycin-resistant callus were transformed, however. Many white, non-transformed escape shoots formed, as well as putatively transgenic green shoots. The kanamycin in the medium may be detoxified by the nptII gene product in the transgenic callus, reducing selective pressure in a localized area. Some transgenic shoots appear to regenerate

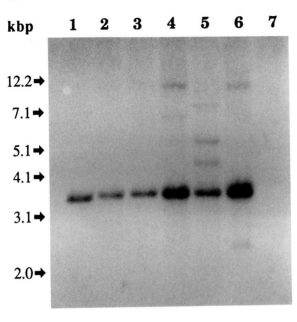

Fig. 3. Southern hybridization analysis of T-DNA integration in six broccoli transformants. Genomic DNA was digested with the restriction enzyme BgIII; a 1.3kb sequence of the *Bt* gene was used as the probe. *Lanes 1–6* independent broccoli transformants; *Lane 7* non-transformed Green Comet broccoli control

directly from the cut ends of the explants 1–2 weeks prior to the development of transformed callus. Such direct shoot regeneration was seen most often from peduncle explants.

Putative transformants that rooted on kanamycin-containing medium were confirmed as transgenic with the nptII expression assay, in which small leaf pieces were placed onto medium containing a higher level of kanamycin. Leaf pieces from escapes turned white after 7–10 days, while leaf pieces from transformants remained green, enlarged significantly and formed callus and roots. All plantlets that rooted on kanamycin-containing medium demonstrated resistance to kanamycin in the nptII expression assay, with the exception of Green Comet broccoli. Several Green Comet shoots regenerated roots on kanamycin-containing medium, but were susceptible in the nptII expression assay. The use of these two steps of antibiotic selection allowed the identification of true transformants with a very high degree of accuracy. All plants that demonstrated both the ability to root in the presence of kanamycin and to remain green during the nptII expression assay showed the presence of the transgene when analyzed by Southern blotting (Fig. 3). All insect-resistant plants analyzed gave the expected 3.6kb band when probed with a 1351bp fragment of the *Bt* gene.

Stable transgene inheritance was demonstrated by segregation of progeny for resistance to kanamycin (Table 3), and by Southern analysis of progeny. All four insect-resistant, selfed transformants tested gave a segregation ratio of 3:1 for kanamycin resistance to kanamycin susceptibility, which is the expected radio for a single integration event. Additionally, twelve of 28 insect-resistant transgenic lines derived from cytoplasmic male-sterile broccoli lines analyzed gave the expected 1:1 ratio of resistance to susceptibility in the

Table 3. Ploidy level of transformants. (Metz et al. 1995)

Explant source	No. plants tested	No. diploid	% diploid
Pedunles (broccoli)			
Fertile no. 1	10	8	80
CMS no. 1	13	12	92
CMS no. 2	48	45	94
CMS no. 3	20	17	85
Green Comet	49	44	90
Seedling explants			
Green Comet broccoli			
Petioles	25	14	56
Hypocotyls	20	14	70
King Cole cabbage			
Petioles	26	19	73
Hypocotyls	24	16	67

Fig. 4. Progeny seedlings of each transgenic broccoli line were screened for resistance to kanamycin. Resistant plants remain green, while susceptible plants show chlorotic spots on leaves

kanamycin spray assay (Fig. 4), consistent with a single T-DNA integration, or one expressed integration event. The remaining transgenic lines analyzed gave segregation ratios indicative of more than one T-DNA integration. Progeny of two insect-susceptible, kanamycin-resistant transgenic lines also showed segregation of progeny for resistance to kanamycin, confirming their transgenic nature. Southern analysis of several progeny from two insect-resistant transgenic plant lines showed segregation for the presence or absence of the expected 3.6-kb band when probed with a 1351-bp fragment of the *Bt* gene (data not shown).

Flowering stalk explants produced more diploid transformants than seedling explants (Table 4); the measured DNA content of diploid transformants (2C = 1.27 ± 0.04 pg) agreed with a previously published value (Arumuganathan and Earle 1991a). The remainder of the regenerated transformants were found to be tetraploid. Flowering stalk explants regenerated diploid shoots on average 90% of the time whereas only 56–73% of shoots from all seedling explants tested were diploid. Similar differences were

Table 4. Segregation of resistance to kanamycin in progeny of selfed diploid transformants. (Metz et al. 1995)

Plant line[a]	No. of seedlings	Seedling type		χ^2 value (3:1 ratio)
		Resistant	Susceptible	
GC	82	0	82	
N59.1	9	7	2	0.04
Q7.1	74	53	21	0.45
Q13.4	63	47	16	0.01
Q23	71	58	13	1.69

[a] *GC* non-transformed Green Comet broccoli control; all others are independent broccoli transformants.

seen in the direct comparison of transformants derived from Green Comet broccoli peduncle and seedling explants. Peduncle explants regenerated 90% diploid transformants, whereas petioles and hypocotyls regenerated only 56% and 70% diploid plants respectively. This difference may be correlated with the higher percentage of direct shoot regeneration from peduncle explants, without an intermediate callus phase.

Insect resistant progeny of several Green Comet transformants have been used in greenhouse experiments designed to study various pest resistance management strategies (Metz et al. 1995). Several homozygous transgenic broccoli lines (Green Comet) have been generated through two generations of self pollination followed by segregation analysis of progeny utilizing the kanamycin spray assay and insect bioassay. Progeny of selected homozygous transgenic lines crossed with a cytoplasmic male sterile non-transformed broccoli cultivar showed excellent insect resistance in field tests conducted during the summer of 1996 (data not shown).

2.3.2 Factors Affecting Transformation

Development of an effective transformation procedure involves the investigation of many interacting factors. Numerous regeneration and transformation experiments were performed to investigate those factors that might influence the frequency of shoot regeneration. The plant cultivars and explant types chosen for *Agrobacterium* inoculation were those that exhibited the ability to regenerate shoots in large numbers and at high frequencies on non-selective medium (Christey and Earle 1991). Optimization of shoot regeneration from each explant type was conducted in the absence of *Agrobacterium* and antibiotic. Several parameters that may be involved in shoot regeneration and cell susceptibility to *Agrobacterium* infection were examined in an attempt to optimize the efficiency of transformation. Among the variables tested were length of inoculation with *Agrobacterium* (1s to 24h), length of co-cultivation (1–5 days), type of nurse cell layer (tobacco, *B. napus, B. campestris*), concentration of *A. tumefaciens* inoculum (undiluted, 1:10, 1:100 dilutions), petri dish sealant (Parafilm, porous tape), use of $AgNO_3$ (1–10mg/l)

Table 5. Effects of tobacco nurse cell layer on broccoli (var. Green Comet) regeneration and transformation frequencies. (Metz et al. 1995)

Explant[a]	No nurse cell layer		Nurse cell layer	
	Regen. frequency (%)	Transform. frequency (%)	Regen. frequency (%)	Transform. frequency (%)
Peduncle	94	0.4	83	5.1
Petiole	81	<0.1	23	1.8
Hypocotyl	97	<0.1	61	9.8

[a] All data for Green Comet broccoli.

as an anti-ethylene agent, initial application of kanamycin selection (0–14 days after co-cultivation), kanamycin concentration (15–50 mg/l), BA concentration (1–10 mg/l), and explant type (leaves, peduncles, seedling cotyledons, hypocotyls, petioles). The transformation procedures described utilize the parameters yielding the best results.

Use of a tobacco nurse cell layer during the co-cultivation with *Agrobacterium* was essential to obtaining transformed plants, although it reduced the frequency of shoot regeneration (Table 5). In the absence of a tobacco feeder layer, all three types of explants regenerated shoots at over an 80% frequency but the transformation frequencies were less than 1%. Use of the tobacco feeder layer reduced shoot regeneration from seedling petiole explants significantly, which is reflected in a lower transformation efficiency than that for either seedling hypocotyls or peduncles. When *B. napus* or *B. campestris* cell suspensions were used in place of tobacco cells, regeneration frequency was also reduced (data not shown), but no transformants were recovered. Optimization of shoot regeneration was conducted in the absence of a nurse cell layer. Enhanced transformation efficiencies may be possible through improvement of shoot regeneration frequency in the presence of the nurse cell layer.

Use of porous surgical tape to seal petri dishes in place of parafilm was essential to promote good shoot regeneration and to obtain a high frequency of transgenic plants. Parafilm provides an air-tight seal and can result in an accumulation of ethylene, high relative humidity, shoot inhibition and hyperhydricity (De Block et al. 1989). The high humidity level in Parafilm-sealed plates resulted in overgrowth of *Agrobacterium*, causing necrosis of the explants. Much less condensation was visible inside the plates sealed with porous tape, resulting in better control of *Agrobacterium* after inoculation and improved shoot regeneration. We attempted to inhibit ethylene action by adding $AgNO_3$ (1–10 mg/l) to the growth medium (Purnhauser et al. 1987), based on encouraging results with *B. oleracea* regeneration (Sethi et al. 1990) and transformation (De Block et al. 1989; Mukhopadhyay et al. 1992), but we found that $AgNO_3$ inhibited shoot regeneration both in controls and in transformation experiments. Porous tape was used as an alternate method for control of ethylene build-up inside plates. Parafilm seals were used during the preculture of seedling explants to maintain high humidity and prevent drying.

A delay in the addition of kanamycin to the medium substantially increased the number of regenerating shoot. Kanamycin selection applied immediately after inoculation with *Agrobacterium* resulted in very poor shoot regeneration and transformation frequencies, also reported by Ovesná et al. (1993). When selection was delayed for 1 week after *Agrobacterium* inoculation (Toriyama et al. 1991), most of the early regenerating shoots were nontransformed escapes, but many more transgenic plants were recovered. Escape shoots initially appeared green and healthy but turned white after 1–2 weeks and then stopped growing. The level of kanamycin used in these experiments inhibited continued development of such escapes, while allowing transformed shoots to develop further. Higher levels of kanamycin (up to 50 mg/l) resulted in fewer transformants.

The peduncle transformation procedure was modified to improve the transformation efficiency of the seedling explant transformation procedure. With seedling explants, a preculture period prior to inoculation with *Agrobacterium* was required. Seedling explants exhibit a greater hypersensitive response upon inoculation with *Agrobacterium*. Without the preculture period, most inoculated explants develop brown, necrotic cut ends, showing little to no regenerative potential. Reduction of the co-cultivation time with *Agrobacterium* from 4 days to 3 days was also necessary in order to obtain a high frequency of transformants from seedling explants (data not shown). Although the original peduncle transformation procedure called for use of a reduced level of MS salts in the regeneration medium (Fry et al. 1987), full strength MS salts were included in the preculture and co-cultivation medium for the seedling explants, as they regenerated poorly on medium containing $1/10 \times$ MS salts.

2.3.3 Comparison of Peduncle and Seedling Explant Transformation Procedures

Each transformation method used has some advantages and disadvantages. Although peduncle explants of several broccoli lines can regenerate shoots at a very high frequency (Christey and Earle 1991) and we were able to produce many transgenic broccoli plants using this method, the procedure also has some disadvantages. The explants are from flowering stalks of mature plants, so it is necessary to plan the transformation experiments several months in advance; even then it is difficult to determine exactly when suitable explant material will be available. Experiments with a biennial crop such as cabbage require even more advanced planning. Variation in plant growth rates can make it difficult to obtain a large number of explants at any one time unless one has the space to maintain many full-grown plants. Using flowering stalks from plants grown in the greenhouse or field also increases the likelihood of fungal or bacterial contamination and can lead to numerous delays in obtaining transformants, including the need to repeat entire experiments. Several transformation experiments utilizing flowering stalk explants had to be entirely discarded due to contamination. If experiments lost to contamination were included in the calculation of transformation efficiency for the flowering stalk procedure, the efficiency would be significantly lower.

Using explants from sterile plant material maintained in culture avoids the frequent problem of fungal or bacterial contamination that is common when using flowering stalk segments. Use of seedling explants eliminates the need to maintain full-grown plants in a greenhouse and reduces the time needed to plan experiments from several months to a few weeks. Most of the effort in these transformation experiments focused on seedling explants of King Cole Cabbage and Green Comet broccoli. Transformation efficiencies, as well as regeneration frequencies of non-inoculated explants varied significantly (greater than tenfold differences in results – data not shown) during the course of these experiments, and seemed to coincide with the transition from one purchased seed lot to another. This suggests that significant genetic heterogeneity exists, even within a particular cultivar. It has been recommended that inbred lines or double haploid stocks be utilized to provide the genetic uniformity necessary to achieve reproducible results (Puddephat et al. 1996).

The flowering stalk transformation procedure may be useful in situations where only a limited amount of seed from the genotype targeted for transformation is available since it is possible to obtain several hundred explants from one flowering broccoli head. This also serves to minimize the problems associated with the lack of homogeneity experienced when using large numbers of seed, or different seed lots. Furthermore, numerous explants can be obtained nondestructively, which allows one to analyze the source material if needed. Moreover, a higher percentage of plants regenerated from peduncle explants are diploid. This might be explained by the fact that a larger proportion of transgenic shoots regenerate directly from the cut ends of the explants without a callus phase. Seedling explants regenerated a higher percentage of shoots from callus, allowing an increased chance of changes in ploidy. With both procedures, it was possible to produce several clonal transformants from each transformation event by removing the regenerating green shoots to rooting medium and placing the remaining green callus back onto regeneration medium. Regenerating several plants from the same transformation event can help compensate for the relatively large percentage of tetraploid plants produced from the seedling explant transformation, since it was possible to recover both diploid and tetraploid plants from the same transgenic callus.

3 Summary and Conclusions

Several published reports demonstrate genetic transformation in broccoli and cabbage. All accounts of transformation within these crops utilize gene transfer through *Agrobacterium* species. In general, most published accounts of transformation have indicated somewhat limited success. Presently, there is not single universally accepted method of genetic transformation for either of these crop plants. The success of the procedure described here with five different broccoli varieties and one cabbage variety suggests the possibility of a general transformation protocol. However, there were also several varieties tested from which no transformants were recovered. Given the lack of genetic uniformity within and between these varieties, it seems unlikely that the current approaches towards achieving genetic transformation will result in a reliable and reproducible procedure that can be widely used for these vegetable crops. However, seedling explants, especially hypocotyls and cotyledonary petioles, seem to currently be the most amenable source material to use, as several groups have developed successful transformation methods using these explants (Lee and Keller 1993; Metz et al. 1995; Christey et al. 1997; Cao et al. 1999; A. Mora and E.D. Earle, pers. comm.). Development of a transformation procedure for a recalcitrant crop requires a great deal of attention to the many factors that may play a role in the gene transfer process and in shoot regeneration from those few cells that possess the transgene (van Wordragen and Dons 1992).

References

Arumuganathan K, Earle ED (1991a) Nuclear DNA content of some important plant species. Plant Mol Biol Rep 9:208–218

Arumuganathan K, Earle ED (1991b) Estimation of nuclear DNA content of plants by flow cytometry. Plant Mol Biol Rep 9:229–241

Ayub R, Guis M, Amor MB, Gillot L, Oustan JP, Latché A, Bouzayen M, Pech J-C (1996) Expression of ACC oxidase antisense gene inhibits ripening of cantaloupe melon fruits. Nat Biotechnol 14:862–866

Beachy RN, Loesch-Fies S, Tumer NE (1990) Coat protein-mediated resistance against viral infection. Annu Rev Phytopathol 28:451–474

Beecher CW (1994) Cancer preventative properties of varieties of Brassica oleracea: a review. Am J Clin Nutr 59:1166–1170

Berthomieu P, Jouanin L (1992) Transformation of rapid cycling cabbage (Brassica oleracea var. capitata) with Agrobacterium rhizogenes. Plant Cell Rep 11:334–338

Berthomieu P, Béclin C, Charlot F, Doré C, Jouanin L (1994) Routine transformation of rapid cycling cabbage (Brassica oleracea) – molecular evidence for regeneration of chimeras. Plant Sci 96:223–235

Birot AM, Bouchez D, Casse-Delbart F, Durand-Tardif M, Jouanin L, Pautot V, Robaglia C, Tepfer D, Tepfer M, Tourneur J, Vilaine F (1987) Studies and uses of the Ri plasmids of Agrobacterium rhizogenes. Plant Physiol Biochem 25:323–335

Bottrell DG, Aguda RM, Gould FL, Theunis W, Demayo CG, Magalit VF (1992) Potential strategies for prolonging the usefulness of Bacillus thuringiensis in engineered rice. Korean J Appl Entomol 31:247–255

Cao J, Tang JD, Strizhov N, Shelton AM, Earle ED (1999) Transgenic broccoli with high levels of Bacillus thuringiensis Cry1C protein control diamondback moth larvae resistant to Cry1A or Cry1C. Mol Breed 5:131–141

Christey MC, Earle ED (1991) Regeneration of Brassica oleracea from peduncle explants. HortScience 26:1069–1072

Christey MC, Sinclair BK (1992) Regeneration of transgenic kale (Brassica oleracea var. acephala), rape (B. napus) and turnip (B. campestris var. rapifera) plants via Agrobacterium rhizogenes mediated transformation. Plant Sci 87:161–169

Christey MC, Sinclair BK, Braun RH, Wyke K (1997) Regeneration of transgenic vegetable brassicas (*Brassica oleracea* and *B. campestris*) via Ri-mediated transformation. Plant Cell Rep 16:587–593

Church GM, Gilbert W (1984) Genomic sequencing. Proc Natl Acad Sci USA 81:1991

Cooper B, Lapidot M, Heick JA, Beachy RN (1995) A defective movement protein of TMV in transgenic plants confers resistance to multiple viruses whereas the functional analogue increases susceptibility. Virology 206:307–313

David C, Tempé J (1988) Genetic transformation of cauliflower (*Brassica oleracea* L. var. *botrytis*) by *Agrobacterium rhizogenes*. Plant Cell Rep 7:88–91

De Block M, De Brouwer D, Tenning P (1989) Transformation of *Brassica napus* and *Brassica oleracea* using *Agrobacterium tumefaciens* and the expression of the *bar* and *neo* genes in transgenic plants. Plant Physiol 91:694–701

Eimert K, Siegemund F (1992) Transformation of cauliflower (*Brassica oleracea* L. var. *botrytis*) – an experimental survey. Plant Mol Biol 19:485–490

Fischhoff DA, Bowdish KS, Perlak FJ, Marrone PG, McCormick SH, Niedermeyer JG, Dean DA, Kusano-Kretzmer K, Mayer EJ, Rochester DE, Rogers SG, Fraley RT (1987) Insect tolerant transgenic tomato plants. Bio/Technology 5:807–813

Fromm M, Armstrong C, Blasingame A, Brown S, Duncan D, Deboer D, Hairston B, Howe A, McCaul S, Neher M, Pajeau M, Parker G, Pershing J, Petersen B, Santino C, Sanders P, Sato S, Sims S, Thorton T (1994) Production of insect resistant corn. J Biol Chem Suppl 18A:77

Fry J, Barnason A, Horsch RB (1987) Transformation of *Brassica napus* with *Agrobacterium tumefaciens* based vectors. Plant Cell Rep 6:321–325

Fujimoto H, Itoh K, Yamamoto M, Kyozuka J, Shimamoto K (1993) Insect resistant rice generated by introduction of a modified ∂-endotoxin gene of *Bacillus thuringiensis*. Bio/Technology 11:1151–1155

Hamilton AJ, Lycett GW, Grierson D (1990) Antisense gene that inhibits synthesis of the hormone ethylene in transgenic plants. Nature 346:284–287

Henzi MX, Christey MC, McNeil DL, Davies KM (1999) *Agrobacterium rhizogenes*-mediated transformation of broccoli (*Brassica oleracea* L. var. *italica*) with an antisense 1-aminocyclopropane-1-carboxylic acid oxidase gene. Plant Sci 143:55–62

Hoekema A, Hirsch PR, Hooykaas JJ, Schilperoort RA (1983) A binary plant vector strategy based on the separation of vir- and T- region of the *Agrobacterium tumefaciens* Ti-plasmid. Nature 303:179–180

Höfte H, Whiteley HR (1991) Insecticidal crystal proteins of *Bacillus thuringiensis*. Microbiol Rev 53:242–255

Hosoki T, Kigo T, Shiraishi K (1989) Transformation and regeneration of broccoli (*Brassica oleracea* var. *italica*) mediated by *Agrobacterium rhizogenes*. J Jpn Soc Hortic Sci 60:71–75

Jianping Y, Qiquan S (1988) Transformation of *Brassica oleracea* with a Ti-plasmid – derived vector. Genet Manipul Crops Newsl 4:75–80

Lee S, Keller W (1993) *Agrobacterium*-mediated transformation of broccoli (*Brassica oleracea* var. *italica*) cotyledons and regeneration of transgenic plants. 8th Crucifer Genetics Workshop, 21–24 July 1993, Saskatoon, Saskatchewan, 94 pp

Lomonossof GP (1995) Pathogen-derived resistance to plant viruses. Annu Rev Phytopathol 33:323–343

Metz TD, Dixit R, Earle ED (1995) *Agrobacterium tumefaciens*-mediated transformation of broccoli (*Brassica oleracea* var. *italica*) and cabbage (*B. oleracea* var *capitata*). Plant Cell Rep 15:287–292

Moloney MM, Walker JM, Sharma KK (1989) High efficiency transformation of *Brassica napus* using *Agrobacterium* vectors. Plant Cell Rep 8:238–242

Monteiro AA, Gabelman WH, Williams PH (1988) The use of sodium chloride solution to overcome self-incompatibility in *Brassica campestris*. Eucarpia Cruciferae Newsl 13:122–123

Mukhopadhyay A, Töpfer R, Pradhan AK, Sodhi YS, Steinbiß H-H, Schell J, Pental D (1991) Efficient regeneration of *Brassica oleracea* hypocotyl protoplasts and high frequency genetic transformation by direct DNA uptake. Plant Cell Rep 10:375–379

Mukhopadhyay A, Arumugam PB, Nandakumar PBA, Pradhan AK, Gupta V, Pental D (1992) *Agrobacterium*-mediated genetic transformation of oilseed *Brassica campestris*: transformation frequency is strongly influenced by the mode of shoot regeneration. Plant Cell Rep 11:506–513

Oeller PW, Wong LM, Taylor LP, Pike DA, Theologis A (1991) Reversible inhibition of tomato fruit senescence by antisense RNA. Science 254:437–439

Ovesná J, Ptáček L, Opatrny Z (1993) Factors influencing the regeneration capacity of oilseed rape and cauliflower in transformation experiments. Biol Plant 35:107–112

Perlak FJ, Deaton RW, Armstrong TA, Fuchs RL, Sims SR, Greenplate JT, Fischhoff DA (1990) Insect resistant cotton plants. Bio/Technology 8:939–943

Perlak FJ, Fuchs RL, Dean DA, McPherson SL, Fischhoff DA (1991) Modification of the coding sequence enhances plant expression of insect control protein genes. Proc Natl Acad Sci USA 88:3324–3328

Pogson BJ, Downs CG, Davies KM (1995a) Differential expression of two 1-aminocyclopropane-1-caboxylic acid oxidase genes in broccoli after harvest. Plant Physiol 108:651–657

Pogson BJ, Downs CG, Davies KM, Morris SC (1995b) Nucleotide sequence of a cDNA clone encoding 1-aminocyclopropane-1-carboxylic acid synthase from broccoli. Plant Physiol 108:857–858

Puddephat IJ, Riggs TJ, Fenning TM (1996) Transformation of *Brassica oleracea* L.: a critical review. Mol Breed 2:185–210

Purnhauser L, Medgyesy P, Czako M, Dix PJ, Marton L (1987) Stimulation of shoot regeneration in *Triticum aestivum* and *Nicotiana plumbaginifolia* Viv. tissue cultures using the ethylene inhibitor AgNO₃. Plant Cell Rep 6:1–4

Sethi U, Basu A, Guha-Mukherjee S (1990) Control of cell proliferation and differentiation by modulators of ethylene biosynthesis and action in *Brassica* hypocotyl explants. Plant Sci 69:225–229

Shure W, Wessler S, Federoff N (1983) Molecular identification and isolation of the Waxy locus in maize. Cell 35:235–242

Siemonsma JS, Piluek K (eds) (1993) Plant Resources of Southeast Asia No 8: Vegetables. Pudoc Scientific Publishers, Wageningen, The Netherlands

Srivastava V, Reddy AS, Guha-Mukherjee S (1988) Transformation and regeneration of *Brassica oleracea* mediated by an oncogenic *Agrobacterium tumefaciens*. Plant Cell Rep 7:504–507

Tabashnik BE, Finson N, Johnson MW (1991) Managing resistance to *Bacillus thuringiensis*: lessons from the diamondback moth (Lepidoptera: Plutellidae). J Econ Entomol 84:49–55

Tacke E, Salamini F, Rohde W (1996) Genetic engineering of potato for broad-spectrum protection against virus infection. Nat Biotechnol 14:1597–1601

Tepfer D (1989) Ri T-DNA from *Agrobacterium rhizogenes*: a source of genes having applications in rhizosphere biology and plant development, ecology and evolution. In: Kosuge T, Nester EW (eds) Plant-microbe interactions. Mol Genet Perspect 3:294–342

Toriyama K (1992) Transformation of *Brassica* species with self-incompatibility gene. In: Oono K, Hirabayashi T, Kikuchi S, Handa H, Kajiwara K (eds) Plant tissue culture and gene manipulation for breeding and the formation of phytochemicals. NIAR, Japan, pp 165–171

Toriyama K, Stein JC, Nasrallah ME, Nasrallah JB (1991) Transformation of *Brassica oleracea* with an *S*-locus gene from *B. campestris* changes the self-incompatibility phenotype. Theor Appl Genet 81:769–776

USDA – National Statistics Service (1994) Agricultural statistics. United States Government Printing Office, Washington, DC

Van Wordragen MF, Dons HJM (1992) *Agrobacterium tumefaciens*-mediated transformation of recalcitrant crops. Plant Mol Biol Rep 10:12–36

Weide R, Koornneef M, Zabel P (1989) A simple, nondestructive spraying assay for the detection of an active kanamycin resistance gene in transgenic tomato plants. Theor Appl Genet 78:169–172

7 Transgenic Vegetable and Forage *Brassica* Species: Rape, Kale, Turnip and Rutabaga (Swede)

M.C. CHRISTEY and R.H. BRAUN

1 Introduction

The economically important *Brassica* genus contains about 85 species. The six cultivated *Brassica* species are interrelated with the three amphidiploids (*B. napus, B. juncea* and *B. carinata*) arising from interspecific hybridization between the three diploid species (*B. oleracea, B. nigra* and *B. campestris*; U 1935). In addition to many important vegetable species, *Brassica* includes oilseed (rape and turnip), vegetables and forage (turnip, swede, rape, kale) species. Turnips and swedes provide winter fodder for sheep and cattle (McNaughton 1976). In New Zealand, arable *Brassica* crops have been important for animal production since 1870. They are used to supplement pasture for sheep and cattle when pasture growth is not sufficient (Palmer 1983).

Forage crops are found in three *Brassica* species: *B. oleracea, B. napus* and *B. campestris. B. oleracea* is a polymorphic species which provides forage for animals as well as a range of vegetables for human consumption (Thompson 1976). *B. oleracea* originated in southern Europe, where it grows along the coasts of the Mediterranean Sea (Nieuwhof 1993). In marrow stem kale (*B. oleracea* L. var. *acephala*) the fleshy stem is utilised for animal forage, and in 1000 head kale the leaves are consumed. *B. napus* provides two forms of forage crops. The cultivated swede (*B. napus* var. *rapifera*), or rutabaga as it is known in North America, originated in Europe and is an important stock feed in northern Europe, Russia and New Zealand. Forage rape (*B. napus* var. *biennis*) produces leafy fodder for sheep in northern Europe and New Zealand (McNaughton 1976). *B. napus* is an amphidiploid and its spontaneous formation from *B. oleracea* and *B. campestris* is likely to have been a rare event, with multiple origin from different parental combinations possible. *B. napus* is generally self-fertile and tolerant of inbreeding. Its domestication is recent, with swedes and rapes probably only a few 100 years old (McNaughton 1976). Turnip is widely cultivated throughout the world for the development of a swollen hypocotyl as the storage organ.

Forage brassicas have many traits which require improvement to enhance their agricultural potential. All of these crops have several insect and disease

New Zealand Institute for Crop & Food Research Limited, Private Bag 4704, Christchurch, New Zealand

Biotechnology in Agriculture and Forestry, Vol. 47
Transgenic Crops II (ed. by Y.P.S. Bajaj)
© Springer-Verlag Berlin Heidelberg 2001

problems. In New Zealand, turnips are prone to infection by beet western yellows virus, which occurs throughout the turnip growing area. No resistant cultivars are currently available. In addition, improvements in levels of toxic factors are needed, particularly in rape and kale. Decreases in glucosinolates and S-methyl cysteine sulphoxide (SMCO) levels are required because the breakdown products of these compounds are harmful to livestock when large amounts are consumed (Nieuwhof 1993).

In recent years the development of plant transformation procedures has enabled the genetic modification of many crop plants, including *Brassica* species. *Agrobacterium rhizogenes*-mediated transformation has been used to produce transgenic plants, including brassicas (reviewed in Christey 1997), but is less widely used than *A. tumefaciens*. However, *A. rhizogenes*-mediated transformation is the most common method of transformation for forage brassicas (e.g. Christey and Sinclair 1992; Christey et al. 1999a,c). A recent review (Christey 1997) indicates that transgenic plants have been regenerated from hairy roots of 60 different taxa, representing 51 species from 23 families; including 14 examples from the Brassicaceae. Agronomically useful traits have been introduced into various brassicas (see reviews by Earle et al. 1996; Puddephat et al. 1996; Christey 1997; Earle and Knauf 1999). This review summarises published transformation research in forage rape, forage kale, swede and turnip, and outlines current transformation research being conducted in New Zealand on these crops.

2 Genetic Transformation

2.1 Susceptibility to *Agrobacterium* Strains

Published reports on the susceptibility of forage brassicas to *Agrobacterium* strains are rare. Research in our laboratory demonstrated that a range of New Zealand cultivars of *Brassica* species, including kale, rape, swede and turnip, were susceptible to *A. tumefaciens* and *A. rhizogenes* and therefore were potential candidates for *Agrobacterium*-mediated transformation. *A. tumefaciens* inoculations were conducted on the hypocotyl and stem of 6- to 8-week-old greenhouse-grown seedlings of 16 *Brassica* cultivars and lines. Five to seven *A. tumefaciens* strains, including nopaline (C58, A208) and octopine (A6, B6, B6S3, ACH5, A722) strains, were tested per cultivar. Results after 3 months indicated that all cultivars were susceptible to at least three and up to seven *A. tumefaciens* strains, but with differences between lines in frequency of infection, size and number of tumours and response to different *Agrobacterium* strains. Tumour morphology was generally similar across different strains and cultivars. In addition, in vitro co-cultivation experiments were conducted with an *Agrobacterium* strain (A4T) harbouring a Ri plasmid. Results demonstrated that all 19 lines of rape, kale, turnip and swede tested were susceptible to infection, with hairy root cultures obtained (Table 1). To date, shoot

Table 1. In vitro response to *Agrobacterium* strain A4T. +Achieved, –not achieved

Brassica species and variety		Response to *Agrobacterium* strain A4T			
Cultivar	Selection of hairy roots	Shoot regeneration from hairy roots	No. of plants to greenhouse[a]	Survival on transfer to soil (%)	Phenotype
B. campestris var. *rapifera* (turnip)					
Green Globe	+	+	1 (1)	100	Ri, died before flowering
Manga	+	–	–		
Red Globe	+	+	17 (4)	59	Ri, fertile
B. napus var. *biennis* (forage rape)					
95A2/5	+	+	18 (3)	100	Ri, fertile
Giant	+	+	469 (56)	92	mix Ri and normal, fertile
H103d	+	+	5 (1)	100	Ri, fertile
Rangi	+	–	–		
Striker	+	+	15 (2)	80	Ri, fertile
B. napus var. *rapifera* (swede)					
Doon Major	+	–	–		
HYf8	+	–			
RGab2	+	+	7 (2)	86	Ri, in progress flowering
Tina	+	+	1 (1)	100	Ri, not yet flowering
TC22e9	+	+	18 (2)	67	Ri, in progress flowering
TNG15be	+	–	–		
B. oleracea var. *acephala* (forage kale)					
CN95	+	+	68 (7)	19	Ri, fertile
Kapeti	+	+	24 (5)	96	Ri, fertile
Medium Stem	+	+	64 (2)	38	Ri, fertile
Midas	+	+	26 (5)	61	Ri, fertile
Rawara	+	+	37 (1)	54	Ri, fertile

[a] Number in parentheses is the number of independent hairy root lines from which the shoots were obtained.

regeneration has been achieved from hairy root cultures of 14 of these lines (Table 1).

2.2 Explant Regeneration

In some forage brassicas, lack of an efficient explant regeneration system is limiting the development of a transformation system. Forage brassicas, especially turnip, are somewhat recalcitrant to shoot regeneration. Recent results obtained with oilseed *B. campestris* are likely to provide key parameters for obtaining transgenic forage brassicas. In oilseed *B. campestris*, use of ethylene inhibitors has been beneficial in the development of efficient shoot regeneration systems and production of transgenic plants. Radke et al. (1992) found silver nitrate increased regeneration frequencies. They also noted sealing Petri dishes with porous tape instead of Parafilm increased transformation frequencies. Others have also observed beneficial effects of reducing ethylene; Mukhopadhyay et al. (1992) reported increased shoot regeneration in oilseed *B. campestris* with use of silver nitrate. In mustard (*B. juncea*) plants expressing an antisense 1-aminocyclopropane-1-carboxylate oxidase gene the regeneration capacity of tissues was greatly enhanced (Pua and Lee 1995).

2.3 Forage *Brassica* Transformation

Use of *A. tumefaciens* to produce transgenic plants of forage brassicas is limited to one paper by Li et al. (1995) who co-cultivated cotyledonary explants of swede with *A. tumefaciens* strain ASE 367 harbouring the binary vector pBS21, which contains NPTII and *cryIA* genes. Some transgenic swede plants were lethal to *Pieris rapae* larvae (cabbage caterpillar) and others caused significant decreases in larval weight. Fertile plants were obtained and transmission of kanamycin resistance to progeny was demonstrated.

In contrast, there are several papers describing the use of *A. rhizogenes* to produce transgenic forage brassica plants (Table 2). We have developed *A. rhizogenes*-mediated transformation systems for commercially important NZ cultivars of forage rape, forage kale, swede and turnip (Tables 1, 2). This involves co-cultivation of leaf and/or cotyledonary petiole explants with *Agrobacterium* strain A4T containing a binary vector. Transgenic hairy roots are selected by their ability to grow on hormone-free media supplemented with selective levels of kanamycin. Transgenic hairy roots are transferred to a callus induction and shoot regeneration medium for regeneration of transgenic shoots. Using this method, we have successfully transferred to a greenhouse over 550 transgenic plants from 14 cultivars of rape, kale, swede and turnip (Table 1). Most of these plants were transformed using the binary vector pKIWI110 (Janssen and Gardner 1989), which contains genes for in planta expression of kanamycin resistance, β-glucuronidase (GUS) activity, and chlorsulfuron resistance. Transgenic hairy roots grew prolifically on hormone-free medium containing kanamycin (50 or 100 mg/l) or chlorsulfuron (10 µg/l).

Table 2. Summary of published transformation studies in forage brassicas[a]

Protoplast infection		
Variety Cultivar	Method and results	Reference
Turnip		
Just Right	uptake of liposome-packaged CaMV into evacuolated protoplasts	Hussain et al. (1985)
Boule d'Or	electroporation of CAT DNA	Rouan et al. (1991)
Just Right	PEG uptake, obtained G418 and kanamycin resistant cell lines	Paszkowski et al. (1986)

In planta transfection		
Variety Cultivar	Method and results	Reference
Turnip		
Just Right	CaMV containing MMT applied to leaves decreased glucosinolates	Lefebvre (1990)

Tumour induction		
Variety Cultivar	*A. tumefaciens* strain and results	Reference
Turnip		
Snowball	Tumours with N2/73 and A281, none Ach5[b]	Godwin et al. (1991)
NR	Tumours with 1D135, data not shown	Kado et al. (1972)
Purple Top White Globe	Tumours with strain isolated from hops	Hoerner (1945)
Swede		
Atow	Tumours with nopaline strains (C58, T37), none octopine[c]	Holbrook and Miki (1985)
Laurentian	Tumours with nopaline strains (C58, T37), none octopine[c]	Holbrook and Miki (1985)
Doon Major	Tumours with A6, A722, B6, none ACH5 and B6S3[c]	Christey et al. (1999a)
HYa3/2	Tumours with ACH5, A6, B6, none A722 and B6S3[c]	
Kiri	Tumours with A6, A722, B6, B6S3, C58, none A281[c]	
Tina	Tumours with ACH5, A6, A722, A208, B6, B6S3, C58, none A281[c]	
TNG15be	Tumours with ACH5, A6, A722, B6, B6S3[c]	

Hairy root induction			
Variety Cultivar	*A. rhizogenes* strain	Foreign genes introduced	Reference
Turnip			
NR	A4	WT, produced opines	Tanaka et al. (1985)
Manga	A4T	WT	Christey and Sinclair (1992)
Red Globe	A4T	GUS, NPTII, ALS	
Forage rape			
Giant	A4T	GUS, NPTII, ALS	Christey and Sinclair (1992)
Giant	A4T	WT	Downs et al. (1994a,b)
Giant	A4T	NPTII, GS	Downs et al. (1994b)

Table 2. *Continued*

Rangi	A4T	WT	Christey and Sinclair (1992)
Striker	A4T	BAR	Christey et al. (1999c)
95A2/5	A4T	BAR	
H103d	A4T	BAR	
Swede			
Doon Major	A4T	WT	Christey and Sinclair (1992)
Tina	A4T	WT	
Hyf8	A4T	BAR	Christey et al. (1999a)
Kiri	A4T	BAR	
RGab2	A4T	BAR	
TC22e9	A4T	BAR	
Tina	A4T	BAR	
TNG15be	A4T	BAR	
Forage kale			
CN95	A4T	BAR	Christey et al. (1999c)
Kapeti	A4T	GUS, NPTII, ALS	Christey and Sinclair (1992)
Medium Stem	A4T	GUS, NPTII, ALS	
Midas	A4T	GUS, NPTII, ALS	
Rawara	A4T	GUS, NPTII, ALS	

Transgenic plants

Variety Cultivar	*Agrobacterium* strain[d]	Genes introduced	No. of plants[e]	Reference
Turnip				
Red Globe	A4T	GUS, NPTII, ALS	3 (1)	Christey and Sinclair (1992)
Forage rape				
Giant	A4T	GUS, NPTII, ALS	133 (6)	Christey and Sinclair (1992)
Giant	A4T	NPTII, GS	NR	Downs et al. (1994b)
Striker	A4T	BAR	15 (2)	Christey et al. (1999c)
95A2/5	A4T	BAR	18 (3)	
H103d	A4T	BAR	5 (1)	
Swede				
Lion Head	ASE 367	NPTII, CRY1A	36 (36)	Li et al. (1995)
RGab2	A4T	BAR	7 (2)	Christey et al. (1999a)
TC22e9	A4T	BAR	18 (2)	
Tina	A4T	BAR	1 (1)	
Forage kale				
CN95	A4T	BAR	68 (7)	Christey et al. (1999c)
Kapeti	A4T	WT	21 (1)	Christey and Sinclair (1993)
Kapeti	A4T	GUS, NPTII, ALS	5 (2)	Christey and Sinclair (1992)
Medium Stem	A4T	GUS, NPTII, ALS	64 (2)	
Midas	A4T	GUS, NPTII, ALS	21 (5)	
Rawara	A4T	GUS, NPTII, ALS	37 (1)	

[a] *ALS* mutant acetolactate synthase, *BAR* phosphinothricin acetyl transferase, *CaMV* cauliflower mosaic virus, *CAT* chloramphenicol acetyltransferase, *CRY1A Bacillus thuringiensis* crystal protein gene cry1A, *GS* glutamine synthetase, *GUS* β-glucuronidase, *MMT* mammalian metallothionein, *NPTII* neomycin phosphotransferase, *NR* not reported, *WT* wildtype strain.

[b] In vitro co-cultivation.

[c] In planta inoculation.

[d] A4T is *A. rhizogenes*, ASE 367 is *A. tumefaciens*.

[e] Number in parentheses is the number of independent lines from which the shoots were obtained.

In addition, hairy root lines were positive for the expression of β-glucuronidase (GUS) after histochemical GUS assays.

Plants regenerated from nine of the above transgenic hairy root cultures were transferred to a containment greenhouse, and detailed observations and measurements were made to determine the extent of the hairy root phenotype. Measurements included height, leaf number, leaf dimensions and seed number for 20–25 plants from each line. These results showed all kale lines exhibited characteristic Ri phenotypic effects. In contrast, two lines of Giant forage rape were barely distinguishable from the non-transgenic control, in both vegetative and reproductive characteristics (Christey and Sinclair 1993; Christey et al. 1994). Segregation analysis in rape and kale seedlings demonstrated Mendelian inheritance in five lines with GUS expression ratios consistent with a single insertion site. Four lines had irregular inheritance patterns including one line of Kapeti kale, the progeny of which had no GUS expression.

2.4 Current Research in New Zealand

Our current research involves introducing Basta resistance into forage rape, forage kale and swede cultivars and breeding lines. Using the method outlined above, we have obtained transgenic Basta-resistant plants of CN95 kale, Striker, H103d and 95A2/5 forage rape and Tina, RGab2, TC22e9 swede using the binary vector pMOA4. The T-DNA of pMOA4 (Barrell and Conner 1995) contains NOS-BAR-pAg7; a chimeric gene conferring resistance to phosphinothricin, the active ingredient in the herbicide Basta. This binary vector is a minimal T-DNA vector and contains the minimum features for efficient plant transformation. There is no extraneous DNA in the T-DNA region, therefore the *bar* gene is tightly flanked by the T-DNA left and right borders.

Hairy root morphology is very similar between all lines even though they are from two different species. Six weeks after subculture, there is rapid growth with prolific dense hairy root growth over the media surface. Roots are thin and white and covered with root hairs. In these experiments, selection for Basta resistance was generally not applied until shoot regeneration. Initial selection is based primarily on hairy root morphology.

Shoot regeneration was achieved for all six CN95 kale transgenic hairy root lines tested. For each hairy root line generally two media from the thidiazuron-containing RCC media of Christey et al. (1997) were tested. In every case shoot regeneration was readily achieved after 3 weeks and was followed by transfer of small shoot buds to hormone-free media. In contrast to kale, only rare shoot regeneration has been achieved with forage rape and swede. However, spontaneous regeneration of Basta-resistant shoots direct from hairy root cultures has been noted with Striker forage rape.

Due to problems with rooting in vitro hairy root-derived shoots of CN95 kale, use of increased levels of Phytagel (Sigma) were investigated. Increase in Phytagel from the usual 0.3 to 0.45% or 0.6%, gave a dramatic increase in the number of in vitro cuttings initiating roots (Fig. 1a). In one experiment,

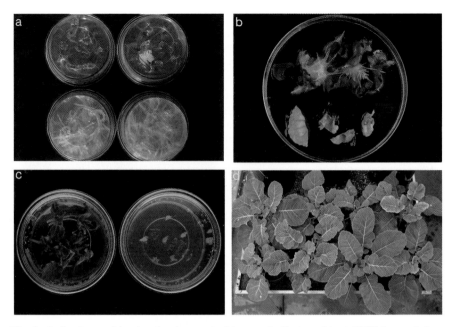

Fig. 1a–d. In vitro and in vivo development of transgenic Basta-resistant CN95 forage kale. **a** Root initiation on hairy root-derived CN95, 3 weeks after transfer of apical cuttings to media containing 0.3% (*top*) or 0.6% (*bottom*) Phytagel. **b** Leaf pieces of transgenic Basta-resistant CN95 (*top*) and non-transgenic CN95 (*bottom*), 2 weeks after transfer to media containing 10 mg/l Basta. **c** Transgenic Basta-resistant CN95(*left*) and non-transgenic CN95 (*right*), 5 weeks after transfer of apical cuttings to media containing 10 mg/l Basta. **d** Transgenic Basta-resistant CN95 2 months after transfer to soil

differences in root initiation were already apparent after 9 days, with no root initiation on 0.3% Phytagel, but on 0.6% Phytagel 49% of cuttings had roots. After 5 weeks on 0.6% Phytagel, excellent root development was apparent throughout the media with 100% root initiation, compared with rare production of roots on 0.3% Phytagel. In a second experiment, root initiation on 0.45% Phytagel was comparable to 0.6% Phytagel. After 2 weeks there was no root initiation on 0.3% Phytagel, but roots on 61 and 53% of cuttings on 0.45 and 0.6% Phytagel, respectively. After 4 weeks, the bases of cuttings on 0.3% Phytagel had small amounts of callus, with roots being rare. On 0.45 and 0.6% Phytagel, cuttings also had small amounts of friable callus but with excellent development of roots on 84% of cuttings. There was no difference in the phenotypic appearance of shoots grown at different Phytagel strengths.

Prior to transfer to the greenhouse, putative transgenic shoots were checked for Basta resistance by conducting a leaf-piece assay. One small leaf per plant was removed, cut into two to three pieces and placed on media containing 10 mg/l Basta (Hoechst Australia Ltd.). After 7 days, successful transformation was very evident, with enlargement of resistant explants, often followed by callus and root development. In contrast, sensitive explants remained small, became pale brown and failed to initiate callus or roots (Fig.

1b). This level of Basta was also selective for in vitro shoots. One week after transfer of apical cuttings to 10 mg/l Basta, browning of sensitive cuttings was noted with no further shoot development. In contrast, resistant shoots remained green and healthy and showed good root and shoot development (Fig. 1c).

Shoots were selected for transfer to soil when there was sufficient root and shoot development, usually 4–5 weeks after transfer of the apical cutting to fresh media. To maintain humidity, a plastic bag was placed over plants for the first 2 weeks and plants were placed under a bench out of direct light. Twenty Basta-resistant plants of CN95 kale, H103d and Striker forage rape grew well in a containment greenhouse. Plants showed some evidence of the hairy root phenotype with leaf wrinkling visible (Fig. 1d). Plants were fertile and a good seed set was obtained. Survival of CN95 kale on transfer to the soil was lower than usual due to callus on the base of cuttings. PCR analysis on 13 of the greenhouse kale and rape plants confirmed the presence of the BAR gene. Nineteen Basta-resistant plants of Tina, RGab2, TC22e9 swede are growing well in a containment greenhouse with slight evidence of the hairy root phenotype with leaf wrinkling apparent (Fig. 2). Some plants have initiated flowers. PCR analysis on nine of these plants has confirmed the presence of the *bar* gene (Fig. 3). Future work will involve Southern analysis to determine copy number insertion. In addition, further phenotype studies and controlled pollinations will be conducted to enable progeny analysis. Due to the breeding systems of kales, intermating of several independent trans-

Fig. 2. Transgenic RGab2 and TC22e9 swede 5 or 9 months after transfer to soil

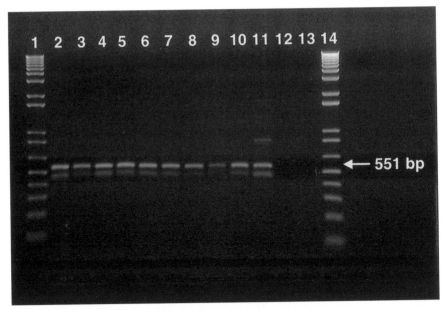

Fig. 3. PCR for the *bar* gene in transgenic RGab2 (RG), TC22e9 (TC) and Tina (T) greenhouse-derived swede. *Lanes 1* and *14* 1kb Plus DNA ladder; *2* TC#4b; *3* TC#4f; *4* TC#14a; *5* TC#14b; *6* TC#14c; *7* TC#14e; *8* RG2#6a; *9* RG2#9a; *10* T#2a; *11* pMOA4; *12* non transgenic cauliflower; *13* control with no DNA

genic plants will be required to maintain heterozygosity and prevent inbreeding depression (Conner and Christey 1994).

2.5 Field Evaluation of Transgenic Forage Brassicas

Field testing of transgenic forage brassicas is limited to two small-scale field trials in New Zealand. Kapeti kale plants regenerated from hairy roots from *Agrobacterium* wild-type strain A4T have been field tested (Christey and Sinclair 1993). At all stages in the development of these plants, both in vivo and in vitro, hairy root-derived plants were clearly distinguishable from control seed-derived plants. Hairy root-derived plants were shorter with smaller, more wrinkled leaves and total biomass was reduced. Both sources of plants had the same phenology. All plants were fertile but the seed set was considerably reduced on the hairy root-derived plants. In contrast, a field trial of several vegetable brassicas containing foreign genes included four lines (20%) which were indistinguishable from the controls, in addition to eight (40%) lines with moderate or slight hairy root (Ri) phenotype and eight (40%) lines with severe Ri phenotype (Christey et al. 1999b).

In addition, Basta-resistant forage rape and kale have been field tested. A total of 329 seedlings were planted, including three independent transgenic lines of CN95 forage kale, and one line each of Striker, H103d and 95A2/5

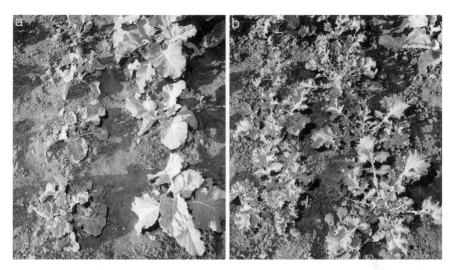

Fig. 4 a–b. Transgenic Basta-resistant plants 2 months after transplanting 8-week-old seedlings to the field. **a** CN95 forage kale; **b** Striker forage rape. In both cases control plants are in the *right-hand* row

forage rape. Selfed seed was collected from the original transgenic plants and planted in a contained greenhouse. As seedlings were segregating, 0.05% Basta was applied with a paint brush to one leaf per plant to select transgenic plants for field testing. As a control, non-transgenic plants were also planted. Eight-week-old seedlings were transplanted to the field. Two months after transplanting, Ri phenotype effects were obvious in CN95 forage kale (Fig. 4a) and H103d forage rape with plants being smaller and having smaller wrinkled leaves. In contrast, transgenic Striker rape (Fig. 4b) showed less effect of transformation. However, by the end of the trial Ri phenotype effects were obvious in all lines with four (66%) lines showing moderate or slight Ri phenotype effects and two (33%) lines severe Ri effects. Plant height was reduced for all transgenic lines by 1.5- to 2.2-fold and final biomass was 3- to 10-fold lower for all lines (Christey et al. 1999c).

2.6 Potential Uses

Numerous genes of agronomic interest have been introduced into *Brassica* species, generally via *A. tumefaciens*-mediated transformation of oilseed rape (reviewed in Poulsen 1996; Earle and Knauf 1999). In contrast to research with oilseed and vegetable brassicas, transformation research in forage brassicas is less extensive, with only two groups reporting the regeneration of transgenic plants (see Table 2). Once more efficient transformation systems are developed for forage brassicas, numerous traits could be improved via genetic engineering. In addition to the wide range of traits listed by Dale et al.

(1993) and Earle and Knauf (1999) already introduced into crop plants, including brassicas, other traits are likely to be of particular benefit in forage brassicas. For example, due to the use of self-incompatibility in forage brassicas, alteration of the self-incompatibility response may be useful, as has been achieved in broccoli by Toriyama et al. (1991).

Several herbicide-resistant genes are available that could be introduced into forage brassicas. These would enable better weed control and are likely to be of particular benefit in kale and swede. For example, wild turnip is a major problem in New Zealand and there is no selective herbicide available to control it in *Brassica* crops. In addition, alterations in glucosinolates are likely to be of major benefit in rape and kale. Use of transformation technology has enabled decreased levels of indole glucosinolates in oilseed *B. napus* via transformation with a tryptophan decarboxylase gene (Chavadej et al. 1994).

There are several insect and disease problems in forage brassicas of economic importance where biotechnology approaches could be of benefit. Palaniswamy (1996) lists 13 different types of insects as important cruciferous pests. Borges and Sequeira (1996) list numerous fungi, bacteria and viruses that are pathogenic on brassicas. In particular, swedes suffer from the fungal disease dry rot (or black-leg in rape caused by *Leptospheria maculans/Phoma lingam*), and conventional breeding for resistance is difficult. Clubroot caused by *Plasmodiophora brassicae* is a serious problem in many brassicas. Genetic manipulation in swede has already produced plants with increased insect resistance through use of *Bt* genes (Li et al. 1995). In addition, expression of a chitinase gene in oilseed rape has given field tolerance to the fungal pathogens *Leptospheria maculans*, *Sclerotinia sclerotiorum* and *Cylindrosporium concentricum* (Grison et al. 1996). Use of proteins and peptides with antibacterial activity like attacins and cecropins may enable control of bacterial diseases, as has been demonstrated in potato by Arce et al. (1999). In addition, extensive genetic mapping projects are well underway in various *Brassica* species and the closely related *Arabidopsis thaliana* to map the chromosomal location of useful genes (Quiros 1998). Progress towards mapping a gene involved in disease resistance has been made by Fuchs and Sacristan (1996), who used morphological markers to determine the chromosomal location of a gene for *P. brassicae* resistance in *A. thaliana*. Other mapping projects have located genes for blackleg resistance, glucosinolate and erucic acid content, oil seed quality and cytoplasmic male sterility restorers (Hu and Quiros 1996). Eventually, access to these genes will enable genetic manipulation of many traits in forage brassicas.

3 Conclusion

The demonstration of susceptibility to *Agrobacterium* spp. and regeneration of transgenic plants in important cultivars and elite breeding lines suggests that future research directed towards developing transgenic forage brassicas is promising. Results obtained to date suggest that forage kale and forage rape are more amenable to transformation than turnips, members of *B. campestris*, which is often recalcitrant to in vitro manipulation. As discussed above, numerous traits and genes could be introduced to improve forage brassicas. In particular, incorporation of herbicide resistance and genes to confer resistance to various diseases is likely to be of most benefit. Further field trials of transgenic forage brassicas containing such agronomically useful foreign genes are likely to be conducted soon.

Acknowledgments. The authors thank Jill Reader for greenhouse and field assistance, Robert Lamberts for the photos and Tony Conner and Stuart Gowers for comments on the manuscript.

References

Arce P, Moreno M, Gutierrez M, Gebauer M, DellOrto P, Torres H, Acuna I, Oliger P, Venegas A, Jordana X, Kalazich J, Holuigue L (1999) Enhanced resistance to bacterial infection by *Erwinia carotovora* subsp. *atroseptica* in transgenic potato plants expressing the attacin or the cecropin SB-37 genes. Am J Potato Res 76:169–177

Barrell P, Conner T (1995) Minimal T-DNA vectors for plant transformation. 6th Australasian Gene Mapping Workshop and New Zealand Genetical Society Conference, 27 Nov–1 Dec, University of Otago, Dunedin, New Zealand, Abstr 43

Borges MLV, Sequeira JC (1996) Soil solarization and phytosanitary problems of Brassica. Acta Hortic 407:461–468

Chavadej S, Brisson N, McNeil JN, de Luca V (1994) Redirection of tryptophan leads to production of low indole glucosinolate canola. Proc Natl Acad Sci USA 91:2166–2170

Christey MC (1997) Transgenic crop plants using *Agrobacterium rhizogenes*-mediated transformation. In: Doran PM (ed) Hairy roots: culture and applications. Harwood Academic Publishers, Amsterdam, pp 99–111

Christey MC, Sinclair BK (1992) Regeneration of transgenic kale (*Brassica oleracea* var. *acephala*), rape (*B. napus*) and turnip (*B. campestris* var. *rapifera*) plants via *Agrobacterium rhizogenes* mediated transformation. Plant Sci 87:161–169

Christey MC, Sinclair BK (1993) Field-testing of Kapeti kale regenerated from *Agrobacterium*-induced hairy roots. NZ J Agric Res 36:389–392

Christey M, Sinclair BK, Braun RH (1994) Phenotype of transgenic *Brassica napus* and *B. oleracea* plants obtained from *Agrobacterium rhizogenes* mediated transformation. VIII Int Congr of Plant Tissue and Cell Culture, 12–17 June, Florence, Italy, p 157

Christey MC, Sinclair BK, Braun RH, Wyke L (1997) Regeneration of transgenic vegetable Brassicas (*Brassica oleracea and B. campestris*) via Ri-mediated transformation. Plant Cell Rep 16:587–593

Christey MC, Braun RH, Kenel FO, Podivinsky E (1999a) *Agrobacterium rhizogenes*-mediated transformation of swede. Proc 10th Int Rapeseed Congr, 26–29 Sept, Canberra, Australia, CDROM

Christey MC, Braun RH, Reader J (1999b) Field performance of transgenic vegetable brassicas (*Brassica oleracea* and *B. rapa*) transformed with *Agrobacterium rhizogenes*. SABRAO J Breed Gen 31(2):93–108

Christey MC, Braun RH, Reader JK, Lambie JS, Forbes ME (1999c) Field testing transgenic Basta resistant forage kale and forage rape. Proc 10th Int Rapeseed Congr, 26–29 Sept, Canberra, Australia, CDROM

Conner AJ, Christey MC (1994) Plant breeding and seed marketing options for the introduction of transgenic insect-resistant crops. Biocontrol Sci Technol 4:463–473

Dale PJ, Irwin JA, Scheffler JA (1993) The experimental and commercial release of transgenic crop plants. Plant Breed 111:1–22

Downs CG, Christey MC, Maddocks D, Seelye JF, Stevenson DG (1994a) Hairy roots of *Brassica napus*: I. Applied glutamine overcomes the effect of phosphinothricin treatment. Plant Cell Rep 14:37–40

Downs CG, Christey MC, Davies KM, King GA, Seelye JF, Sinclair BK, Stevenson DG (1994b) Hairy roots of *Brassica napus*: II. Glutamine synthetase overexpression alters ammonia assimilation and the response to phosphinothricin. Plant Cell Rep 14:41–46

Earle ED, Knauf VC (1999) Genetic engineering in *Brassica*. In: Gomez-Campo C (ed) Biology of *Brassica* cenospecies. Elsevier, Amsterdam, pp 287–313

Earle ED, Metz TD, Roush RT, Shelton AM (1996) Advances in transformation technology for vegetable *Brassica*. Acta Hortic 407:161–168

Fuchs H, MD Sacristan (1996) Identification of a gene in *Arabidopsis thaliana* controlling resistance to clubroot (*Plasmodiophora brassicae*) and characterization of the resistance response. Mol Plant-Microbe Interact 9:91–97

Godwin I, Todd G, Ford-Lloyd B, Newbury HJ (1991) The effects of acetosyringone and pH on *Agrobacterium*-mediated transformation vary according to plant species. Plant Cell Rep 9:671–675

Grison R, Grezes-Besset B, Schneider M, Lucante N, Olsen L, Leguay J-J, Toppan A (1996) Field tolerance to fungal pathogens of *Brassica napus* constitutively expressing a chimeric chitinase gene. Nat Biotechnol 14:643–646

Hoerner GR (1945) Crowngall of hops. Plant Dis Rep 29:98–110

Holbrook LA, Miki BL (1985) *Brassica* grown gall tumourigenesis and in vitro of transformed tissue. Plant Cell Rep 4:329–332

Hu J, Quiros CF (1996) Application of molecular markers and cytogenetic stocks to *Brassica* genetics, breeding and evolution. Acta Hortic 407:79–85

Hussain MM, Melcher U, Essenberg RC (1985) Infection of evacuolated turnip protoplasts with liposome-packaged cauliflower mosaic virus. Plant Cell Rep 4:58–62

Janssen B-J, Gardner RC (1989) Localised transient expression of GUS in leaf discs following co-cultivation with *Agrobacterium*. Plant Mol Biol 14:61–72

Kado CI, Heskett MG, Langley RA (1972) Studies on *Agrobacterium tumefaciens*: characterization of strains 1D135 and B6, and analysis of the bacterial chromosome, transfer RNA and ribosomes for tumor-inducing ability. Physiol Plant Pathol 2:47–57

Lefebvre DD (1990) Expression of mammalian metallothionein suppresses glucosinolate synthesis in *Brassica campestris*. Plant Physiol 93:522–524

Li X-B, Mao H-Z, Bai Y-Y (1995) Transgenic plants of rutabaga (*Brassica napobrassica*) tolerant to pest insects. Plant Cell Rep 15:97–101

McNaughton IH (1976) Swedes and rapes *Brassica napus* (Cruciferae). In: Simmonds NW (ed) Evolution of crop plants. Longman, London, pp 53–56

Mukhopadhyay A, Arumugam N, Nandakumar PBA, Pradhan AK, Gupta V, Pental D (1992) *Agrobacterium*-mediated genetic transformation of oilseed *Brassica campestris*: transformation frequency is strongly influenced by the mode of shoot regeneration. Plant Cell Rep 11:506–513

Nieuwhof M (1993) Cole crops. In: Traditional crop breeding practices: an historical review to serve as a baseline for assessing the role of modern biotechnology. OECD, Paris, pp 159–171

Palaniswamy P (1996) Host plant resistance to insect pests of cruciferous crops with special reference to flea beetles feeding on canola – a review. Acta Hortic 407:469–481

Palmer TP (1983) Forage brassicas. In: Wratt GS, Smith HC (eds) Plant breeding in New Zealand. Butterworths, New Zealand, pp 63–70

Paszkowski J, Pisan B, Shillito RD, Hohn T, Hohn B, Potrykus I (1986) Genetic transformation of *Brassica campestris var. rapa* protoplasts with an engineered cauliflower mosaic virus genome. Plant Mol Biol 6:303–312

Poulsen GB (1996) Genetic transformation of *Brassica*. Plant Breed 115:209–225

Pua E-C, Lee JEE (1995) Enhanced de novo shoot morphogenesis in vitro by expression of anti-sense 1-aminocyclopropane-1-carboxylate oxidase in transgenic mustard plants. Planta 196:69–76

Puddephat IJ, Riggs TJ, Fenning TM (1996) Transformation of *Brassica oleracea* L.: a critical review. Mol Breed 2:185–210

Quiros CF (1998) Molecular markers and their application to genetics, breeding and evolution of *Brassica*. J Jpn Soc Hortic Sci 67(6):1180–1185

Radke SE, Turner JC, Facciotti D (1992) Transformation and regeneration of *Brassica rapa* using *Agrobacterium tumefaciens*. Plant Cell Rep 11:499–505

Rouan D, Montané M-H, Alibert G, Teissié J (1991) Relationship between protoplast size and critical field strength in protoplast electropulsing and application to reliable DNA uptake in *Brassica*. Plant Cell Rep 10:139–143

Tanaka N, Hayakawa M, Mano Y, Ohkawa H, Matsui C (1985) Infection of turnip and radish storage roots with *Agrobacterium rhizogenes*. Plant Cell Rep 4:74–77

Thompson KF (1976) Cabbages, kales etc. *Brassica oleracea* (Cruciferae) In: Simmonds NW (ed) Evolution of crop plants. Longman, London, pp 49–52

Toriyama K, Stein JC, Nasrallah ME, Nasrallah JB (1991) Transformation of *Brassica oleracea* with an *S*-locus gene from *B. campestris* changes the self-incompatibility phenotype. Theor Appl Genet 81:769–776

U N (1935) Genome-analysis in *Brassica* with special reference to the experimental formation of *B. napus* and peculiar mode of fertilization. Jpn J Bot 7:389–453

8 Transgenic Chicory (*Cichorium intybus* L.)

M. ABID, B. HUSS, and S. RAMBOUR

1 Introduction

Chicory, a member of the *Asteraceae* family, is a dioecious species ($2x = 2n = 18$). In Europe three species are usually grown. *Cichorium endivia* L. and *C. spinosum* L. are annuals predominantly grown in southern Europe and northern Africa and consumed as salads, whereas *C. intybus* L. comprises annual, biennial or perennial varieties which are widespread in Europe and Asia, and have been naturalised in America, Australia and South Africa. For a long time, *C. intybus* was used as a medicine, or as a foodstuff by humans and cattle. Besides endive and curly endive, some other varieties with large and more or less red leaves are predominantly grown in Italy (cv. Verona, Trevisia, Chioggia), and consumed as salads as well. The varieties sativum (= Magdeburg) and Witloof are usually grown as biennials and are mainly cultivated in the Netherlands, Belgium and North of France.

 C. intybus has a stout tap-root and large, slightly hairy serrated leaves with pointed lobes. The stems grow to a height of about 1–1.5 m and bear clusters of small capitula composed of ligulated flowers in the axils of their leaves. The corolla is blue in colour, pollination is by insects and the achenes are about 3 mm long, grey-brown, angular and surmounted by a pappus of finely divided scales.

 Cultivars of the variety Magdeburg produce deep and well-developed roots storing fructans of the inulin type. They are grown as root crops in much the same way as sugar beet. The roots are roasted to a powder that closely resembles ground coffee in both flavour and appearence. Nowadays there is growing interest in these cultivars as a source of fructose and fructans which represent 80 to 85% of the dry weight (Fiala and Jolivet 1980). The Witloof cultivars arose in Belgium in 1845 and were derived from the Magdeburg variety. A gardener of the botanical garden of Brussels is said to have obtained this new variety which is sometimes called "chicory of Brussels". Seeds are sown in May, it is harvested in September–October and the defoliated roots are grown in forcing pits or chambers in complete darkness at 18 °C and 90% relative humidity so that they produce an etiolated bud, commonly named

Laboratoire de Physiologie et Génétique Moléculaire Végétales, Université des Science et Techniques de Lille, 59655 Villeneure d'Ascq cedex, France

Biotechnology in Agriculture and Forestry, Vol. 47
Transgenic Crops II (ed. by Y.P.S. Bajaj)
© Springer-Verlag Berlin Heidelberg 2001

"chicon". It is built of white tightly imbricated leaves which are consumed as a salad.

In France, the first in vitro cultures of *C. intybus* were initiated by Gautheret (1941). Chicory is a particularly suitable species for in vitro regeneration and genetic engineering.

The growth rate of the seedlings of *C. intybus* is low and fast developing weeds compete with them, so that resistance to non-selective herbicides is of interest. Both the quality of the bud and the susceptibility of the tap-root to bacterial or fungal pathogens during the hydroponic process depend upon the cultural conditions encountered in the field, and particularly from both the quantity and the quality of the nitrogen source. Thus, genetic engineering, which could introduce specific resistance to herbicides or to pathogens by modifying the nitrogen metabolism, may be of use to chicory producers. Similarly, altering the fructan metabolism in the Magdeburg varieties is gaining interest.

In chicory, promoters driving genes expressed in an organ-specific fashion have so far not been characterised. Thus, before starting transformation of chicory with genes providing traits of agronomic interest, we used more conventional constructs to allow analysis of gene expression in tissues at different developmental stages. A summary of the genetic transformation methods used with chicory by various workers is given in Table 1.

The aim of our work was: (1) to compare the ability of genetic transformation of different organs by *Agrobacterium tumefaciens* strains containing the neomycine phosphotransferase II gene as a selectable marker; (2) to analyse how *uidA* coding for β-glucuronidase and used as a reporter gene is controlled by either constitutive (35 S of Cauliflower Mosaic Virus) or inducible (mas2′ of mannopine synthase) promoters; (3) to analyse how *uidA* is expressed in different organs at different developmental stages of R1-transformed chicory. Activation of mas2′ and CaMV 35S promoters by wounding and by some growth regulators and the stability of the transgenes in R1 progeny were analysed as well.

2 Genetic Transformation

2.1 Review of Work (Table 1)

Genetic transformation of *C. intybus* was first reported by Sun et al. (1991). Two different strains of *Agrobacterium rhizogenes* were used and regeneration of whole plants was performed from transformed roots. The transgenic plants showed the characteristic phenotype of wrinkled leaves, increased branching, sterility and annual flowering, which may be of interest since chicory is biennial.

An acetolactate synthase gene from *Arabidopsis thaliana*, conferring resistance to sulfonylurea herbicides, was introduced by Vermeulen et al. (1992).

Table 1. Characteristics of and references on transformation of chicory. *csrl-1* is a mutant aceto-lactate synthase gene from *Arabidopsis thaliana*; *nptII* neomycinphosphotransferase II conferring kanamycin resistance; *uidA* β-glucuronidase encoding gene. *pGUSIN* contains an intron in the *uidA* gene that is correctly spliced in eukaryotic cells. *6G-FFT* encodes 6G-fructosyltransferase from onion

Varieties	Agrobacterium strain/plasmid	Explant source	Selective agent	Genes transferred	Reference
1. *C.* intybus L. Hybrid Flash	*A. rhizogenes* A4RSII and 8196	Roots, floral stems	–	T_L and T_R-DNA	Sun et al. (1991)
2. *C. intybus* L.	*A. tumefaciens* LBA4404/ pGH6	Leaf	Kanamycin	csrl-1/nptII	Vermeulen et al. (1992)
3. *C. intybus* cv. Rossa di Chioggia	pGV3850/ pKU2 LBA4404/ pBI121 pGV2260/ pGUSIN	Leaf	Kanamycin	nptII uidA/nptII uidA	Genga et al. (1994)
4. *C. intybus* L. Hybrid Flash	pGV2260/ pGSGLUC1 pGV2260/ pTDE4	Tap-root, leaf, Cotyledon	Kanamycin	nptII/uidA nptII/uidA	Abid et al. (1995)
5. *C. intybus* L. cv. *sativum*	pGV2260/ 35SGUSIN pGV1531	Shoot buds	Kanamycin	nptII/uidA	Frulleux et al. (1997)
6. *C. intybus*	*A. tumefaciens*	Leaf	Kanamycin	6G-FFT	Vijn et al. (1997)

Numerous shoots regenerated from leaf discs infected with *A. tumefaciens*. The transgenic ones were selected once shoots were on rooting medium.

Histochemical analysis of *uidA* encoding β-glucuronidase has been performed in *C. intybus* by Abid et al. (1995) and showed differential expression of the transgene when controlled either by the mas2′ or CaMV35S promoter. Tap-root, leaf and cotyledon explants were assayed, the last being best adapted for efficient transformation (this work). Frulleux et al. (1997) used micropropagated shoot buds for transformation, but the efficiency was no better.

Recently, sucrose: fructan 6-fructosyltransferase (SST) from onion has been introduced into chicory (Vijn et al. 1997). SST is the key enzyme in the formation of fructan of the inulin neoseries. In addition to the endogenous linear inulin, transgenic chicories accumulated new types of fructan with a higher degree of polymerisation.

2.2 Materials and Methods

2.2.1 Plant Material

Tap-root explants, cotyledons and leaves from *Cichorium intybus* L. var. Witloof, cv. Flash were used for transformation. The tap-roots were surface sterilised for 20 min in a calcium hypochloride solution ($70 \, g \, l^{-1}$) and rinsed three times in sterile water. The explants (6 mm diameter and 2 mm thickness) were cut out of the vascular cambium with a sharpened cork-borer.

Seeds were surface sterilised for 15 min in a 0.1% mercuric chloride aqueous solution, rinsed and germinated for 2 days on H2O medium. Some of the seedlings were grown in the dark for 5 days and etiolated cotyledons were cut off and used for agroinfection. The remainder were grown for 15 to 17 days at a light irradiance of $14 \, \mu mol \, m^{-2} \, s^{-1}$ with a photoperiod of 16 h/8 h light/dark and the young leaves were used for *Agrobacterium* infection.

2.2.2 Agrobacterium *Strains*

Transformation was carried out with *Agrobacterium tumefaciens* C58C1 harbouring pGV2260 (Deblaere et al. 1985) and the binary plasmid pGSGLUC1 (Mendel et al. 1989) or pTDE4 (kindly provided by Plant Genetic Systems, Gent-Belgium). The pGSGLUC1 vector contained *nptII* and *uidA* genes under the control of the mas1′ and mas2′ promoter respectively (Fig. 1A). The vector pTDE4 contained the same genes except *uidA* was under the control of the CaMV 35S promoter and *nptII* under the *nos* promoter of the nopaline synthase (Fig. 1B). The *nptII* gene was used for the selection of transformed plant cells which acquired kanamycin resistance, while *uidA* expression was assayed in different tissues and organs by means of histochemical and fluorimetric analysis.

The bacterium strains were grown at 28 °C in YEB medium (bacto-yeast extract $1 \, g \, l^{-1}$, bacto-beef extract $5 \, g \, l^{-1}$, peptone $5 \, g \, l^{-1}$, sucrose $5 \, g \, l^{-1}$, pH 7.2) containing appropriate antibiotics. When the OD value of the suspension at 600 nm reached 1, bacteria were pelleted and resuspended in 10 ml of liquid Murashige and Skoog medium (1962) supplemented with 100 µM acetosyringone.

2.2.3 Genetic Transformation

The tap-root explants were floated for 10 min in 10 ml of infection medium H5-1 containing 1 ml of *Agrobacterium* suspension. The explants were slightly dried on filter paper, transferred onto H5 medium and incubated at low light intensity. After 2 days of co-culture, the explants were washed with H5 liquid medium containing $500 \, mg \, l^{-1}$ cefotaxime, patted dry on sterile filter paper and transferred onto H5 solid medium supplemented with $75 \, mg \, l^{-1}$ kanamycin.

A

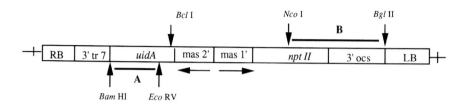

B

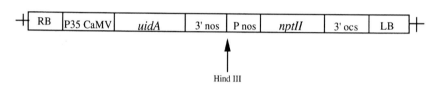

Fig. 1. Structure of the T-DNA of the binary vectors pGSGLUC1 and pTDE4 used for transforming chicory explants (Abid et al. 1995). In pGSGLUC1, *nptII* and *uidA* genes were under the dual mas1'2' promoter respectively (*horizontal arrows*). *3' ocs 3'* region of the octopine synthase gene. *3'tr 7 3'* region of gene 7. *Arrows* define the restriction fragments used as probes. In pTDE4, the *uidA* gene was behind the CaMV 35S promoter and the *nptII* gene was controlled by the nopaline synthase gene (nos) promoter. *3'nos*: 3' region of *nos* gene

The explants were then kept at 24 °C at high light intensity with a day length of 16 h.

The same transformation technique was used with wounded etiolated cotyledons and leaves. Infection was carried out on M1 medium, co-culture on M2 medium and M3 was used as the selection medium.

Small shoots which differentiated on the calli 2 or 3 weeks after the onset of selection were transferred onto the rooting medium (H10).

2.2.4 Growth Media

H5: salts of Heller's medium (Heller 1953) containing $5 \mathrm{g} \mathrm{l}^{-1}$ sucrose, $5 \mathrm{g} \mathrm{l}^{-1}$ agar.

H5-1: liquid H5 medium supplemented with 100 µM acetosyringone.

H10: solid H5 medium except it contained $10 \mathrm{g} \mathrm{l}^{-1}$ sucrose and $75 \mathrm{mg} \mathrm{l}^{-1}$ kanamycin sulphate.

H20: solid H5 medium except it contained $20 \mathrm{g} \mathrm{l}^{-1}$ sucrose.

M: Murashige and Skoog's medium (1962) except macroelements were used at their half concentration, $20 \mathrm{g} \mathrm{l}^{-1}$ sucrose, $5 \mathrm{g} \mathrm{l}^{-1}$ agar.

M1: liquid M medium supplemented with 100 μM Acetosyringone.
M2: solid M medium supplemented with 0.1 mg l^{-1} BAP.
M3: solid M2 medium supplemented with 500 mg l^{-1} cefotaxime and 75 mg l^{-1}
kanamycin sulphate.

2.2.5 Electron Microscopy

Tap-roots, leaves and cotyledons were harvested 2 days after infection with
the pTDE4-containing strain. After a brief wash in 0.1 M cacodylate pH 7.4,
samples were fixed in 1% glutaraldehyde in 0.1 M cacodylate pH 7.4 and
further treated as described by McCowan et al. (1978).

2.2.6 GUS Assays

Histochemical localisation of GUS activity was performed with the chro-
mogenic substrate 5-bromo-4-chloro-3-indolyl-β-D-glucuronic acid (X-Gluc)
according to Jefferson et al. (1987) with some modifications. Freshly harvested
tissues from transformed and untransformed chicories were immersed into a
0.1 M sodium phosphate buffer pH 7.0, containing 1 mM X-Gluc and then
incubated at 37 °C for various times (1 to 12 h). Sectioning was performed on
a microtome using a glass knife. After fixation and dehydration, 2 μm-thick
sections were observed under the microscope. Coloration appears pink in sections observed under dark field optics. Flower buds were sectioned with a
vibroslicer microtome. Cross sections 100 μm thick were incubated for several
hours with the reactives, then dehydrated in 95% ethanol and observed with
an Olympus BH-2 microscope.

Fluorimetric GUS assays were performed with protein extracts of plant
tissues of the R1 issue containing a single T-DNA (lines 9 and 4 transformed
respectively with pGSGLUC1 and pTDE4). Leaves and roots were collected
from the same plant. Transformed tissues were wounded or treated with plant
growth regulators (indole-3-acetic acid, 2,4-dichlorophenoxyacetic acid, 6-
benzylaminopurine, abscisic acid, gibberellic acid A3 used at 10 mg l^{-1}) in MS
solid medium for 24 h. Proteins were extracted according to Jefferson et al.
(1987). GUS activity was assayed and expressed as nmol of 4-methylumbel-
liferone per min and mg protein. Fluorescence was measured with an excita-
tion length of 365 nm and emission at 455 nm on a Hitachi spectrofluorimeter.
Fluorescence of 100 and 1000 nM 4-methylumbelliferone in 0.2 M sodium carbonate was used for calibration. Protein levels were measured according to
Bradford (1976).

2.2.7 Evaluation of Seed Progeny from Transgenic Plants

To induce flowering, the transformants were vernalised at 4 °C for 7 weeks and
subsequently transferred to the greenhouse for autopollination. Seeds were

harvested and evaluated for kanamycin resistance by aseptically germinating them on H20 media supplemented with 75 mg l^{-1} kanamycin.

2.2.8 Southern Analysis

Genomic DNA was extracted from leaves of either transformed parents and the R1 generation or untransformed plants according to Dellaporta et al. (1983) and digested with HindIII. Ten micrograms of DNA was elec-trophoresed on 0.8% agarose gel, blotted, and fixed to Hybond-N$^+$ (Amer-sham) filters. Hybridisation was performed at 42 °C in the presence of 50% formamide and 10% dextran sulfate (Sambrook et al. 1989). ^{32}P-labelled DNA fragments containing either the coding sequence of *uidA* or *nptII* were sepa-rately used on the same blot. For dehybridation, membranes were incubated 30 min in 0.4 M NaOH at 45 °C and 30 min in 0.1 SSC, 0.1 SDS, 0.2 M Tris-HCl, pH 7.5 at 45 °C.

3 Results

3.1 Genetic Transformation

Chicory explants were first subjected to antibiotic selection: the level of kanamycin suitable for selection of the transformed tissues was determined at 75 mg l^{-1} regardless of the origin of the explants. At this concentration, untrans-formed tissues did not survive whilst the putative transformed calli showed little if any growth inhibition. Tap-roots contain numerous cambial cells which can differentiate numerous, vigorous, fast growing shoots. Thus, in order to prevent a very rapid differentiation of putative chimaeric shoots, the Heller's medium, which is relatively poor in overall mineral nutriments, was used. Con-versely, cotyledon and leaf explants which are unable to produce organs on simple media, were grown on MS medium supplemented with BAP. Under these conditions, leaf explants produced more shoots than did root or cotyle-donary explants (Table 2). Transformation efficiency was not related to regen-eration capability. In our experiments, acetosyringone was used during the co-cultivation period in order to increase transformation efficiency. The trans-formation frequencies were calculated as a percentage expressing the number of explants forming transgenic shoots under selective pressure to the total number of explants grown under selective pressure. It reached 5% in root explants regardless of the strain used (Table 3). Moreover, the transformants showed a transient GUS activity which was detectable in the first leaves, sub-sequently decreased, and became undetectable in the further developing leaves. Thus, tissues excised from the tap-root were not suitable for genetic transformation.

Table 2. Frequency of callus formation and shoot regeneration from chicory explants. Tap-root explants were grown on H5 medium, leaf and cotyledon explants on M2 medium

Explant origin	Number of explants tested	Calluses → shoots	Percent	Shoots/callus (mean + SE)
Tap-root	70	68	97	4 ± 0.2
Leaf	40	40	100	9 ± 0.5
Cotyledon	40	36	90	5 ± 0.4

Table 3. Efficiency of the transformation of chicory explants with pTDE4 and pGSGLUC1 vectors; km, kanamycin

Origin of the explant	Plasmids used	Number of explants tested	Explants with km-resistant shoots	Total number of transgenic shoots	Frequency of transformation (%)
Tap-root	pTDE4	39	2	9	5
	pGSGLUC1	57	3	5	5
Leaves	pTDE4	45	4	11	9
	pGSGLUC1	84	5	7	6
Cotyledon	pTDE4	60	9	21	15
	pGSGLUC1	71	7	13	10

With leaf and cotyledon explants, the transformation process varied according to the strain used. More regenerants expressing both GUS and NPTII activities were obtained when co-cultivated with the strain containing pTDE4, whereas far fewer transformants were obtained with the strain containing pGSGLUC1. Transformation efficiency has been increased by performing the antibiotic resistance selection in two steps, first using 50 mg l^{-1} of the selective agent for 5 days, then 75 mg l^{-1}, since tissues tended to desiccate when immediately placed to the lethal dose. This selection process did not succeed with tap-roots because of a high escape rate.

Although leaf tissues possessed a higher regenerative capacity than cotyledonary tissues, their transformation efficiency was lower. Using the pGSGLUC1 strain, the transformation frequency only reached 6% while it rose up to 9% with the pTDE4 strain. In cotyledons, the transformation frequencies attained 10% with pGSGLUC1 and 15% with pTDE4 (Table 3). Transformation efficiency has recently been enhanced to 15% when leaf samples were incubated for 3 days, prior to *Agrobacterium* infection, on M medium containing 2,4-D at 1 mg l^{-1}. Hormonal pretreatment of cotyledons did not result in an increased shoot formation since numerous vitrous calluses, unable to regenerate, were obtained. Nevertheless, cotyledons proved to be the most convenient organs for efficient transformation of chicory since twice as many shoots which were able to root were obtained in less time. Regenerated shoots were transferred onto the H10 medium after 3–4 weeks. Attempts to improve rooting of the shoots showed that 0.1 mg l^{-1} α-naphtaleneacetamide is efficient, nevertheless rooting has also been obtained without hormonal supply. The rooted plants (up to 75%) were grown to maturity in a greenhouse.

3.2 Electron Microscopy

The attachment of *Agrobacterium* was documented using scanning electron microscopy. Massive aggregates of agrobacteria occurred in the wounded regions of the different explants. However, the density of the bacteria was higher on cotyledons than on leaves and tap-root explants (Fig. 2). Pretreatment with 2,4-D seemed to attract a high density of agrobacteria over the whole leaf surface, thus increasing their adhesion and leading to a higher transformation frequency. Cotyledon pretreatment did not result in a better bacterial colonization (data not shown).

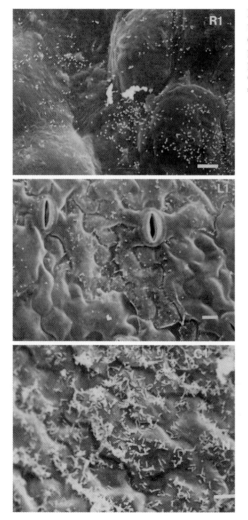

Fig. 2. Scanning electron micrographs of tap-root (*R1*), leaf (*L1*) and cotyledon (*C1*) explants, 24 h after infection with the pTDE4-containing strain. Note the high presence of bacteria on the cotyledonary explant (Abid et al. 1995). *Bar* = 10 µm, *R1* = ×400, *L1* = ×600, *C1* = ×1000

3.3 Analysis of β-Glucuronidase Activity

Tissue-specific expression of *uidA* was analysed by histochemical staining with X-Gluc as the substrate. GUS activity measured in leaves and roots of plants transformed by *Agrobacterium* containing either pGSGLUC1 or pTDE4 displayed distinct patterns of activity. Within the roots, the mas2′-*uidA* expression was intense in the apices and staining was particularly dense in the cap, meristem, outer cortex and epidermis. Expression was less intense within the inner cortex and the vascular stele. In the elongating zone, outside the meristem, staining was restricted to the outer cortex. The further the outer cortex was away from the tip, the fewer cellular layers were stained (Fig. 3A). Conversely, in leaves, all cellular types expressed β-glucuronidase (Fig. 3B). In the flower bud, GUS activity was detected in vascular tissues of the receptacle (Fig. 3C) and in those of the anthers and the ovaries. *uidA* was rarely expressed in pollen grains. High activity was mainly found in cotyledonary areas of zygotic embryos (Fig. 3D). In the young seedlings, expression was observed only as one or two spots in the cotyledons (Fig. 3E), but was not present in mature plants (2- to 3-month-old).

In chicory transformed with *uidA* under the control of the CaMV 35S promoter, GUS activity in the leaves was detected within all cellular types, however, it was higher in vascular tissues than in mesophyll cells (Fig. 4A). In the roots, the expression pattern varied according to the developmental stage and among individual transformants: activity was detected in the tip, in the meristematic region, in the whole cortex and in the vascular tissues of very young organs, for instance in 3-week-old roots (Fig. 4B). At older stages (2-month-old), most of the plants examined lacked detectable expression inside the stele. Similarly, staining was undetected at the emergence of lateral roots (Fig. 4C). Thin histochemical sections clearly showed that GUS activity was essentially localised within the cortex and endodermis whereas it was absent in the vascular tissues (Fig. 4D). In flower buds, activity was high in the receptacle at the location where female organs differentiate (Fig. 4E), but was never detected within ligulate florets. In more detail, expression was observed in the vascular tissues of the ovule, in the nucellus and in the embryo sac (Fig. 4F). In the male reproductive organs, staining was found in the vascular bundles of the anthers and the style (Fig. 4G). Expression was further found in the five individualised anthers (Fig. 4H). The percentage of X-Gluc positive pollen grains reached only 1%, suggesting that the CaMV 35S promoter was not active. A heterogenous distribution pattern was observed in embryos where coloration was detected in the whole of the embryo or only in cotyledonary or root tissues (Fig. 4I). After the seeds germinated, the pattern of GUS activity changed: activity was higher in the cotyledons than in the leaves. Nevertheless, during the plant growth, promoter activity became substantially higher in mature leaves than in young developing ones (Fig. 4J). A gradual increase of GUS expression from young leaves towards the mature ones was observed.

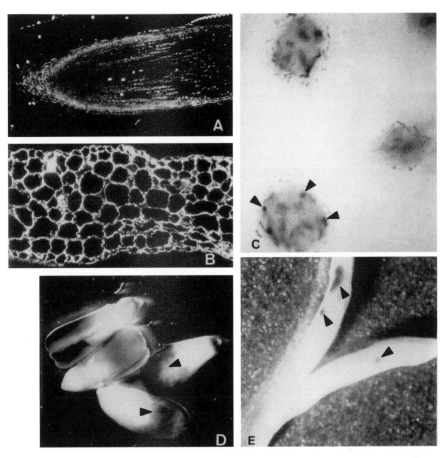

Fig. 3A–E. Histochemical localisation of β-glucuronidase under the control of the mas2′ promoter in transformed chicory. *Arrows* indicate GUS activity. **A** Longitudinal section of the root tip; **B** thin cross section of a leaf; **C** transverse section of flower bud; vascular tissues are stained. **D** Zygotic embryos; **E** 3-day-old seedling

3.4 Enhancing Effects of Wounding and Plant Growth Regulators on *uidA* Expression Controlled by mas2′ or CaMV 35S Promoters

The histochemical staining was extended to fluorimetric analysis of GUS activity measured in roots and leaves of R1-transformed chicory plants belonging to lines 4 and 9 and treated with different plant growth regulators. Roots and leaves from the same transformed or untransformed plants were wounded or treated for 24 h with IAA, 2,4-D, BAP, ABA or GA₃ as described in materials and methods.

It should be noted that under the control of mas2′, GUS activity was eight times more important in transformed versus untransformed leaves, whereas activities were similar in roots (Table 4). This result may be due to the whole root being used in the fluorimetric analysis which weakened the signal mainly

Table 4. Fluorimetric assays of GUS activity in leaf and root tissues from transformed chicory containing one copy of T-DNA of either pGSGLUC1 or pTDE4. Measurements were taken 24h after wounding or incubation with the plant growth regulators used at a concentration of 10mg l⁻¹. Results correspond to two different tests done on the same sample

| | GUS activity (nmole 4-MU per min⁻¹ per mg⁻¹ protein) | | | |
| | mas2′ promoter | | CaMV 35S promoter | |
	Leaf	Root	Leaf	Root
Untransformed	174 ± 8	104 ± 10	33 ± 2	15 ± 6
Transformed	1 390 ± 251	131 ± 22	710 ± 113	425 ± 96
Wounded transformed	8 518 ± 709	3 500 ± 183	815 ± 88	350 ± 76
2,4-D	16 684 ± 5 419	1 756 ± 435	1095 ± 102	630 ± 72
AIA	2 500 ± 928	1 470 ± 366	–	–
BAP	1 622 ± 758	–	–	–
GA₃	1 418 ± 153	1 341 ± 294	–	–
ABA	3 550 ± 240	17 708 ± 3 580	657 ± 45	470 ± 42

– Not determined.

localized at the apices. Driven by the CaMV 35S promoter, GUS activity was considerably increased, confirming this promoter is strong in chicory tissues.

When wounded by laceration, GUS activity controlled by the mas2′ promoter was 27 times greater in roots and six times greater in leaves compared with the transformed controls (Table 4). On the other hand, the CaMV 35S promoter did not seem to be inducible by wounding since activities measured in transformed or untransformed tissues were similar.

Regarding plant growth regulators, the synthetic auxin 2,4-D increased the GUS activity 12-fold in mas2′-*uidA*-transformed leaves, whereas the effect of IAA was far weaker. Also, ABA exerted a lesser effect than 2,4-D, since GUS activity was only 2.5 times greater. BAP and GA₃ only promoted a very weak response similar to the transformed control. In the roots, 2,4-D, IAA and GA₃ induced mas2′ promoter to the same extent while ABA increased the response 135-fold. CaMV 35S promoter cannot be described as inducible by plant growth regulators as an increase of only 1.5-fold has been determined in 2,4-D treated leaves or roots.

3.5 Southern Hybridisation Analysis of Transformants

Independently transformed plants were analysed for the presence of the transgenes and for the pattern of T-DNA integration. Plant DNA was digested by *Hind*III and two probes prepared from pGSGLUC1 were used: an *Eco*RV/*Bam*HI fragment corresponding to the *uid*A gene (probe A) and an *Nco*I/*Bgl*II fragment corresponding to the *npt*II gene (probe B; Fig. 1A). Results belonging to pTDE4-containing lines are shown in Fig. 5. The number of transgenes present in the transformed plants varied from one to five copies per 2C genome of chicory.

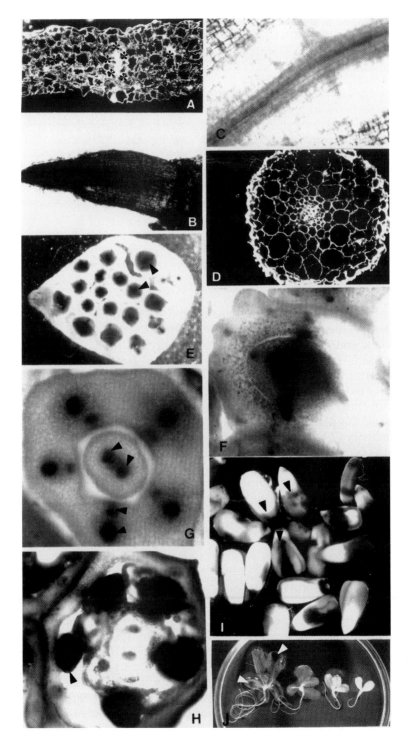

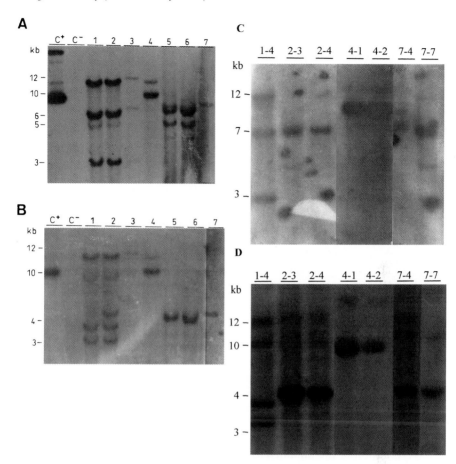

Fig. 5. Southern blot analysis of transgenic R0 and R1 chicories. Hybridisation with the *uidA* probe (**A, C**) and with the *nptII* probe (**B, D**) from the mas1'*nptII*/mas2'*uidA*-containing plants. *Numbers in the margin* refer to molecular weights (in kb). **A, B.** Parental chicories; *lane C⁻* untransformed plant; *lane C⁺* positive control; *lanes 1–7* transformed plants. **C, D** R1 progeny. Corresponding parental plants are indicated by the first number

◄───

Fig. 4A–J. Histochemical localisation of β-glucuronidase under control of CaMV 35S promoter. *Arrows* or *asterisks* indicate GUS activity. **A** Cross section of leaf. Expression is detected in mesophyll cells but is particulary intense in vascular tissues. **B** Young root (5-day-old) showing intense activity over all the organ; **C** longitudinal section of an older root (2-month-old) at the point of lateral root emergence. Expression is present in vascular tissues, but remains undetected in the vascular stele of main and lateral roots. Cortex cells are still stained; **D** thin cross section of a 2-month-old root. Little if any activity is detected in vascular tissues. Cortex and rhizoderm cells express *uidA*. **E** Receptacle of a young flower bud. All the florets are stained. **F** Cross section of an ovary. **G** Particularly high GUS activity was observed in vascular tissues of both the anthers and the stamen (in the middle of the photograph). **H** Transverse section of anthers just before dehiscence. **I** Zygotic embryos. **J** Plantlets (1-month-, 2- and 1-week- and 5-day-old; from *right* to *left*). Weak activity is shown in young compared with older tissue

The hybridisation pattern of plant 7 showed only one band when probed with either *uidA* or *nptII*, indicating that it contained one copy of a complete T-DNA. Seeds from self-pollination of R0 transformants were germinated and analysed to verify the stability of the transgenes. The progeny 7-4 showed a loss of the *uidA* fragment while the progeny 7-7 exhibited the same integration pattern as the parent.

Plant 4 showed single copies of *uidA* and *nptII*. The fragments hybridised with either probe A or B migrated the same distance (10.5 kb) supposing that both *Hin*dIII sites flanking the T-DNA were equidistant or the *Hin*dIII site inside the T-DNA was lost. Two plants, 4-1 and 4-2, of the R1 generation showed that the hybridising *uidA* and *nptII* fragments detected in the R0 plant segregated in the R1 progeny, indicating that the T-DNA was not subjected to recombination events and was stably transmitted.

Plants 5 and 6 (from the same transformation event) and plant 3 contained more copies of *uidA* than of *nptII* while plants 1 and 2 showed more copies of *nptII* than of *uidA*. In these plants the transferred DNA was probably fragmented, yielding incomplete copies of T-DNA. Analysis of plants 1-3 and 1-4 showed, as in the parent plant, three *uidA* fragments. A reprobing filter revealed the presence of four *nptII*-hybridising fragments. Thus four T-DNA copies were integrated, one insertion of *uidA* was lost during the transformation process and the *uidA* and *nptII* sequences segregated independently.

The situation was much more complicated in the R1 generation of plants from line 2 which lost both *uidA* and *nptII* sequences. Analysis of the parent has determined five copies of *nptII* and three copies of *uidA*. Plants 2-3 and 2-4 retained only one copy of *uidA* localised at 8 kb and only one of *nptII* localised at 5 kb. Four copies of *nptII* and two of *uidA* were lost in the progeny.

Segregation of the kanamycin resistant phenotype among the R1 generation was easily detected as seedlings possessed healthy green leaves in contrast to the yellow/white leaves of the sensitive phenotype. The ratio of resistant to sensitive seedlings indicated that for 60% of the plants the *nptII* transgene was transmitted to the R1 progeny in a Mendelian fashion as expected for a dominant nuclear gene and segregated in 3:1 ratio, as shown for plants 4, 7 and 9 (Table 5). Conversely, plants 1 and 2 showed inverted 1:3 ratios indicating that the kanamycin resistance was transmitted as a recessive trait. G analysis confirmed the homogeneity of the statistical data between plants 1 and 2 and between plants 4, 7 and 9. Why a number of seeds did not germinate remains unexplained.

4 Discussion

Chicory can be transformed by different *Agrobacterium* strains, but the efficiency of the process is dependent upon the nature of the organs used.

Table 5. Segregation ratios of some transformants, based on kanamycin resistance during seed germination. R1 seeds were germinated on H2O medium with 75 mg l^{-1} kanamycin. The number of kanamycin-resistant seedlings were screened after 3 weeks of culture at 24 °C. R or S kanamycin-resistant or -sensitive. Probability of segregation was determined by X^2. A X^2 value of ≥3.83 implies significant deviation from the expected ratio: 3/1 at $p = 0.05$. The G-test was used at $p = 0.001$

Plants	Sown seeds	Germinated seeds	R	S	X^2 R/S; 3/1	X^2 S/R; 1/3	G-test
1	173	62	23	39	47.5	3.63	2.29
2	215	72	18	54	96	0.00	
4	133	91	69	22	0.03	–	1.74
7	59	29	19	10	1.39	–	
8	150	59	40	19	1.63	–	

Explants from tap-roots produced transformants at a low frequency. The tap-root explants were isolated from the vascular cambium, which retains meristematic characteristics. The difficulties in obtaining transgenic plants may be due to the low competence often ascribed to meristematic cells. On the other hand, wounded tap-root tissues produce latex, thus the lack of transformation might be due to bacterial attachment being hindered. Scanning electron microscopy showed that this was not the case. Even on very simple media such as the Heller's, explants cut out of tap-roots first undergo an intense proliferation and then rapidly differentiate vigorous shoots. Formation of chimaeric shoots containing both stably or transiently transformed and untransformed cells may thus easily arise. This suggests that shoots obtained from tap-roots present a high frequency of "escapes" which results in transient expression, because the copies of the T-DNA are not integrated into the nucleus. Thus, despite its outstanding regeneration capacities, the vascular cambium of chicory seems to be inappropriate for genetic transformation by *Agrobacterium* strains.

Leaf discs constituted a more suitable material for genetic transformation. Failure in yielding chicory transformants by the leaf disc procedure was reported by Vermeulen et al. (1992): in the presence of kanamycin, leaf explants became necrotic and shoot regeneration failed. We did not encounter such difficulties with our material, whatever organ was used. Although the yield of regenerants was lower in the presence of kanamycin than in control conditions, shoot differentiation occurred. However, reliable transformation was only effective in the presence of acetosyringone. The discrepancy between these results may be explained in terms of the initial physiological state of the plant material: the leaf discs we used originated from 3-week-old seedlings grown in vitro, whilst Vermeulen et al. used leaves from 2-month-old plants. The transformation efficiency of leaf explants was lower than with cotyledonary explants except when pretreated with 2,4-D which is more time consuming. Thus, cotyledons were the most suitable organs for fast and easy transformation. The frequency of transformation correlates with the greater abundance of agrobacteria observed on cotyledonary explants.

Analysis by Southern blot hybridisation suggested that rearrangement of the transferred T-DNAs occurred prior to the transformation event. This may result from rearrangement of T-DNA into the host plant genome or by fragmentation of the transferred DNA before integration. Restriction sites may be lost and modified copies of the T-DNA may be integrated. Analysis of R1 progeny showed a stable transmission of the T-DNA for some plants, whereas in other lines some T-DNA insertions were unstable and were completely or partially lost during segregation. It seems that there is a reasonable agreement between the copy number and transmission of kanamycin resistance. Some plants with one copy transmitted kanamycin resistance to progeny in a Mendelian fashion and the transgene was conserved. Plants with a high copy number showed a reduced resistance transmission trait. Analysis of the R2 progeny is underway to test further the stability of the T-DNA integration.

Histochemical analysis of plants transformed with mas2'-*uidA* showed that vascular tissues were the site of specific expression. This is in agreement with data obtained with several plants (Langridge et al. 1989; Leung et al. 1991; Saito et al. 1991; Stefanov et al. 1994). The mas2' promoter was subjected to developmental regulation. Its expression was very weak in germinated plantlets and the first leaf exhibited lower activity than the mature leaves. A gradient of *uidA* expression reported by others seemingly operated in chicory as well. Loss of activity in the young leaves might well be due to some inhibitor produced by the shoot apex which down-regulates the mas2' promoter (Langridge et al. 1989). Tissue-specific expression has been confirmed by Northern analysis in potato. Correlation between the number of *uidA* transcripts and those encoding *mas*-binding factors has been shown (Feltkamp et al. 1995). These factors are able to bind *ocs*-like sequences which act as enhancers in gene expression (Ebert et al. 1987; Lam et al. 1989) and represent binding sites for proteins belonging to the bZip family (Foley et al. 1993).

mas2' promoter was wound-inducible in leaves, roots and zygotic embryos (data not shown). Indeed, mas1'2' promoter contains various *cis*-regulatory sequences which respond to environmental stress (Guevara-Garcia et al. 1993; Ni et al. 1994). However, the *cis*-elements involved in mechanical damage have not yet been characterised. On the other hand, ABA treatments significantly enhanced GUS activity. ABA, like jasmonic acid or methyl jasmonate are essential compounds in the response of plants to wounding and are able to activate wound-inducible genes in the absence of any damage (Pena-Cortes et al. 1989). Involvement of ABA in the gene activation process following mechanical damage is supported by the fact that endogenous ABA concentrations rise upon wounding. ABA was further shown to induce jasmonic acid biosynthesis, which in turn activates wound-responsive genes (Pena-Cortes et al. 1995). Further investigations are necessary to demonstrate if ABA exerts its effects via the same regulatory sequences.

Other hormonal compounds regulate *uidA* when controlled by mas2'. Application of 2,4-D, IAA and GA$_3$ increased β-glucuronidase activity in chicory as well as in other species (Teeri et al. 1989; Langridge et al. 1989; Saito et al. 1991). High expression of GUS was detected within the cap and the meristematic cells of chicory roots. The cap of the root is a place of cytokinin and auxin accumulation. Moreover, the auxin/cytokinin ratio modulated the gene expression, cytokinin prompting an increased expression (Langridge et al. 1989). mas2′ promoter appears to be provided with regulatory sequences that are sensitive to changes in developmental signals and hormonal balances within the host plant.

The auxin is synthesised in the leaf primordia and then transported preferentially to more basal regions of the plant such as the root (Rubery 1987). This may explain the important activity found in leaves treated with 2,4-D. Surprisingly, when IAA was applied to leaf fragments, the response was only enhanced 2.5 times. The effect of IAA was less than that of 2,4-D in tobacco leaves, whereas in roots no enhancement was observed (Saito et al. 1991). The greater effects of 2,4-D compared with IAA may be explained by a better stability of the former, although treatments were done for only 24h. Why 2,4-D and IAA had analogous responses on roots remains unexplained. Leung et al. (1991) suggested that auxin signals act by modifying pre-existing transcription complexes already bound to DNA. This modification in turn enhances the transcription rate of the mannopine synthase genes. Nevertheless, no single sequence was sufficient for maximal auxin response, enhancement resulting from co-operative interactions between multiple responsive elements interspersed in the mas promoter.

In regenerants containing the CaMV 35S-*uidA* fusion, GUS activity was detected within all leaf cell types, while in roots the expression pattern varied among individual transformants. Similar spatial and temporal variations were also described in tobacco plants transformed with other reporter genes fused to the CaMV 35S promoter (Williamson et al. 1989; Schneider et al. 1990) indicating that the activity of this promoter is not expressed equally either in all cells or over time. Although the CaMV 35S promoter was originally considered to provide constitutive expression, it possesses domains that confer developmental and tissue-specific expression and which may be recognised differently depending on the plant species (Benfey and Chua 1989; Benfey et al. 1990a,b). High activity in vascular tissues from reproductive organs seems to be a general feature and found in petunia, tobacco (Benfey and Chua 1989), melon (Dong et al. 1991; Vallès and Lasa 1994), carnation (Lu et al. 1991) and tomatillo (Assad-Garcia et al. 1992). Likewise, expression in ovaries is consistent in all species (Williamson et al. 1989; Terada and Shimamoto 1990; Lu et al. 1991; Assad-Garcia et al. 1992).

Curiously, neither CaMV 35S nor mas2′ promoter exhibit real activity in the pollen of chicory, confirming results found by others (Langridge et al. 1989; McCormick et al. 1989; Twell et al. 1991; Mascarenhas and Hamilton 1992). Inactivity would involve a weak concentration of a specific transcription factor required for promoter activity or it might be due to the structure of chromatin. The generative nucleus possesses a compact chromatin structure which may prevent activation of the promoter (Wagner et al. 1990). Langridge et al. (1989) showed that ungerminated tobacco pollen exerted no luciferase activity under the mas1'2' promoters whereas a high activity has been reported within the first hour of germination. This might coincide with recovered metabolic activity during germination.

5 Greenhouse Performance/Present Status of Transgenic Plants

Regenerated shoots with a well-developed root system were acclimatised for several days under high moisture content before transferring them in pots to the greenhouse. Usually plants were grown on for 2 to 3 months in order to increase robustness and to favour accumulation of stores. Vernalisation was performed for 8 weeks at 5 °C in a cold chamber under 14h/10h light/dark conditions, with a weekly watering. Once returned to the greenhouse, leaves begin to develop and flower spikes differentiated. Autopollination was performed 1–2 months later and seeds harvested.

Behaviour of transgenic plants was analogous to non-transformed ones. Probably plants affected in major metabolic pathways or potentially not viable were eliminated during the in vitro regeneration process. No particular phenotypes were distinguished except that fever seeds are harvested.

6 Summary and Conclusion

Genetic transformation of *Cichorium intybus* L. has been carried out successfully with several *Agrobacterium tumefaciens* strains. Tap-root explants produced transformants at only a low frequency while leaves from 3-week-old plantlets constituted more suitable material, allowing transformation efficiencies up to 15%. Nevertheless, etiolated cotyledons were the most suitable organs, appropriate for fast and easy transformation. Moreover, electron microscopy observations showed the highest density of agrobacteria on cotyledons compared with other organs used. Stable transgene integrations occur in R_0 and R_1 transformed plants whereas in others the transferred DNA was subjected to deletions or rearrangements.

Analyses of transgenic plants concerned mainly the expression of *uidA* coding for β-glucuronidase under the control of either mas2' or CaMV 35S promoters. When driven by mas2', GUS activity was shown in the cap, meristem and outer cortex of the root, in all cellular leaf types and in the vascular tissues of the flower. β-Glucuronidase activity was strongly enhanced both in leaves and roots by wounding and by treatment with 2,4-D or ABA. Under the control of CaMV 35S, *uidA* expression was particularly intense in the vascular tissues of the leaf, the ovule and the style. Gus activity was also high in young roots, while in older ones, staining was absent in the stele and at lateral root emergence, corroborating developmental regulation. Zygotic embryos showed GUS activity when placed under the control of either the inducible mas2' or the constitutive CaMV 35S, while expression remained silent in pollen whatever the promoter used.

Phenotypically normal transgenic regenerants are obtainable within 2 months. The gene transfer system described here will facilitate the study of gene regulation in chicory and may be useful for the introduction of genes responsible for agronomically important traits in different varieties. The taproots of the variety Magdeburg which are traditionally roasted and ground to mix with coffee, are presently considered as interesting potential sources of fructose syrups, and particularly of modified fructans. Thus, a potent transformation process might be a suitable tool for improving both the quality and the amount of stored fructosans. In the absence of homologous promoters specifying tissue-specific expression in chicory, further work is needed in order to find out tags which could target the products in suitable cellular compartments.

Acknowledgments. The authors are indebted to T. Dubois for help in the interpretation of data about flower analysis, and to JP Teissier for electron microscopy manipulations. We wish to thank V. Kimpe for data obtained from further transformation experiments.

References

Abid M, Palms B, Derycke R, Tissier JP, Rambour S (1995) Transformation of chicory and expression of the bacterial *uidA* and *nptII* genes in the transgenic regenerants. J Exp Bot 46:337–346

Assad-Garcia N, Ochoa-Alejo N, Garcia-Hernandez E, Herrera-Estrella L, Simpson J (1992) *Agrobacterium*-mediated transformation of tomatillo (*Physalis ixocarpa*) and tissue specific and developmental expression of the CaMV 35S promoter in transgenic tomatillo plants. Plant Cell Rep 11:558–562

Benfey PN, Chua N-H (1989) Regulated genes in transgenic plants. Science 244:174–181

Benfey PN, Ren L, Chua N-H (1990a) Tissue-specific expression from CaMV 35S enhancer subdomains in early stages of plant development. EMBO J 9:1677–1684

Benfey PN, Ren L, Chua N-H (1990b) Combinatorial and synergistic properties of CaMV 35S enhancer subdomains. EMBO J 9:1685–1696

Bradford MM (1976) A rapid and sensitive method for quantitation of microgram quantities of protein utilizing the principle of protein-dye binding. Ann Biochem 72:248–254

Deblaere R, Byteber B, De Greve H, Deboeck F, Schell J, Van Montagu M, Leemans J (1985) Efficient atropine Ti-plasmid derived vectors for *Agrobacterium*-mediated gene transfer to plants. Nucleic Acids Res 13:4777–4788

Dellaporta SL, Wood J, Hicks JB (1983) A plant DNA minipreparation; version II. Plant Mol Biol Rep 1:19–21

Dong J-Z, Yang M-Z, Jia S-R, Chua N-H (1991) Transformation of melon (*Cucumis melo* L.) and expression from the Cauliflower Mosaic Virus 35S promoter in transgenic melon plants. Biotechnology 9:858–863

Ebert PR, Ha SB, An G (1987) Identification of an essential upstream element in the nopaline synthase promoter by stable and transient assays. Proc Natl Acad Sci USA 84:5745–5749

Feltkamp D, Baumann E, Schmalenbach W, Masterson R, Rosahl S (1995) Expression of the mannopine synthase promoter in roots is dependent on the *mas* elements and correlates with high transcript levels of *mas*-binding factor. Plant Sci 109:57–65

Fiala V, Jolivet E (1984) Mise en évidence d'une nouvelle fraction glucidique dans la racine de chicorée et son évolution au cours de la formation des réserves. Physiol Vég 22:315–321

Foley RC, Grossmann C, Ellis JG, Llewellyn DJ, Dennis ES (1993) Isolation of a maize bZIP protein subfamily: candidates for the *ocs*-element transcription factor. Plant J 3:669–679

Frulleux F, Weyens G, Jacobs M (1997) *Agrobacterium tumefaciens*-mediated transformation of shoot-buds of chicory. Plant Cell, Tissue and Organ Culture, 50:107–112

Gautheret RJ (1941) Recherches expérimentales sur la polarité des tissus de la racine d'Endive. C R Acad Sci SerIII-Vie 213:37–39

Genga A, Giansante L, Bernacchia G, Allavena A (1994) Plant regeneration from *Cichorium intybus* L. leaf explants transformed by *Agrobacterium tumefaciens*. J Genet Breed 48:25–32

Guevara-Garcia A, Mosqueda-Cano G, Argüello-Astorga G, Simpson J, Herrera-Estrella L (1993) Tissue-specific and wound-inducible pattern of expression of the mannopine synthase promoter is determined by the interaction between positive and negative *cis*-regulatory elements. Plant J 4:495–505

Heller R (1953) Recherche sur la nutrition minérale des tissus végétaux cultivés in vitro. Ann Sci Nat Bot 14:1–223

Jefferson RA, Kavanagh TA, Bevan MW (1987) GUS fusions: β-glucuronidase as a sensitive and versatile gene fusion marker in higher plants. EMBO J 6:3901–3907

Lam E, Benfey PN, Gilmartin PM, Fang R-X, Chua N-H (1989) Site-specific mutations alter in vitro factor binding and change promoter expression pattern in transgenic plants. Proc Natl Acad Sci USA 86:7890–7894

Langridge WHR, Fitzgerald KJ, Koncz C, Schell J, Szalay AA (1989) Dual promoter of *Agrobacterium tumefaciens* mannopine synthase genes is regulated by plant growth hormones. Proc Natl Acad Sci USA 86:3219–3223

Leung J, Fukuda H, Wing D, Schell J, Masterson R (1991) Functional analysis of *cis*-elements, auxin response and early developmental profiles of the mannopine synthase bidirectional promoter. Mol Gen Genet 230:463–474

Lu C-Y, Nugent G, Wardley-Richardson T, Chandler SF, Young R, Dalling MJ (1991) *Agrobacterium*-mediated transformation of carnation (*Dianthus caryophyllus* L.) Bio/Technology 9:864–868

Mascarenhas JP, Hamilton DA (1992) Artifacts in the localization of GUS activity in anthers of petunia transformed with a CaMV 35S-GUS construct. Plant J 2:405–408

McCormick S, Twell D, Wing R, Ursin V, Yamaguchi J, Larabell S (1989) Anther-specific genes: molecular characterization and promoter analysis in transgenic plants. In: Lord E, Bernier G (eds) Plant reproduction: from floral induction to pollination. Am Soc Plant Physiol, Rockville, MD, pp 128–135

McCowan PP, Chen KJ, Barley CMB, Costerton JW (1978) Adhesion of bacteria to epithelial cell surfaces of the reticulo-rumen of cattle. Appl Environ Microbiol 35:149–155

Mendel RR, Muller B, Schulze Y, Kolesnikov V, Zelenin A (1989) Delivery of foreign genes to intact barley cells by high velocity microprojectiles. Theor Appl Genet 78:31–34

Murashige T, Skoog F (1962) A revised medium for rapid growth and bioassays with tobacco tissue cultures. Physiol Plant 15:473–497

Ni M, Cui D, Gelvin SB (1994) Wound inducibility and tissue specificity of chimaeric promoters derived from the mannopine and octopine synthase genes. 4th Int Congr Plant Mol Biol, 19–24 June 1994, Amsterdam

Pena-Cortes H, Sanchez-Serano JJ, Mertens R, Willmitzer L (1989) Abscisic acid is involved in the wound-induced expression of the proteinase inhibitor II gene in potato and tomato. Proc Natl Acad Sci USA 86:9851–9855

Pena-Cortes H, Fisahn J, Willmitzer L (1995) Signals involved in the wound-induced proteinase inhibitor II gene expression in tomato and potato plants. Proc Natl Acad Sci USA 92:4106–4113

Rubery PH (1987) Auxin transport. In: Davies PJ (ed) Plant hormones and their role in plant growth and development. Martinus Nijhoff, The Hague, pp 341–362

Saito K, Yamazaki M, Kaneko H, Murakoshi I, Fukuda Y, Van Montagu M (1991) Tissue-specific and stress-enhancing expression of the TR promoter for mannopine synthase in transgenic medicinal plants. Planta 184:40–46

Sambrook J, Fritsh EF, Maniatis T (1989) Molecular cloning: a laboratory manual, 2nd edn. Cold Spring Harbor Laboratory Press, Cold Spring Harbor, New York

Schneider M, Ow DW, Howell SH (1990) The in vivo pattern of firefly luciferase expression in transgenic plants. Plant Mol Biol 14:935–947

Stefanov I, Fekete S, Bögre L, Pauk J, Fehér A, Dudits D (1994) Differential activity of the mannopine synthase and the CaMV 35S promoters during development of transgenic rapeseed plants. Plant Sci 95:175–186

Sun LY, Touraud G, Charbonnier C, Tepfer D (1991) Modification of phenotype in Belgian endive (*Cichorium intybus*) through genetic transformation by *Agrobacterium rhizogenes*: conversion from biennial to annual flowering. Transg Res 1:14–22

Teeri TH, Lehvaslaiho H, Franck M, Uotila J, Heino P, Palva ET, Van Montagu M, Herrera-Estrella L (1989) Gene fusions to lacZ reveal new expression patterns of chimeric genes in transgenic plants. EMBO J 8:343–350

Terada R, Shimamoto K (1990) Expression of CaMV-GUS gene in transgenic rice plants. Mol Gen Genet 220:389–392

Twell D, Yamaguchi J, Wing RA, Ushiba J, Mc Cormick S (1991) Promoter analysis of three genes that are coordinately expressed during pollen development reveals pollen-specific enhancer sequences and shared regulatory elements. Gene Dev 5:496–507

Vallès MP, Lasa JM (1994) *Agrobacterium*-mediated transformation of commercial melon (*Cucumis melo* L., cv. Amarillo Oro). Plant Cell Rep 13:145–148

Vermeulen A, Vaucheret H, Pautot V, Chupeau Y (1992) *Agrobacterium*-mediated transfer of a mutant *Arabidopsis* acetolactate synthase gene confers resistance to chlorsulfuron in chicory (*Cichorium inthybus* L.). Plant Cell Rep 11:243–247

Vijn I, Van Dijken A, Sprenger N, Van Dun K, Weisbeek P, Wiemken A, Smeekens S (1997) Fructan of the inulin neoseries is synthesized in transgenic chicory plants (*Cichorium intybus* L.) harbouring onion (*Allium cepa* L.) fructan : fructan 6G fructosyltransferase. Plant J 11:387–398

Wagner VT, Cresti M, Salvatici P, Tiezzi A (1990) Changes in volume, surface area, and frequency of nuclear pores on the vegetative nucleus of tobacco pollen in fresh hydrated and activated conditions. Planta 181:304–309

Williamson JD, Hirsch-Wyncott ME, Lankins BA, Gelvin SB (1989) Differential accumulation of a transcript driven by the CaMV 35S promoter in transgenic tobacco. Plant Physiol 90:1570–1576

9 Transgenic Watermelon (*Citrullus lanatus*)

J.R. Liu, P.S. Choi, and Y.S. Kim

1 Introduction

Watermelon, one of the most important vegetable crops, is eaten chiefly as a fresh fruit. It originated from tropical and subtropical Africa, and now is widely distributed throughout the tropics, Mediterranean region, South Asia, and East Asia including China.

Watermelon is attacked by many fungi, the most serious diseases in the tropics and in the United States being caused by *Collctotrichum lagenarium*, *Phytophthora parasitica*, *Fusarium oxysporum*, and *Mycosphaerella melonis*, etc. (du Cellier and Duke 1993). In addition, bacterial wilt (spread by cucumber beetles), squash bugs, and pickleworms are also common problems with cultivation of watermelon. Bacterial fruit blotch is a major concern in the US. It is thought to be a seedborne disease and is more serious in humid areas than in dry areas. Excess rain, heat, or drought can affect watermelon production. Excess rain during any stage of growth can reduce watermelon yields. Excessive heat and direct sunlight increase the likelihood of yield losses due to sunburn, which causes yellowing of the rind. Drought may reduce watermelon yields by diminishing plant growth, and limiting the development and size of the melons.

By using pathogen-resistant and stress-resistant varieties, watermelon yields can be increased. Several varieties, including Crimson Sweet, Jubilee, Dixilee, etc., have been developed by conventional breeding to have resistance to *Fusarium* wilt and anthranose. However, conventional breeding has limitations in improving the quality and productivity of watermelon. Recent advances in *Agrobacterium*-mediated transformation have made it possible to introduce foreign genes to improve their productivity and quality beyond the limit of conventional breeding. Using genetic transformation, pathogen-resistant and stress-resistant watermelons can be developed in a relatively short time. Furthermore, the sweetness of watermelon can be regulated through the introduction of enzymes which regulate carbohydrate content into the watermelon genome. Genetic transformation of watermelon has been reported by Choi et al. (1994). This chapter describes their transformation system for watermelon by co-culturing cotyledonary explants with *Agrobac-*

Plant Cell Biotechnology Laboratory, Korea Research Institute of Bioscience and Biotechnology, P.O. Box 115, Yusong, Taejon, 305-600, Korea

Biotechnology in Agriculture and Forestry, Vol. 47
Transgenic Crops II (ed. by Y.P.S. Bajaj)
© Springer-Verlag Berlin Heidelberg 2001

terium harboring the pBI121 binary vector carrying the GUS gene of *Escherichia coli* as a reporter gene.

2 Genetic Transformation

Numerous studies have been conducted on various aspects of in vitro culture, i.e., micropropagation, organogenesis, somatic embryogenesis, production of triploids and tetraploids (Anghel and Rosu 1985; Srivastava et al. 1989; Dong and Jia 1991; Compton and Gray 1993a) etc., and the subject has been reviewed (Jelaska 1986; Adelberg et al. 1997). The studies on genetic engineering, however, are recent (Choi et al. 1994).

2.1 Plant Materials and Culture Conditions

Zygotic embryos of F_1 hybrid watermelon (*Citrullus lanatus*; cvs. Sweet Gem and Gold Medal) were dissected out of the mature seeds and surface-disinfected with 70% ethanol for 1 min and 1% sodium hypochlorite for 10 min. They were rinsed three times with sterile deionized-distilled water.

The basal MS medium used throughout the experiments consisted of Murashige and Skoog's (1962) inorganic salts, $100 \, mg \, l^{-1}$ myo-inositol, $0.4 \, mg \, l^{-1}$ thiamine. HCl, 3% sucrose, and 0.4% Gelrite. The pH of all media was adjusted to 5.8 before autoclaving. Twenty-five milliliters of medium was dispensed into 87×15-mm plastic Petri dishes. Nine zygotic embryos were placed in each Petri dish containing the basal medium and incubated at $25\,°C$ in the dark. After 5 days incubation, seedlings 2- to 3-cm-long germinated, and their cotyledons were excised and transversely cut into proximal and distal halves (cotyledonary explants).

2.2 Induction of Adventitious Shoots

To induce adventitious shoots, cotyledonary explants of Sweet Gem and Gold Medal were placed onto medium supplemented with either 0.02, 0.2, 1, 2, 4 or $9 \, mg \, l^{-1}$ 6-benzylaminopurine (BA) in Petri dishes. Two to four Petri dishes were cultured per treatment with four explants per dish. Cultures were maintained at $25\,°C$ in the light (16-h photoperiod, about $7 \, Wm^{-2}$ cool-white fluorescent lamps) or in the dark. After 4 weeks of culture, the number of explants with adventitious shoots and the number of shoots formed per explant were counted under a dissecting microscope.

2.3 Transformation of *Agrobacterium*

pBI121, a binary vector carrying the CaMV 35S promoter-GUS gene-NOS terminator fusion and NOS promoter-neomycin phosphotransferase gene-NOS terminator fusion (Fig. 1; Choi et al. 1994) was transformed into *Agrobacterium tumefaciens* LBA 4404, a disarmed strain, by a heat shock treatment (Jefferson et al. 1987) and used for co-culture with cotyledonary explants.

2.4 Transformation of Cotyledonary Explants

Cotyledonary explants of Sweet Gem were gently punctured 10 to 15 times throughout the adaxial surface with the tip of a scalpel, and precultured on medium with $1\,\mathrm{mg\,l^{-1}}$ BA for 0, 1, 2, 3, 4, 5 or 6 days. *Agrobacterium* was grown in YEP liquid medium. Explants were co-cultured with *Agrobacterium* at the log phase of growth in liquid MS medium with $1\,\mathrm{mg\,l^{-1}}$ BA and $200\,\mu$M β-hydroxyacetosyringone (10^6 bacterial cells per ml) for 48h. (β-Hydroxyacetosyringone was filter-sterilized and added to the medium after autoclaving.) Explants were briefly blotted with sterile Whatman filter paper and placed onto MS medium with $1\,\mathrm{mg\,l^{-1}}$ BA, $250\,\mathrm{mg\,l^{-1}}$ carbenicillin, and $100\,\mathrm{mg\,l^{-1}}$ kanamycin, and cultured in the light as described above. (The antibiotics were added to medium after autoclaving.) Five Petri dishes were cultured per treatment (explant preculture for 0 to 6 days) with nine explants per dish. After 4 weeks of culture, young leaves were excised from 20 to 30 adventitious shoots per treatment and incubated with X-gluc at 37 °C for 24h according to the Clone Tech manual. Data recorded at 4 weeks included the percentage of explants, kanamycin-resistant shoots and the frequency of histochemical GUS-positive shoots. Statistical analysis was conducted using regression analysis.

2.5 Southern Blot Analysis

GUS-positive and GUS-negative adventitious shoots were separately transferred onto MS basal medium to be rooted. The regenerants were transplanted to potting soil and maintained in the phytotron (RH 60–70%, 27 °C day/22 °C

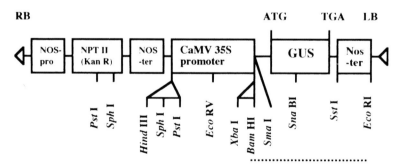

Fig. 1. A binary vector harboring *Escherichia coli* β-glucuronidase gene. *LB* T-DNA *left border, RB* T-DNA *right border, npt* neomycin phosphotransferase gene, *Tet* tetracycline resistance gene, *Pnos nopaline synthase* promoter, *NPTII* neomycin phosphotransferase gene II; *Tnos*, nopaline synthase terminator, *GUS* β-glucuronidase gene. The dotted line indicates the 2.1-kb *Eco*RI/*Bam*HI DNA fragment used as a Southern hybridization probe

night, 16-h photoperiod, about 80 Wm⁻²). The genomic DNA of GUS-positive and GUS-negative regenerants was extracted according to the method of Deblaere et al. (1987), digested with *Eco*RI, *Hind*III, *Bam*HI or *Eco*RI/*Bam*HI for 4 h and subjected to electrophoresis (0.8% agarose gel). The DNA bands were transferred to positive-charged Nylon membrane, and a 2.1 kb GUS NOS poly(A) probe labeled with digoxigenine (DIG; Boehringer Mannheim) was used for Southern hybridization.

3 Results and Discussion

3.1 Induction of Adventitious Shoots

After 1 week of culture, etiolated cotyledonary explants turned green and enlarged two to three times. After 2 weeks of culture, callus formed on the cut edges of the explants (Fig. 2A; Choi et al. 1994). After 3 weeks of culture, callus on the proximal cut edges of both of the proximal and distal halves of the cotyledonary explants gave rise to leaf primordia (Fig. 2B; Choi et al. 1994), indicating that the cotyledon has a polarity for competence of adventitious shoot formation. After 4 weeks of culture, adventitious shoots had grown to about 1 cm in height. The number of adventitious shoots per explant and the

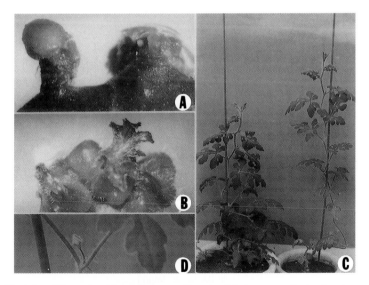

Fig. 2A–D. Adventitious shoot formation on cultured cotyledonary explants of watermelon and transformed plant regeneration following co-culture with *Agrobacterium*. **A** Green callus formed on the distal half cotyledon. **B** Numerous shoots formed on the explant after 3 to 4 weeks of culture. **C** Transformed plants transplanted to potting soil. **D** A transformed plant bearing flower buds

percentage of explants with shoots were greatest when the explants were cultured on medium with $1\,mg\,l^{-1}$ BA for both cultivars (Fig. 3; Choi et al. 1994). The percentage of Sweet Gem explants with shoots was higher than that of Gold Medal explants on the whole. Likewise, Sweet Gem produced about three times more shoots than Gold Medal at the same concentration of BA and was more competent to form shoots over a wider range of BA concentrations. This finding is similar to that of Compton and Gray (1993a) and Srivastava et al. (1989) who reported optimal shoot formation on medium with $5\,\mu M$ BA. The distal half cotyledonary explants produced more shoots than the proximal half (in Sweet Gem, it was about two times higher; in Gold Medal, about 1.25 times higher), which is in contrast to the results of Compton and Gray (1993a), who obtained adventitious shoots only on the proximal region of the cotyledons, and suggested that the competence for adventitious shoot formation in watermelon was restricted to the proximal region of cotyledons. However, our results indicate that cells competent for shoot formation are not localized at one site of the cotyledon in these cultivars. Light was essential for adventitious shoot formation, because no cotyledonary explants of either cultivar formed shoots in the dark (data not shown).

3.2 Transformation of Cotyledonary Explants

The percentage of explants forming kanamycin-resistant shoots after co-culture with *Agrobacterium* was increased by preculture of the explants on

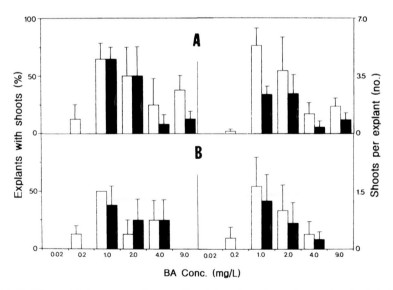

Fig. 3A–B. Effect of BA concentration on adventitious shoot formation on the distal (□) and proximal (■) half of Sweet Gem (**A**) and Gold Medal (**B**) cotyledonary explants. Explants were cultured on MS medium with various concentrations of BA in the light and data were collected after 4 weeks of culture. *Vertical bars* represent standard error of the mean

medium with 1 mg l⁻¹ BA for up to 5 days (Fig. 4; Choi et al. 1994). The proportion of GUS-positive shoots relative to total adventitious shoots subjected to GUS histochemical assay was greatest (16%) for the 5-day preculture treatment. Thus, preculturing explants for 5 days made cells more competent for *Agrobacterium*-mediated transformation. After 2 weeks of culture, all GUS-positive shoots rooted on MS basal medium. Seven regenerants were transplanted to potting soil and grown to maturity in the phytotron (Fig. 2C, D; Choi et al. 1994).

When the genomic DNA of three randomly selected GUS-positive regenerants was digested with *Eco*RI or *Bam*HI and subjected to Southern blot analysis using DIG-labeled GUS NOS poly(A) probe, only one band of about 23 kb greater in length than the intact pBI121 was obtained, indicating that the GUS gene was incorporated into the genomic DNA of the regenerants (pBI121 has one *Eco*RI site and the total size of the vector is 12.7 kb; see Fig. 5A; Lanes 3, 5 and 7; Choi et al. 1994). One band of 2.2 kb obtained by *Eco*RI or *Bam*HI digestion showed, as expected, that the incorporated DNA fragment included the intact GUS gene (Fig. 5A; Lanes 4, 6 and 8; Choi et al. 1994). The genomic DNA of regenerant III digested with *Hind*III yielded one major band of 9.2 kb and several minor bands of smaller sizes (Fig. 5B; Lane 4; Choi et al. 1994). The major band confirmed that the GUS gene was incorporated

Fig. 4A–B. Effect of explant preculture on kanamycin-resistant shoot formation (●) and transformation efficiency (○) of Sweet Gem cotyledonary explants following co-culture with *Agrobacterium*. The explants were precultured on MS medium with 1 mg l⁻¹ BA before co-culture with *Agrobacterium* harboring the pBI121 binary vector. Data were collected after 4 weeks of culture. The transformation efficiency is obtained from the number of histochemical GUS-positive shoots divided by the number of shoots subjected to X-gluc assay. * Significant at $P = 0.05$

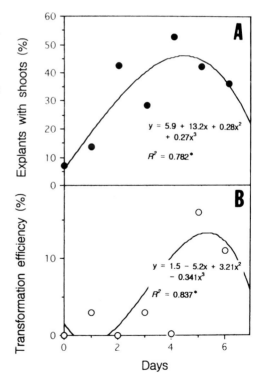

A

$$y = 5.9 + 13.2x + 0.28x^2 + 0.27x^3$$
$$R^2 = 0.782*$$

B

$$y = 1.5 - 5.2x + 3.21x^2 - 0.341x^3$$
$$R^2 = 0.837*$$

Explants with shoots (%)

Transformation efficiency (%)

Days

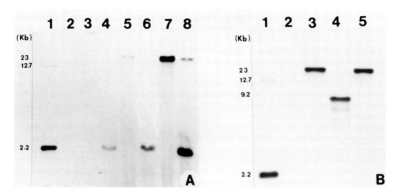

Fig. 5A–B. Southern blot analysis of GUS-positive and GUS-negative regenerants of water-melon. The 2.2 kb GUS NOS poly(A) probe labeled with DIG was hybridized with genomic DNA of the regenerants. Each lane was loaded with the following DNA. **A** *lane 1* pBI121 digested with *Eco*RI/*Bam*HI; *lane 2* a GUS-negative regenerant; *lane 3* GUS-positive regenerant I digested with *Eco*RI; *lane 4* GUS-positive regenerant digested with *Eco*RI/*Bam*HI; *lane 5* GUS-positive regenerant II digested with *Eco*RI; *lane 6* GUS-positive regenerant II digested with *Eco*RI/*Bam*HI; *lane 7* GUS-positive regenerant III digested with *Eco*RI; *lane 8* GUS-positive regenerant III digested with *Eco*RI/*Bam*HI. **B**: *lanes 1 and 2* are the same as in **A**. *Lane 3* GUS-positive regenerant III digested with *Eco*RI; *lane 4* *Hind*III; *lane 5* *Bam*HI

into the genomic DNA of the regenerant. The minor bands were probably due to non-specific binding of the probe with the genomic DNA.

4 Present Status/Field Performance of Transgenic Watermelon

Mature transformed plants (T_0) of watermelon were morphologically normal and bore fruits with viable seeds after self-pollination in a growth chamber. The seeds (T_1) were sown in the field. PCR confirmed that the GUS gene was stably transmitted to the germinated seedlings in a Mendelian pattern. The seedlings were also grown to maturity in the field without noticeable abnormality.

Using the protocol described above, a newly designed gene for the potent sweetner monellin (Kim et al. 1989) was also introduced into watermelon. Results from this experiment are not yet available.

5 Summary and Conclusion

Adventitious shoots formed on the proximal cut edges of different cotyledonary explants of watermelon [*C. lanatus* (Thunb.) Matsum. and Nakai; cvs. Sweet Gem and Gold Medal] cultured on MS medium with

1 mg l^{-1} BA. Light (16-h photoperiod, about 7 wm^{-2} cool-white fluorescent lamps) was essential for shoot formation. To obtain transformed plants, cotyledonary explants of Sweet Gem were co-cultured with *A. tumefaciens* LBA4404, a disarmed strain harboring a binary vector pBI121 carrying the CaMV 35S promoter-β-glucuronidase (GUS) gene fusion used as a reporter gene and NOS promoter-neomycin phosphotransferase gene as a positive selection marker, for 48 h on MS medium with 1 mg l^{-1} BA and 200 mg l^{-1} β-hydroxyacetosyringone. After 48 h of culture, explants were transferred to medium with 1 mg l^{-1} BA, 250 mg l^{-1} carbenicillin, and 100 mg l^{-1} kanamycin and cultured in the light. Adventitious shoots were formed on the explants after 4 weeks of culture. When subjected to GUS histochemical assay, young leaves obtained from the shoots showed a positive response at a frequency of up to 16%. Preculturing cotyledonary explants on MS medium with 1 mg l^{-1} BA for 5 days enhanced the competence of the cells to be transformed by *Agrobacterium*. Southern blot analysis confirmed that the GUS gene was incorporated into the genomic DNA of the GUS-positive regenerants. The transformed plants were grown to maturity.

References

Adelberg JW, Zhang XP, Rhodes BB (1997) Micropropagation of *Citrullus lanatus* (Thunb.) Matsum. and Nakai (Watermelon). In: Bajaj YPS (ed) Biotechnology in agriculture and forestry, vol 39. High-Tech and micropropagation V. Springer, Berlin Heidelberg New York, pp 60–76

Anghel I, Rosu A (1985) In vitro morphogenesis in diploid, triploid and tetraploid genotypes of watermelon – *Citrullus lanatus* (Thunb.) Mansf. Rev Roum Biol-Biol Veg 30:43–55

Choi PS, Sho WY, Kim YS, Yoo OJ, Liu JR (1994) Genetic transformation and plant regeneration of watermelon using *Agrobacterium tumefaciens*. Plant Cell Rep 13:344–348

Compton ME, Gray DJ (1993a) Shoot organogenesis and plant regeneration from cotyledons of diploid, triploid, and tetraploid watermelon. J Am Soc Hortic Sci 118:151–157

Compton ME, Gray DJ (1993b) Somatic embryogenesis and plant regeneration from immature cotyledons of watermelon. Plant Cell Rep 12:61–65

Deblaere R, Reynaerts A, Hofte H, Hernalsteens JP, Leemans J, Van Montagu M (1987) Vectors for cloning in plant cells. Methods Enzymol 153:277–292

Dong J-Z, Jia S-R (1991) High efficiency plant regeneration from cotyledons of watermelon (*Citrullus vulgaris* Schrad.). Plant Cell Rep 9:559–562

duCellier JL, Duke JA (1993) CRC handbook of alternative cash crops, pp 134–137

Jefferson RA, Kavanaugh TA, Bevan MW (1987) GUS fusion: β-glucuronidase as a sensitive and versatile gene fusion marker in higher plants. EMBO J 6:3901–3907

Jelaska S (1986) Cucurbits. In: Bajaj YPS (ed) Biotechnology in agriculture and forestry, vol 2. Crops I. Springer, Berlin Heidelberg New York, pp 371–386

Kim SH, Kang CH, Kim S, Cho JM, Lee TK (1989) Redesigning a sweet protein: increased stability and renaturability. Protein Eng 2:571–575

Murashige T, Skoog F (1962) A revised medium for rapid growth and bioassays with tobacco tissue cultures. Physiol Plant 15:473–497

Srivastava DR, Andrianov VM, Piruzian ES (1989) Tissue culture and plant regeneration of watermelon (*Citrullus vulgaris* Schrad. cv. Melitopolski). Plant Cell Rep 8:300–302

10 Transgenic Cucumber (*Cucumis sativus* L.)

P.P. CHEE

1 Introduction

Cucumber is an economically important crop. It is widely grown in the tropics, subtropics, and milder parts of the temperate zones of both hemispheres. Its total world tonnage production ranks sixth among the vegetable species (Reynolds 1986). Numerous diseases attack the cultivated species and cause tremendous annual losses. In many cases these diseases are the result of infection by viruses or microbes. Although various pest control measures have been attempted, none has yet proven entirely successful. A main objective of breeding programs is thus not only to increase yield and quality, but also to select cultivars resistant to diseases. One method of obtaining these traits is to hybridize many of the wild species with the cultivated species (Deakin et al. 1971). However, this approach is associated with many difficulties such as sterility barriers and the number of backcrosses required to obtain the original cultivar containing the desirable traits. The recent advances in genetic engineering appears to offer a promising alternative for improvement of this crop. The major advantage of genetic transformation is the potential to add a characteristic trait directly to an existing genotype, while avoiding the necessity for a further conventional breeding procedure to stabilize the new gene in a background suitable for sexual propagation. The transformation of a pre-existing genotype would also enable the retention of already proven cultivars and products (Chee 1990b,c, 1993; Slightom et al. 1990; Chee and Slightom 1991, 1992).

The success of any gene transfer method depends on the susceptibility of the target cells and on the availability of a regeneration procedure for the transformed tissue. Many of the species belonging to the family Cucurbitaceae are known to be susceptible to infection by *Agrobacterium* pathogens (Anderson and Moore 1979; Smarrelli et al. 1986). In addition, procedures for regenerating many of these species have already been established (Chee 1990a,b, 1991a,b, 1992).

This chapter describes the use of *Agrobacterium*-mediated and microprojectile bombarment methods to transfer genetic materials into the genome of cucurbit species. Table 1 summarizes various studies on transformation of cucumber.

Pharmacia & Upjohn Inc., 301 Henrietta St., Kalamazoo, Michigan 49008, USA

Biotechnology in Agriculture and Forestry, Vol. 47
Transgenic Crops II (ed. by Y.P.S. Bajaj)
© Springer-Verlag Berlin 2001

Table 1. Summary of various studies on transformation of cucumber

Reference	Explant/culture used	Method/vector used	Observation
1. Trulson et al. (1976)	Inverted hypocotyl sections from 5-day-old seedlings.	*Agrobacterium*-mediated transformation. Vector: Disarmed A4 *A. rhizogenes* harboring a binary plasmid pARC8 containing *Nos-NPTII* selectable marker.	Thirty percent of regenerated roots contained NPTII gene.
2. Chee (1990b)	Cotyledons from 3- to 5-day-old-seedlings.	*Agrobacterium*-mediated transformation. Vector: Disarmed C58Z707 *A. tumefaciens* harboring a binary plasmid pGA482 containing *Nos-NPTII* selectable marker.	Ten percent of regenerated plantlets contained the NPTII gene.
3. Chee and Slightom (1991)	Cotyledons from 3- to 5-day-old-seedlings.	*Agrobacterium*-mediated transformation. Vector: Disarmed C58Z707 *A. tumefaciens* harboring a binary plasmid pGA482GG/cpCMV19 containing GUS, *Nos-NPTII* and CMV-C coat protein gene.	All Regenerated transformed cucumber plants contained NPTII, GUS and CMV-C genes in the plant genome.
4. Chee and Slightom (1992)	Cotyledons from 3- to 5-day-old seedlings.	Microprojectile-mediated transformation. Vector: Plasmid pGA482 containing *Nos-NPTII* selectable marker.	Sixteen percent of regenerated plants contained the NPTII gene.
5. Sarmento et al. (1992)	Petiole and leaf segments from 19- to 21-day-old seedlings.	*Agrobacterium*-mediated transformation. Vector: Disarmed LBA 4404 *A. tumefaciens* harboring a binary plasmid pBIN19 containing 35S-NPTII gene.	Nine percent of regenerated plants contained the NPTII gene.
6. Schulze et al. (1995)	Embryogenic cell suspension culture	Microprojectile-mediated transformation. Vector: Plasmid pRT99 containing NPTII and uidA genes.	Regenerated plants from suspension culture expressed uidA gene 1 year after bombardment.
7. Nishibayashi et al. (1996)	Hypocotyl segments from 7- to 10-day-old seedlings.	*Agrobacterium*-mediated transformation. Vector: Disarmed EHA101 *A. tumefaciens* harboring a binary plasmid pIG121 containing *Nos-NPTII*,	Transformed cells were selected using hygromycin. Twelve of twenty one regenerated plantlets expressed GUS in the very young leaves.

Table 1. *Continued*

Reference	Explant/culture used	Method/vector used	Observation
		CaMV 35S-I-GUS, CaMV 35S-hph genes.	
8. Raharjo et al. (1996)	Pickling cucumber petiole segments from the first and second true leaves of 10- to 21-day-old seedlings.	*Agrobacterium*-mediated transformation. Vectors: Three *A. tumefaciens* (EHA105, MOG301, MOG301), each harboring one of three binary vectors which contain an acidic chitinase gene from petunia (pMOG196), and basic chitinase genes from tobacco (pMOG198) and bean (CH5B)	Regenerated pickling plants expressed three chitinase genes in the transgenic leaves.

2 Gene Transfer Method via *Agrobacterium tumefaciens*

2.1 Materials and Methods

Seeds of cucumber (*Cucumis sativus* L. Poinsett 76, Asgrow Seed Co., Kalamazoo, MI) were soaked in tap water for approximately 15 min. The seed coats were remove manually. The de-coated seeds were surface sterilized with 70% alcohol for 1 min. A 25-min treatment with 25% (v/v) solution of Clorox (commercial bleach, 5.25% sodium hypochlorite) followed. The seeds were then rinsed four times with sterile distilled water. Sterilized seeds were germinated at 26 °C on 0.8% water agar for 3 to 5 days in darkness. Unless otherwise stated, all media were supplemented with 3% sucrose and solidified with 0.8% Phytagar (Gibco). The pH of all media was adjusted to 5.9 before being autoclaved. All media were autoclaved at 121 °C for 20 min.

Transformation was performed according to Chee (1990b). Cotyledons from 3- to 5-day-old in vitro grown seedlings (Poinsett 76) were used as donors of explants. The cotyledons were removed aseptically and were cut into pieces approximately 5 mm^2 in size. The pieces were submerged in a diluted overnight culture (2×10^8 cells/ml) of the disarmed *Agrobacterium* strain C58Z707 (Hepburn et al. 1985). The *Agrobacterium* strain contains the binary plasmid pGA482G (Fig. 1; An 1986) which was grown in LB medium containing 25 mg Km/l (kanamycin). After gentle shaking to ensure that all edges were infected, the cotyledon pieces were blotted dry and cultured abaxial side down in a sterile 100 × 20 Petri plate (20 pieces per plate) containing the initiation medium [MS (murashige and Skoog) basal medium + 2 mg 2, 4-D/l (2,4-D-dichlorophenoxyaceticacid) + 0.5 mg kinetin/l; Chee 1990a]. After 4 days of growth, cucumber cotyledon pieces were transferred to Petri plates containing the same medium supplemented with 500 mg Cb/l (carbenicillin) and

Fig. 1. Structure of the *Agrobacterium* binary plasmid pGA482. It contains the T-DNA border fragments of pTi37 (labeled B_R and B_L), the *cos* site of bacterial phage lamda, a restriction enzyme polylinker, and the *Nos-NPT*II fusion gene, neomycin phosphotransferase II (NPTII), driven by the nopaline synthase (NOS) promoter. Restriction enzyme sites are: *C Cla*I, *H Hin*dIII; *Hp Hpa*, *St* Stu. Not all site locations are shown

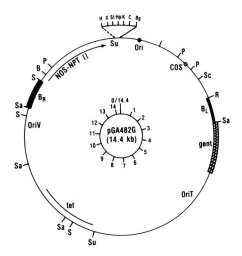

100 mg Km/l and cultured as before for 5 additional weeks in the dark at 26 °C. Regeneration of potentially transformed embryogenic callus was done according to the procedure described by Chee (1990a,b).

NPTII (neomycin phosphotransferase II) enzyme activity was detected by the in situ gel assay procedure described by Reiss et al. (1984). Total nucleic acids were extracted from cucumber leaves using the rapid DNA isolation procedure described by Chee et al. (1991a). Genomic DNA from individual cucumber plants (between 5 and 10 µg) was digested with a fivefold excess of *Bam*HI and *Hin*dIII enzymes followed by electrophoresis through a 0.7% agarose gel and electroblot transferred onto nylon filters (Hybond-N, Amersham) and hybridized as described by Chee et al. (1991a). These filters were hybridized against a 600 bp *Bgl*II-*Nco*I DNA fragment isolated from the bacterial *NPT*II gene of transposon Tn5 (Mazodier et al. 1985), the fragment was ^{32}P-labeled using a nick translation kit obtained from Bethesda Research Laboratories.

2.2 Experimental Results

2.2.1 Transformation of Cucumber Tissue

Cucumber cotyledon pieces were subjected to antibiotic selection by transferring onto initiation medium which was supplemented with different levels (25 to 200 mg/l) of kanamycin. The lowest level of kanamycin found to be useable for selection of the transformed callus tissues without allowing escapes, was determined to be 100 mg/l. At 100 mg/l, non-transformed tissues were not capable of growth while the putative transformed calli showed little if any growth inhibition. After 5 to 6 weeks all putative transformed calli obtained from cotyledon pieces cultured on initiation medium supplemented with 100 mg Km/l were tested for NPTII activity and all of these were found to be positive. These results indicate that, at least, the modified *Nos* (nopaline syntrase)-*NPT*II gene, contained within the T-DNA region of pGA482, had been transferred into the cucumber tissues by using the disarmed *Agrobacterium* strain.

2.2.2 Regeneration of Plantlets from Transformed Embryogenic Callus

After about 6 weeks on initiation medium supplemented with 500 mg Cb/l and 100 mg Km/l, a characteristic gel-like callus was formed on the surface of the explants and at the site of tissue contact with the medium. The upper part of the gelatinous tissue contained small sectors of putative embryogenic tissue which, upon microscopic examination, consisted of cytoplasmically dense, multicellular aggregates, resembling the proembryonic masses. The sectors with proembryonic masses were selected for transfer to the secondary medium supplemented with 100 mg Km/l. Upon transfer, the putative transformed proembryonic structures developed into embryo-like structures. After 2 weeks, the embryos were transferred to hormone-free MS medium supplemented with 50 mg Km/l. Ten percent of the explants produced plantlets and all followed normal development upon transfer to the latter medium. Transgenic plantlets were visually identifiable by their ability to form roots on MS medium supplemented with 50 mg Km/l. In contrast, non-transgenic, control plantlets were inhibited from root development on medium supplemented with 50 mg Km/l. All transgenic plants flowered and set seeds normally.

2.2.3 Detection of NPTII Enzyme Activity in Transformed R_0 Plants

A total of more than 100 transformed Km^r R_0 plants were obtained. In order to determine if this group of plants contained any escapes during the regeneration process, each plant was tested for NPTII enzyme activity. All of the putative transgenic R_0 plants showed the presence of NPTII enzyme activity in their protein extracts which co-migrated with the bacterial-derived NPTII enzyme activity. Figure 2 shows the analysis of NPTII enzyme activity in the protein extracts of some transformed R_0 plants. In contrast, the negative control plants, which resulted from co-cultivation with C58Z707 minus the binary plasmid, showed no co-migrating NPTII enzyme activity in leaf extracts.

2.2.4 Analysis of Transgenic R_0 Plants for the Presence of Nos-NPTII Gene

Total nucleic acids were extracted from leaves of the NPTII positive R_0 plants. This DNA was subjected to restriction endonuclease digestion with both *Bam*HI and *Hind*III. The restriction map of the *Nos-NPT*II fusion gene of pGA482G (Fig. 1; An 1986) predicts a fragment of about 2.0 kb in length with *Hind*III and *Bam*HI digestion. Digested cucumber DNAs were attached to

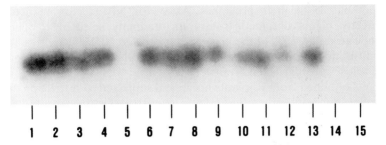

Fig. 2. Analysis of NPTII activity in transgenic R_0 cucumber plants obtained via the *Agrobacterium*-mediated method. *Lane 1* contains the bacterially derived NPTII protein control; *lanes 5, 14, and 15* contain extracts from leaf material of non-transformed cucumber plants. *Lanes 2 to 4 and 6 to 13* contain extracts from leaf material of independent transgenic R_0 cucumber plants

Fig. 3. Detection of the NPTII gene in R_0 cucumber plants obtained via the *Agrobacterium*-mediated method. Total DNA was isolated from plants, and digested with *Bam*HI and *Hind*III. *Lanes 1 to 7* contain a digest of DNA isolated from independently transformed cucumber plants. *Lane 8* contains a digest of pGA482 DNA. *Lane 9* contains a digest of DNA isolated from a non-transformed cucumber plant

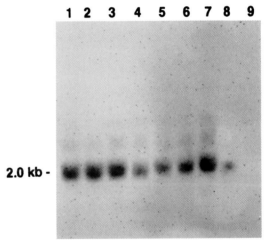

nylon filters which were hybridized against a nick-translated ^{32}P-labeled *Bgl*II-*Nco*I 600 bp fragment isolated from the bacterial *NPT*II gene. This fragment contains only the bacterial *NPT*II gene portion of Tn5 (Beck et al. 1982) and was selected because of its low background hybridization against various plant genomic DNA (Chee et al. 1989). The predicted 2 kb *Nos-NPT*II hybridizing fragment was observed for DNAs isolated from all NPTII positive plants. Figure 3 shows the genomic blots for *Hind*III and *Bam*HI enzyme digestion on transgenic R_0 plants. Copy number reconstruction suggests that each of these plants contain one gene copy of the *NPT*II gene.

2.2.5 Analysis of Transgenic R_1 Plants for the Nos-NPTII Gene

The most convincing evidence for the integration of the *Nos-NPT*II gene of pGA482 would be the transfer of the transgenic phenotype from the R_0 plants (self pollinated) to their progenies. Assay for NPTII enzyme activity in 40 progeny plants obtained from four randomly selected R_0 cucumber lines indicated that 75% of the progenies expressed NPTII activities. In order to confirm that the progenies have inherited the *Nos-NPT*II gene, total nuleic acids were extracted from leaves of five NPTII positive R_1 cucumber plants obtained from R_0 plants no. 36 and subjected to restriction endonuclease digestion with either *BamHI* and *Hind*III (which is expected to show the characteristic 2.0 kb hybridizing fragment) or with only *Hind*III which should yield a hybridizing fragment characteristic of the chromosomal location of the integrated T-DNA within the genome of R_0 cucumber plants. The results of the genomic blots for both enzyme digestions are shown in Fig. 4. The digestion with *Hind*III and *BamHI* shows the expected 2.0 kb band, while digestion with *Hind*III alone yields a hybridizing band of about 5 kb for each of the progeny plants. The presence of the identically sized band for the *Hind*III digest suggests, as expected, that all of these progeny plants were derived from a single transformation event which gave rise to the parent R_0 plant.

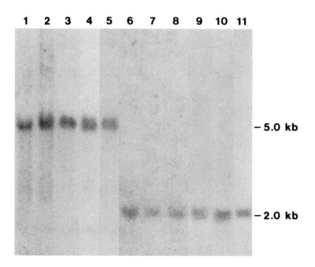

Fig. 4. Genomic blot analysis of DNAs isolated from progenies (R_1) of transformed R_0 cucumber plant (no. 36) obtained via the *Agrobacterium*-mediated method. DNAs from progeny plants, numbered 36-1, 36-2, 36-3, 36-4, 36-5, were subjected to restriction enzymme digests using both *Hind*III and *Bam*HI (*lanes 6 to 10*), and only *Hind*III (*lanes 1 to 5*). *Lane 11* contains a digest (*Hind*III and *Bam*HI) of DNA isolated from a non-transformed cucumber plant, to which pGA482 DNA had been added at a concentration which represents a reconstruction of one copy equivalence per haploid cucumber genome. DNA fragment sizes indicated on the *right* correspond with two fragments from the BRL DNA size standard mix. The observed hybridizing bands were measured to be 2 kb for the *Hind*III and *Bam*HI digests and about 5.0 kb when only the *Hind*III enzyme was used

3 Gene Transfer via Microprojectile Bombardment

3.1 Materials and Methods

The target tissues for microprojectile bombardment were cucumber embryogenic callus cultures. Embryogenic callus cultures were initiated from cotyledon explants (cv. Poinsett 76) as described by Chee (1990a). Each plate was bombarded three times with tungsten particles coated with the plasmid pGA482G (Fig. 1; An 1986), which were prepared according to the procedure described by Klein et al. (1988), and using aseptic conditions. Control tissues were bombarded using identical conditions with microprojectiles coated with pUC19 DNA (Yanisch-Perron et al. 1985). Plasmid DNAs were adsorbed onto 2 mg of surface sterilized tungsten particles (median size 1 μm) by adding 2.5 μg of DNA (1 μg/μl in 0.1 M Tris-EDTA buffer, pH 7.7) to 25 μl of tungsten particles in a 1.5-ml Eppendorf tube. $CaCl_2$ (25 μl of a 2.5 M solution) and spermidine-free base (10 μl of a 0.1 M solution) were then added to the microprojectile-suspension. Microprojectile bombardment was done using the Biolistics Particle Accelerator Apparatus model BPG that was originally leased from Biolistics Inc., Ithaca, NY (this lease was subsequently transferred to DuPont, Wilmington, Delaware). Each bombardment used 2.5 μl of the microprojectile-suspension, loaded onto a macroprojectile, which was propelled under partial vaccuum using a gun powder charge (no. 1 gray exra light; Speed Fasteners, St. Louis). Embryogenic callus tissues were positioned 13 cm from the macroprojectile stopping plate. The microprojectile bombarded *Cucumis sativus* tissues were regenerated using the procedure described by Chee (1990a), except that no kanamycin was added to any of the culture media, after bombardment, embryogenic callus tissue was transferred to fresh initiation medium and cultured in the dark. After 5 to 6 weeks on initiation medium, the cultures were transferred to MS medium + 1 mg NAA/l (naphthaleneacetic acid) + 0.5 mg kinetin/l for somatic embryo development. These cultures were incubated for an additional 2 weeks at 26 °C under diffuse cool-white fluorescent lamps (4 klx) with a 16-h photoperiod. These tissues were then transferred to MS medium with no growth regulators for development of plantlets.

3.2 Identification of Transgenic R_0 Cucumber Plants Regenerated from Bombarded Emryogenic Tissues

After microprojectile bombardment, the cucumber embryogenic callus tissues were allowed to grow 5–6 weeks on induction medium during which time the surface of the callus formed either separate or clusters of yellowish glossy embryos. Single cucumber plantlets were regenerated from embryos that were bombarded with microprojectiles coated with either pGA482 or pUC19 plasmid DNAs using the procedure described by Chee (1990a,b). A total of 107 phenotypically normal cucumber plants were regenerated from ten different batches of embryogenic callus bombarded with pGA482 and 25 cucumber plants were regenerated from three different batches of embryogenic callus bombarded with pUC19. The chances that each plantlet resulted from an independent transformation event was increased by obtaining only one regenerated plantlet from a cluster of microprojectile bombarded embryos.

Total leaf proteins were extracted from 107 cucumber plants obtained from pGA482 bombarded tissues and seven control plants obtained from pUC19 bombarded tissues and NPTII enzyme activities were determined (Chee et al. 1989). Co-migrating NPTII enzyme activity was found in only four of the regenerated plants obtained from callus cultures bombarded with pGA482. The simplest explanation for finding such a low number of plantlets with NPTII enzyme activity is that the NPTII gene was not efficiently transferred into the cucumber embryo tissues. However, there are reasons for not finding NPTII enzyme activity even if the gene is present. The *Nos* promoter can be inactivated by methylation (Gelvin et al. 1983), especially if its activity is not required for growth (positive selection), or the *Nos-NPT*II gene may not be intact due to gross rearrangement that occurred during the miroprojectile-mediated transfer process (McCabe et al. 1988). The presence or absence of the *Nos-NPT*II gene in the 107 R_0 cucumber plants was determined by genomic blot hybridization (Southern 1975). Previous analyses of the *Nos-NPT*II gene of pGA482 showed that its coding region is contained within a 2.0 kb *Bam*HI-*Hin*dIII fragment (Chee et al. 1989; Chee 1990b; Chee and Slightom 1991). Southern blot analysis of the *Bam*HI-*Hin*dIII digested DNAs isolated from the 107 regenerated R_0 cucumber plants found that 17 (16%) were transformed with at least a portion of the *Nos-NPT*II coding region. This finding suggests that the microprojectile-mediated transfer system is more efficient than suggested by the NPTII enzyme activity assay, and secondly, it indicated that for some reason the transferred *Nos-NPT*II gene is not functional in many of these transformed cucumber plants.

The Southern blot hybridization results for the *Hin*dIII-*Bam*HI digest of DNAs isolated from 12 transgenic R_0 is shown in Fig. 5. The presence of the 2 kb *Hin*dIII-*Bam*HI hybridization fragment suggests that all of these plants contain a major portion of the NPTII coding region; however, this analysis does not reveal the status of the *Nos* promoter, nor whether the *NPT*II gene has been subjected to any minor mutations (small insertions or deletions) that would render it non-functional. The number of *NPT*II gene copies integrated into the genome of these plants appears to vary considerably, as indicated by the presence of more than one band (lanes 7 to 13) and by the intensity of the hybridization signals (Fig. 5). It appears that most of the plants contain multiple copies of the *NPT*II gene, a common result for genes transferred using microprojectile bombardment (McCabe et al. 1988).

Copy number reconstruction suggests that only six R_0 plants contain a single copy of the *Nos-NPT*II gene; these are R_0 plants nos. 28, 34, 35, 36, 55, and 58, five of which are included in Fig. 5. Interestingly, the four plants that express the *Nos-NPT*II gene are among these six R_0 plants that contain a single copy of the *Nos-NPT*II gene (plant nos. 28, 34, 55, and 58), it appears

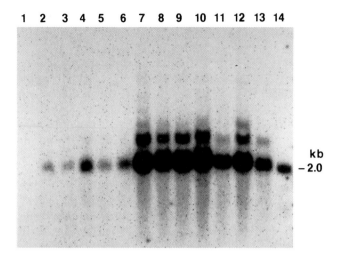

Fig. 5. Genomic DNA blot hybridization of DNAs isolated from R_0 cucumber plants obtained via microprojectile bombardment method. The numbered lanes contain DNAs from the following R_0 cucumber plants: *lane 1* control, *lane 2* no. 36, *lane 3* no. 28, *lane 4* no. 34, *lane 5* no. 55, *lane 6* no. 58, *lane 7* no. 6, *lane 8* no. 8, *lane 9* no. 11, *lane 10* no. 15, *lane 11,* no. 20, *lane 12* no. 24, *lane 13* no. 80, and *lane 14* contains 10 ng of pGA482 DNA that was added to 10 µg of DNA isolated from a non-transformed cucumber plant. The band in *Lane 14* shows the appropriate size hybridization fragment (about 2 Kb) expected from a *Bam*HI-*Hind*III digest of an intact *NPT*II gene

that plants with a single copy of the *Nos-NPT*II gene have a higher probability of its expression. That is, the lack of *Nos-NPT*II expression could be due to its presence in multiple gene copies in the remaining 14 transgenic cucumber plants. Such a result is consistent with data that shows that the *Nos* promoter is subject to methylation inactivation and that this inactivation is increased if multiple copies of the promoter is present (Gelvin et al. 1983). In addition, the microprojectile-mediated gene transfer process is subject to many different types of rearrangement, which include partial gene transfer processes and the interruption of transferred genes (McCabe et al. 1988). Goring et al. (1991) have shown that the presence of a partial copy of the *Nos* gene can inhibit the expression of the wild-type gene. Similar inactivation has also been observed for the cauliflower mosaic virus 35S promoter (Hobbs et al. 1990). Thus, the fact that the *Nos-NPT*II is non-functional or functions at a very low level in transgenic cucumber plants which contain multiple gene copies is not surprising. We have not yet determined if any of the 14 transgenic cucumber plants that do not express the *Nos-NPT*II gene contain a complete gene that could be re-activated. That is, if *Nos* promoter inactivity is due to methylation it is possible to restore its activity by treatment with 5-azacytidine or by backcrossing to reduce its gene copy number (Gelvin et al. 1983). However, 5-azacytidine treatment is not practical for cucumber plants and backcrossing may not be effective, because multiple copies transferred by microprojectile bombardment are usually closely linked (McCabe et al. 1988).

3.3 PCR and Genomic Blot Analysis of the Nos-NPTII Gene in R_1 Cucumber Plants

PCR amplification analysis was done as described by Chee et al. (1991a). Genomic cucumber DNAs were subjected to PCR amplification using the 5' oligo, 5'-CCCCTCGGTATCCAATTAGAG-3', which shares identity with the *Nos* promoter region 33 bp 5' of its start codon and the 3' (antisense) oligo, 5'-CGGGGGGTGGGCGAAGAACTCCAG-3', which shares identity with the 3'-flanking region of the *NPT*II gene, 150 bp 5' of its translation termination codon (Mazodier et al. 1985). PCR amplification cycles were controlled by a Perkin-Elmer-Cetus (Norwalk, CT) DNA Thermal Cycler and were subjected to 30 cycles, involved the following steps: denaturation of duplex DNA at 94 °C for 2 min, followed by annealing at 55 °C for 2 min, DNA synthesis at 72 °C for 3 min. PCR amplified samples were analyzed by removing 30 μl of each sample and electrophoresing it through a 0.7% agarose gel, after which the gel was stained with ethidium-bromide and photographed.

The six R_0 plants that contained a single copy of the *Nos-NPT*II gene were allowed to mature and set seed (self pollination); however, plants nos. 35 and 36 (*Nos-NPT*II non-expressing) were male sterile and were not advanced. Genomic DNAs from 20 progenies of each fertile R_0 plant were subjected to PCR amplification of the *Nos-NPT*II gene. This analysis indicated that the *Nos-NPT*II gene was present in many of the progeny plants from each R_0 plant; for example, PCR analysis of the 20 progeny plants derived from R_0 plant no. 58 showed that 14 contained the *Nos-NPT*II gene (70%). This 3:1 segregation ratio is expected for a single copy gene, which is further evidence that a single *Nos-NPT*II gene was transferred into R_0 plant no. 58. Figure 6 shows the PCR amplification results for DNAs isolated from ten plants of R_0 plant no. 58, the DNAs from five progeny plants show the expected 1 kb amplified DNA fragment.

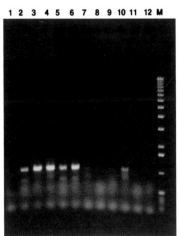

Fig. 6. PCR identification of the *Nos-NPT*II gene in R_1 cucumber plants obtained via microprojectile bombardment method. The lanes contain the PCR amplification results for DNAs isolated from the following plants: *lane 1* non-transformed plant, negative control; *lane 2*, 10 ng of pGA482, positive control; *lane 3*, no. 58-1; *lane 4*, no. 58-2; *lane 5*, no. 58-3; *lane 6*, no. 58-7; *lane 10*, no. 58-8; *lane 11*, no. 58-9; *lane 12*, no. 58-10; and lane *M* contains the 1 Kb ladder from Bethesda Research Laboratories

Fig. 7. Genomic blot analysis of DNAs isolated from progenies of cucumber plant (no. 58) obtained via the microprojectile bombardment method. DNA samples were digested with either *Hin*dIII and *Bam*HI (*lanes 4 to 7*) or only *Hin*dIII (*lanes 8 to 11*). The numbered lanes contain the following DNA samples; *lane 1*, nontransformed cucumber DNA digested with *Hin*dIII + *Bam*HI; *lane 2*, no. 58-7, digested with *Hin*dIII + *Bam*HI; *lane 3*, no. 58-7, digested with *Hin*dIII; *lane 4*, no. 58-1; *lane 5*, no. 58-2; *lane 6*, no. 58-3, *lane 7*, no. 58-4; *lane 8*, no. 58-1; *lane 9*, no. 58-2; *lane 10*, no. 58-3; *lane 11*, no. 58-4

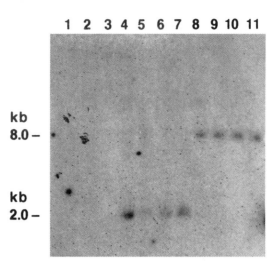

Genomic integration of the *Nos-NPT*II gene in the transformed R_1 progeny of plant no. 58 was confirmed by genomic blot hybridization. Total genomic DNAs from four of the PCR positive progeny plants were digested with either *Hin*dIII and *Bam*HI or with only *Hin*dIII. The *Hin*dIII digestion results should yield a hybridizing *NPT*II gene fragment that is characteristic of its integration into the genome of cucumber plant no. 58 and its progeny. The results of this genomic blot hybridization are shown in Fig. 7, where DNAs digested with *Hin*dIII and *Bam*HI show the expected 2.0 Kb hybridizing band, while the *Hin*dIII digest yielded a 8.0 Kb hybridizing band. The presence of the identical 8.0 Kb *Hin*dIII hybridizing band in DNAs isolated from these R_1 plants provides further evidence to support the transfer of a single copy of the *Nos-NPT*II gene into the genome of plant no. 58, since additional integration events would be expected to give rise to different size hybridizing bands.

4 Present Status of Transgenic Plants

4.1 Greenhouse Evaluation

The performance of transformed cucumber plants was determined by mechanical inoculation of CMV. The results showed that all non-transformed cucumber plants showed mosaic and chlorotic mottle symptoms on the first true leaf 6 days after inoculation and symptoms appeared on all subsequent true leaves that developed (Gonsalves et al. 1992). However, 36% of the inoculated plants of transgenic lines were symptomless, 50% showed symptom delays and were confined only to the lowest two leaves, and 14% showed symptoms on all leaves at the same time as the controls.

4.2 Field Evaluation

Three field tests were completed with transgenic cucumber plants expressing the *cp* gene of CMV (Gonsalves et al. 1992). The primary goals were to determine resistance of transgenic cucumbers to inoculations by aphid vectors under field conditions and to compare the resistance of transgenic plants with a commercially available CMV resistant cultivar. Preliminary mechanical inoculations showed that the transgenic cucumbers were not completely resistant to CMV. Instead, a percentage of the plants exhibited delay of and milder symptoms, and new leaves that developed were often devoid of symptoms. In contrast, infected non-transformed plants showed symptoms on all newly developed leaves.

Several factors needed to be considered in the cucumber field test. Some of these included: (1) ensuring that sufficient inocula would be available to start an epidemic; (2) isolating the field so that other viruses would not unduly interfere with the tests; and (3) establishing the test at a time when there would be sufficient aphid populations. The initial field test in 1989 suggested that a field plot isolated from cucurbits and surrounded by trees or grasses would effectively screen out zucchini yellow mosaic virus and papaya ringspot virus, but not watermelon virus 2 (viruses that commonly infect cucurbits). Furthermore, the 1989 test indicated that interspersing a small percentage (10%) of infected cucumber plants in the field would be sufficient to start the epidemic. Test plots were established in mid and late June because aphid flights in upstate New York generally start in late July and peak in August. The selected planting date would allow the cucumber to bear a full crop before frost and still be growing vigorously at the time when aphid flights are predicted to be high. Results from the 1990 and 1991 field tests clearly showed that the transgenic cucumbers showed much less CMV infection than non-transformed plants. Furthermore, the level of resistance was comparable to the commercial CMV tolerant cultivar, Marketmore 76. Eight weeks after transplanting, the critical period when plants were bearing fruit, the percentage of infection in transgenic plants and Marketmore 76 plants averaged less than 5%, in contrast to about 72% in non-transformed plants. Trials were terminated 13 weeks after field transplanting, when plants had entered the senescent stage. At this time, ELISA demonstrated an appreciable level of CMV infection in transgenic lines (35%) and a much higher amount in Marketmore 76 (62%), while an average of 85% of the non-transformed plants tested positive. Fruit yields and vegetable growth of transgenic lines averaged better than those of non-transformed plants of the same cultivar.

These results indicate that transgenic cucumbers which are subjected to high disease pressure by their natural vectors exhibit significant protection against CMV. They should perform even better in fields with only transgenic cucumbers. Such conditions will tend to lessen the disease pressure caused by the presence of infected susceptible plants.

5 Concluding Remarks

The available data strongly support that foreign genes can be transferred to cucumber and be stably inherited in the progeny without detrimental effects on the host plants. Most interestingly, high levels of virus protection are maintained in transgenic plants under field conditions. Thus, breeding of crop varieties by insertion of specific single or a few genes could be a good supplement to conventional crop improvement programs. This approach could be helpful in overcoming the specific defects of otherwise high yielding and well-adapted commercial cultivars. In the next few years, more cases of coat protein-mediated protection with different viruses will be reported and various virus resistant transgenic cucumber plants should become commercially available.

References

An G (1986) Development of plant promoter expression vectors and their use for analysis of differential activity of nopaline synthase promoter in transformed tobacco cell. Plant Physiol 81:86–91

Anderson AR, Moore LW (1979) Host specificity in the genome *Agrobacterium*. Phytopathology 69:320–323

Beck E, Ludwig G, Auerswald EA, Reiss B, Schaller H (1982) Nucleotide sequence and exact localization of the neomycin phosphotransferase gene from transposon Tn5. Gene 19:327–336

Chee PP (1990a) High frequency of somatic embryogenesis and recovery of fertile cucumber plants. HortScience 25:792–793

Chee PP (1990b) Transformation of *Cucumis sativus* tissue by *Agrobacterium tumefaciens* and the regeneration of transformed plants. Plant Cell Rep 9:245–248

Chee PP (1990c) Transformation of cucumber via *Agrobacterium tumefaciens*. In: Nijkamp HJJ, van der Plass LHW, van Aartrikj J (eds) Progress in plant cellular and molecular biology. Kluwer, Dordrecht, pp 201–206

Chee PP (1991a) Somatic embryogenesis and plant regeneration of squash *Cucurbita pepo* L. cv. YC60. Plant Cell Rep 9:620–622

Chee PP (1991b) Plant regeneration from cotyledons of *Cucumis melo* "Topmark". HortScience 26:908–910

Chee PP (1992) Initiation and maturation of somatic embryos of squash *Cucurbita pepo*. HortScience 27:59–60

Chee PP (1993) Transformation in cucumber (*Cucumis sativus*) In: Bajaj YPS (ed) Biotechnology in agriculture and forestry, vol 23. Plant protoplasts and genetic engineering IV. Springer Berlin Heidelberg New York, pp 215–227

Chee PP, Slightom JL (1991) Transfer and expression of cucumber mosaic virus coat protein gene in the genome of *Cucumis sativus*. J Am Soc Hortic Sci 116:1098–1102

Chee PP, Slightom JL (1992) Transformation of cucumber tissues by microprojectile bombardment: identification of plants containing functional and non-functional transferred genes. Gene 118:255–260

Chee PP, Fober KA, Slightom JL (1989) Transformation of soybean (*Glycine max*) by infecting germinating seeds with *Agrobacterium tumefaciens*. Plant Physiol 91:1212–1218

Chee PP, Drong RF, Slightom JL (1991a) Using polymerase chain reaction to identify transgenic plants. In: Gelvin SB, Schilperoort RA (eds) Plant molecular biology manual. Kluwer Dordrecht

Chee PP, Jones JM, Slightom JL (1991b) Expression of bean storage protein minigene in tobacco seeds: introns are not required for seed specific expression. J Plant Physiol 137:402–408

Deakin JR, Bohn GW, Whitaker TW (1971) Interspecific hybridization in *Cucumis*. Econ Bot 25:195–211

Gelvin SB, Karcher SJ, DiRita VJ (1983) Methylation of T-DNA in *Agrobacterium tumefaciens* and in several crown gall tumors. Nucleic Acids Res 1:159–174

Gonsalves D, Chee P, Provvidenti SR, Slightom JL (1992) Comparison of coat protein-mediated and genetically-derived resistance in cucumbers to infection by cucumber mosaic virus under field conditions with natural challenge inoculations by vectors. Bio/Technology 10:1562–1570

Goring DR, Thomson L, Rothsein SJ (1991) Transformation of a partial nopaline synthase gene into tobacco suppresses the expression of a resident wild-type gene. Proc Natl Acad Sci USA 88:1770–1774

Hepburn AG, White J, Pearson L, Maunders MJ, Clarke LE, Prescott AG, Blundy KS (1985) The use of Pnj5000 as an intermediate vector for genetic manipulation of *Agrobacterium* Ti-plasmid. J Gen Microbiol 131:2961–2969

Hobbs SLA, Kapodar P, Delong CMO (1990) The effect of T-DNA copy number, position and methylation on reporter gene expression in tobacco transformants. Plant Mol Biol 15:851–864

Klein TM, Harper EC, Swab Z, Sanford JC, Fromm ME, Maliga P (1988) Stable genetic transformation of intact *Nicotiana* cells by the particle bombardment process. Proc Natl Acad Sci USA 85:8502–8505

Mazodier O, Cossart P, Giraud E, Gasser (1985) Completion of the nucleotide sequence of the central region of Tn5 confirms the process of three resistance genes. Nucleic Acids Res 13:195–205

McCabe DE, Swain WF, Martinell BJ, Christou P (1988) Stable transformation of soybean (*Glycine max*) by particle acceleration. Bio/Technology 6:923–926

Nishibayashi S, Kaneko H, Hayakawa T (1996) Transformation of cucumber (*Cucumis sativus* L.) plants using *Agrobacterium tumefaciens* and regeneration from hypocotyl explants. Plant Cell Rep 15:809–814

Raharjo SHT, Hernandez MO, Zhang YY, Punja ZK (1996) Transformation of pickling cucumber with chitinase-encoding genes using *Agrobacterium tumefaciens*. Plant Cell Rep 15:591–596

Reiss B, Sprengel R, Will H, Schaller H (1984) A new sensitive method for qualitative and quantitative assay of neomycin phosphotransferase in crude cell exracts. Gene 30:211–218

Reynolds JF (1986) Regeneration in vegetable species. In: Vasil IK (ed) Cell culture and somatic cell genetics of plants, vol 3. Academic Press, Orlando, pp 151–178

Sarmento GG, Alpert K, Tang FA, Punja ZK (1992) Factors influencing *Agrobacterium tumefaciens* mediated transformation and expression of kanamycin resistance in pickling cucumber. Plant Cell Tissue Organ Cult 31:185–193

Schulze J, Balko C, Zellner B, Koprek T, Hänsh R, Nerlich A, Mendel RR (1995) Biolistic transformation of cucumber using embryogenic suspension cultures: long-term expression of reporter genes. Plant Sci 112:197–206

Slightom JL, Chee PP, Gonzalves D (1990) Field testing of cucumber plants which express the CMV coat protein gene: field plot design to test natural infection pressures. In: Nijkamp HJJ, van der Plas LHW, van Aartrikj (eds) Progress in plant cellular and molecular biology. Kluwer, Dordrecht, pp 201–206

Smarrelli J Jr, Watters MT, Diba L (1986) Response of various cucurbits to infection by plasmid-harboring strains of *Agrobacterium*. Plant Physiol 82:622–624

Southern EM (1975) Detection of specific sequences among DNA fragments separated by gel electrophoresis. J Mol Biol 98:503–517

Trulson AJ, Simpson RB, Shahin EA (1986) Transformation of cucumber (*Cucumis ativus* L.) plants with *Agrobacterium rhizogenes*. Theor Appl Genet 73:11–15

Yanisch-Perron C, Vieira J, Messing J (1985) Improve M13 phage cloning vectors and host strains: nucleotide sequence of the M13 mp18 and pUC18 vectors. Gene 33:103–119

11 Transgenic Carrot (*Daucus carota* L.)

D.R. GALLIE

1 Introduction

1.1 Distribution and Importance of Carrot

Carrot is a member of the Umbelliferae, a large family that includes parsley, parsnip, celery, fennel, anise, dill, cumin, and coriander as just some of the better known vegetables and spices that make up this economically important family. The *Daucus* genus comprises over 60 species. Carrot is a biennial with only a few annual forms known and is diploid with nine pairs of chromosomes. Carrot is thought to have originated from the area now known as Afghanistan (Mackevic 1929) from where it spread to the Mediterranean and Northern Europe during the tenth to fifteenth centuries, to China in the fourteenth century, Japan in the seventeenth century, and brought to the New World with the first English settlers (Banga 1957, 1963). It is now grown throughout the world and is one of the ten most important vegetables in the US. The commercial carrot has an orange-colored flesh resulting from its high carotene content but its selection originated in the Netherlands in the seventeenth century from the anthocyanin-containing carrot of the Near East (Ammirato 1986). The present day commercial varieties, of which there are over 400, all derive from this narrow genetic base.

1.2 Importance as a Model for Embryogenesis

In addition to its economic importance, carrot has become a model species for studying the molecular and organizational events during somatic embryogenesis in dicots. This is largely due to the ease with which it can be cultured in vitro and its high potential for regeneration. Its growth in vitro and subsequent regeneration was one of the first species to be described in detail (Steward et al. 1958, 1964). Much of what is known about the stages of somatic embryogenesis stems largely from work with carrot (Ammirato 1987).

Department of Biochemistry University of California Riverside California 92521 USA

Biotechnology in Agriculture and Forestry, Vol. 47
Transgenic Crops II (ed. by Y.P.S. Bajaj)
© Springer-Verlag Berlin Heidelberg 2001

2 Direct Gene Transfer in Carrot

2.1 Factors Affecting Transient Gene Expression

Because of its capacity for regeneration, carrot was one of the first species to be used in direct gene transfer studies and for transformation by *Agrobacterium*. Because of its excellent susceptibility to infection by *Agrobacterium*, this has become the procedure of choice for stable transformation in most studies. Direct gene transfer is used for the analysis of transient gene expression or to transiently produce a gene product in regeneration studies. Due to the expense, difficulty, and inherent variability associated with particle bombardment, this method for transient or stable transformation has not been used to any extent with carrot.

Carrot has proven to be quite easy to create protoplasts from and undergoes cell wall regeneration quickly and efficiently. Carrot protoplasts have been used in a number of transfection studies because DNA can be introduced with high efficiency and the DNA is expressed at high levels. These qualitites make carrot an ideal model system for investigations into gene expression. Carrot can be readily transfected using either electroporation or using polyethylene glycol (PEG). Electroporation of plasmid DNA into carrot was first described by Fromm et al. (1985). Electroporation of linear DNA and mRNA has also been described (Bates et al. 1990; Gallie 1993). PEG transformation of carrot using DNA was first described by Ballas et al. (1987) and many of the parameters for optimizing expression have been examined (Rasmussen and Rasmussen 1993). The level of transient expression is sufficiently high in carrot that the resulting RNA transcripts can be analyzed directly without having to resort to more sensitive methods such as RNase protection assays (Murray et al. 1990). The promoter from nopaline synthase (NOS) was more than three times as active than the 35S promoter from cauliflower mosaic virus (Rathus et al. 1993). Two copies of the 35S promoter in tandem, which has resulted in an increased level of expression in other dicot species, was equivalent in strength to just one copy, but introduction of the octopine synthase enhancer immediately upstream of the 35S promoter increased expression approximately twofold (Rathus et al. 1993). The presence of an intron in gene constructs has been shown to enhance expression in plants, particularly in monocots, but also in some dicot species (Callis et al. 1987). The inclusion of intron 1 from the maize alcohol dehydrogenase (*Adh1*) gene in a 35S promoter-based construct had little effect on expression (Rathus et al. 1993). This was probably due to the fact that the monocot *Adh1* intron is not efficiently spliced in dicot species (Kieth and Chua 1986; Leon et al. 1991). The Emu promoter, which contains six copies of the anaerobic regulatory element from the maize *Adh1* promoter, four copies of the OCS enhancer, and the *Adh1* intron, has been reported as a highly active promoter for monocot species (Last et al. 1991) but resulted in a similar level of expression as the 35S promoter in carrot. In summary, although the use of various promoters, enhancers, and introns can improve gene expression in many monocot species, beyond the requirement for a strong promoter such as the 35S or NOS promoters, little else appears to make a significant difference in the resulting level of expression in carrot. This may be due to a naturally high level of gene transcription in this species that obviates the need for optimizing other nuclear steps during gene expression.

In addition to efficient transcription, optimizing translation is essential to achieve a high level of expression, whether in the context of transient expression analysis or in stable transformants. All known plant cellular mRNAs contain a cap structure (m^{7}GpppN, where N represents any nucleotide) at the 5' terminus and a poly(A) tail at the 3' terminus. The cap and poly(A) tail function synergistically to promote a high level of translation initiation in plants (Gallie 1991), including carrot (Fig. 1). These regulatory elements also increase mRNA stability in carrot (Gallie et al. 1989). Transcripts produced from RNA polymerase II promoters are capped in the nucleus and the presence of a polyadenylation sequence positioned downstream of the gene to be introduced will ensure that the pre-mRNA produced will be correctly processed and polyadenylated before transport from the nucleus. Although these are the most critical elements needed for efficient translation in plants, there are other features of the mRNA that can be optimized for a high level of protein synthesis. The immediate context of the initiation codon can influence start site

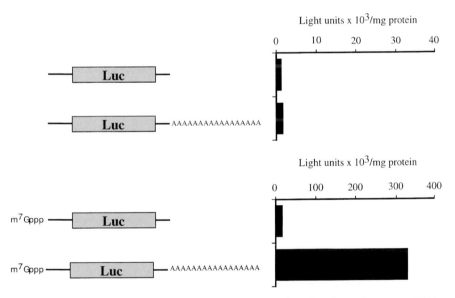

Fig. 1. The effect of the cap and poly(A) tail on expression of introduced or endogenous mRNAs. Luciferase mRNA was synthesized in vitro in the different forms shown and delivered to carrot protoplasts using electroporation. The protoplasts were allowed to translate the mRNA overnight, which was sufficient time to allow the mRNA to be fully translated and degraded. The amount of luciferase enzyme activity is shown as histograms and is a measure of the amount of protein synthesis from each mRNA. Note that the scale of the *bottom* graph is ten times greater than the scale of the *top* graph

selection. (A/G)NN*AUG*GC represents the consensus sequence found most often in plant genes (Gallie 1996). In addition to the sequences surrounding the start codon, the 5′ leader sequence can exert a considerable positive or negative influence on protein synthesis. The 5′ leader from tobacco mosaic virus (called Ω) enhances expression up to 25-fold (Fig. 2) and the leader from tobacco etch virus enhances translation more than 30-fold in carrot (Gallie et al. 1989, 1995; Pitto et al. 1992; Gallie 1993).

In addition to DNA, mRNA has been used for direct delivery to carrot protoplasts using either electroporation (Gallie et al. 1989) or PEG (Gallie 1993). This approach is particularly useful for the analysis of regulatory elements involved in controlling translational efficiency or mRNA stability as it delivers the mRNA directly to the cytoplasm and thereby avoids any complications associated with transcription or pre-mRNA processing. It is also a useful way of producing a short burst of a gene product to examine its subsequent effect on cell physiology.

2.2 Isolation of Protoplasts

A number of carrot varieties has been used to produce protoplasts for transient expression analysis. Protoplasts can be readily isolated from cultured cells and are extremely uniform in size and viability. The highest levels of

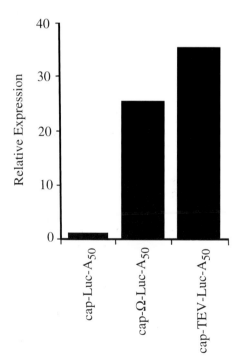

Fig. 2. The effect of 5′ leader sequences on protein synthesis. Luciferase mRNAs with either a control, 17-base polylinker leader (*cap-Luc-A$_{50}$*), the 5′ leader from tobacco mosaic virus (*cap-Ω-Luc-A$_{50}$*), or the 5′ leader from tobacco etch virus (*cap-TEV-Luc-A$_{50}$*), were synthesized in vitro and delivered to carrot protoplasts using electroporation. The protoplasts were allowed to translate the mRNA overnight. The amount of luciferase enzyme activity is shown in the histograms and is a measure of the amount of protein synthesis from each mRNA. Expression is shown relative to the level from the control luciferase mRNA which is set at a value of one

expression are achieved from cells that are in the rapid growth phase of their culture cycle. Typically, for a 7-day culture cycle, cells are harvested for protoplasting 3–4 days following subculture. One packed cell volume is mixed with four volumes of protoplast isolation buffer (12 mM sodium acetate pH 5.8, 50 mM $CaCl_2$, 0.25 M mannitol) containing 0.25% CELF cellulase (Worthington Biochemicals), 1% Cytolase M103S (Genencor), 0.5% BSA, and 7 mM β-mercaptoethanol. Protoplasts are usually obtained within 90–120 min of digestion and their state of digestion is determined with the use of a microscope. It is important that the cells be fully spherical as this indicates that most of the cell wall has been removed, a necessary prerequisite for efficient transfection. If the cells are to be used for electroporation, the protoplasts are then washed once with protoplast isolation buffer, once with electroporation buffer (10 mM HEPES pH 7.2, 130 mM KCl, 10 mM NaCl, 4 mM $CaCl_2$, 0.2 M mannitol), and resuspended in electroporation buffer to a final concentration of 1.0×10^6 cells/ml. If the cells are to transformed using PEG, following cell wall digestion, the protoplasts are washed once with protoplast isolation buffer, once with MaMg buffer (0.1% MES pH 5.6, 15 mM $MgCl_2$, 0.45 M mannitol), and are resuspended in the same buffer to a final concentration of 0.5×10^6 cells/ml.

2.2 Parameters Affecting DNA or mRNA Delivery

For electroporation, the DNA (or mRNA) is mixed with 0.8 ml of protoplasts immediately before electroporation. The electroporation parameters vary with the type of electroporator and must be empirically determined for each instrument. This is due to the internal resistance of the instrument itself, the shape of the wave, i.e., square or exponential, and the shape and size of the electrodes. Also, it is necessary to balance the percentage of cells that survive electroporation versus the level of expression that is desired. To a large extent, one is optimized at the expense of the other. Moreover, the composition of the electroporation buffer, the volume of the cells, and the gap between the electrodes influence electroporation efficiency. Using the IBI GeneZapper with a 4 mm electrode gap and 0.8 ml of cells in the electroporation buffer described above, the electroporation parameters for optimal viability and expression are 500 μF capacitance and 350 volts. Following the electrical discharge, the sample is transferred to a Petri dish containing carrot protoplast incubation medium (Murashige and Skoog macro and micro salts adjusted to pH 5.8, 30 g/l sucrose, 100 mg/l myo-inositol, 0.1 mg/l 2,4-D, 1.3 mg/l niacin, 0.25 mg/l thiamine-HCl, 0.25 mg/l pyridoxine phosphate, 0.25 mg/l calcium panthotenate, 0.2 M mannitol) and is incubated in the dark at room temperature.

For PEG transformation, 100 μg of sheared and denatured salmon sperm DNA is added to 0.4 ml of protoplasts in MaMg solution in a microfuge tube. The test DNA (or mRNA) is added immediately and mixed briefly before the addition of 0.5 ml PEG solution (40% PEG-4000, 100 mM $Ca(NO_3)_2$, 0.4 M mannitol) which is subsequently mixed by inverting the tube several times until the protoplasts are thoroughly mixed with the PEG solution. Following a 15 min incubation at room temperature, 0.6 ml of 0.2 M $CaCl_2$ is added and the tube inverted several times until completely mixed. As PEG is toxic, the cells are then harvested by low-speed centrifugation, the PEG-$CaCl_2$ solution removed, and the protoplasts resuspended in the carrot protoplast incubation medium in a Petri dish. The cells are then incubated in the dark at room temperature.

Nucleic acid delivery is equally efficient whether electroporation or PEG is used for the transfection. However, transformation using PEG has the advantage that no expensive equipment is required. Several parameters can affect the efficiency of the PEG-mediated delivery. The concentration of the PEG significantly affects delivery and expression (Fig. 3). Although a final PEG concentration of 15% is typically used, expression is six times higher when the final concentration of PEG is 24%. The higher PEG concentrations will result in a greater degree of water loss from the protoplast, which may compromise recovery. However, this does not appear to be a significant problem for carrot as viability is not greatly affected by this concentration of PEG.

Both Mg^{2+} and Ca^{2+} ions are present during transfections which are essential for delivery of nucleic acid. Removal of Ca^{2+} from the PEG solution reduces expression from delivered mRNA by 88%. Although the removal of Mg^{2+} from the MaMg solution did not significantly reduce expression when Ca^{2+} was present in the PEG solution, the removal of Mg^{2+} when the PEG solution lacked Ca^{2+} reduced expression a further 96% (Fig. 4). Therefore, the

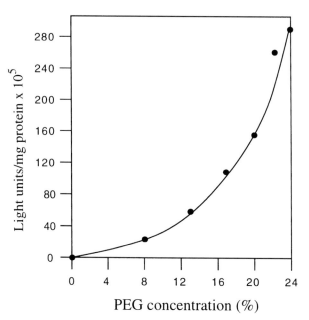

Fig. 3. The effect of the concentration of PEG on mRNA delivery and expression in carrot protoplasts. Luciferase mRNA was used as the reporter mRNA and the resulting level of expression is indicated on the Y-axis. The concentration indicated on the X-axis represents the final concentration of PEG during transfection

Fig. 4. The effect of Mg^{2+} and Ca^{2+} ions during PEG-mediated mRNA delivery in carrot protoplasts. Luciferase mRNA was used as the reporter mRNA and the resulting level of expression is shown relative to the control containing both Mg^{2+} and Ca^{2+}

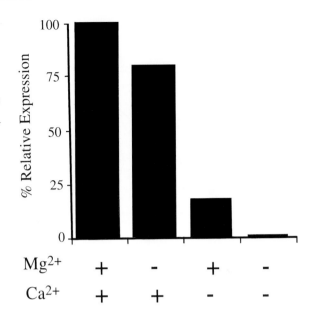

presence of divalent cations, particularly Ca^{2+} is required for PEG transformation. A 0.2 M CaCl$_2$ solution is also used to dilute the PEG-treated protoplasts and is thought an essential step for efficient transformation. However, neither the CaCl$_2$ solution nor dilution of the PEG-treated protoplasts is essential. Nevertheless, dilution is recommended as it can be difficult to pellet protoplasts from a dense PEG solution. Protoplast concentration or the pH of the protoplast solution (from pH 5.5 to 7.0) had little effect on transformation efficiency (Gallie 1993).

Addition of carrier DNA (50–100 µg) increases expression from the transfected test DNA but carrier RNA does not significantly increase expression from introduced mRNA (Gallie et al. 1989). The expression from mRNA delivered to carrot protoplasts using electroporation is linear up to at least 30 µg of input mRNA, whereas expression from mRNA PEG-mediated delivery is linear up to approximately 5 µg of input mRNA (Fig. 5).

3 Stable Transformation of Carrot

3.1 History of Carrot as a Model System for Transformation

Carrot has been used frequently as the species of choice for transformation by *Agrobacterium*. Studies using *Agrobacterium*-mediated transformation of carrot fall into two types: those that use carrot to elucidate the infection process by both *A. tumefaciens* and *A. rhizogenes* (Tepfer 1984; Cardarelli

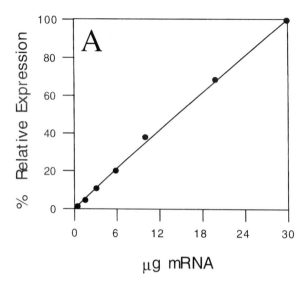

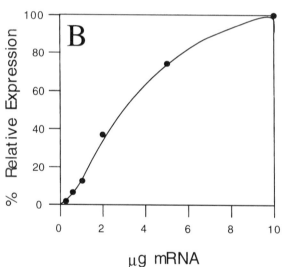

Fig. 5. Dose response of mRNA delivered by electroporation (**A**) and by PEG (**B**) in carrot protoplasts. Luciferase mRNA was used as the reporter mRNA and the resulting level of expression is shown relative to the highest level of mRNA used for each type of delivery

et al. 1985; Ryder et al. 1985; Jen and Chilton 1986; Guivarc'h et al. 1993) and those that focus on optimizing the efficiency of transformation (Scott and Draper 1987; Thomas et al. 1989; Wurtele and Bulka 1989; Liu et al. 1992; Pawlicki et al. 1992) or study the expression of introduced promoters or genes in carrot (Van Sluys et al. 1987; Mattsson et al. 1992; Gilbert et al. 1996). Initially, carrot root discs were used for many of the early studies on *Agrobacterium* infection. The appearance of galls on the apical surface of a root disc, in the case of infection with virulent strains of *A. tumefaciens*, or roots, in the case of *A. rhizogenes*, was used as the means by which virulence was determined. Once the introduction of foreign genes by *Agrobacterium* had been demonstrated, carrot transformation studies focused on determining the optimal conditions for efficient transformation. Early work used embryogenic carrot cultures from which transgenic plants were regenerated (Scott and Draper 1987; Wurtele and Bulka 1989). Subsequent methods using carrot hypocotyls or petioles obviated the need to establish embryogenic carrot cultures (Thomas et al. 1989; Liu et al. 1992; Pawlicki et al. 1992). Attempts have been made to optimize even further the frequency of transformation through the presence of multiple copies of *virG*, the transcriptional activator that induces expression of the *Agrobacterium* virulence genes required for infection, which increased transformation efficiency for some *Agrobacterium* strains, particularly of the nopaline type (Liu et al. 1992).

3.2 Transformation Using Embryogenic Carrot Cultures

Callus cells were initiated from root cortex on MS medium and vitamins supplemented with 30 g/l sucrose, 7 g/l agar and 0.5 mg/l 2,4-D (Wurtele and Bulka 1989) or 1-cm petiole segments from 7-day-old seedlings placed on similar MS medium with 2 mg/l 2,4-D (Scott and Draper 1987). Transformants have also been obtained from cell cultures derived from hypocotyls (Liu et al. 1992). Liquid cultures were initiated by placing callus pieces into liquid medium. Cell suspension cells or callus clumps were incubated with overnight cultures of *Agrobacterium*, incubated in the absence of antibiotic selection for 3 days in the case of the callus (Wurtele and Bulka 1989) or 5–7 days in the case of suspension cells (Scott and Draper 1987). The callus was then transferred to media containing both kanamycin (typically 100–300 µg/ml) to select for successful transformants, as resistance to this antibiotic is the method of choice for carrot, and 200–500 µg/ml cephotaxime to kill the *Agrobacterium*. Regeneration is carried out by transferring the callus to medium containing kanamycin and no hormones. Somatic embryos readily develop within 3–4 weeks on hormone-free medium and under light will regenerate into complete plants that can be transferred to pots.

3.3 Transformation Using Carrot Petioles, Cotyledons, or Hypocotyls

In this procedure, 10 mm sections from carrot petioles, cotyledons, or hypocotyls are taken from aseptic seedlings (typically 3- to 4-weeks-old) and are incubated in a suspension of *Agrobacterium* containing the binary construct of interest for 5 min before placing the explants on MS medium supplemented with 1 mg/ml 2,4-D, 20 g/l sucrose, and 0.8% agar (Pawlicki et al. 1992). Following a 2–3 day co-cultivation period in the absence of antibiotic selection, the explants are rinsed in MS medium containing 500 µg/ml cephotaxime, blotted dry, and cultured on the same medium supplemented with 100 µg/ml kanamycin, and 500 µg/ml cephotaxime (or 400 µg/ml carbinicillin) for a 2-week period, and then transferred to fresh medium in which the cephotaxime has been reduced to 250 µg/ml for a further 2 weeks (Pawlicki et al. 1992). Regeneration is carried out by transferring the embryogenic calli to the same medium without cephotaxime or hormones. Rooted plantlets were obtained within 8–10 weeks (Pawlicki et al. 1992).

The use of petioles resulted in the highest yield of transformants, with cotyledons yielding approximately half as many and hypocotyls even less (Pawlicki et al. 1992). Roots proved to be quite poor. The age of the seedlings from which explants are obtained is critical, with 3- to 4-week-old seedlings proving to be the optimal age. When hypocotyls were taken from 1-week-old seedlings, a significantly lower rate of transformation was reported (Thomas et al. 1989). The presence of acetosyringone during transformation had little effect (Pawlicki et al. 1992) but at least a 2-day preculture on hormone-containing medium proved to be essential for transformation using hypocotyls (Thomas et al. 1989). Transformation rates vary widely with the variety of carrot used and will depend greatly on the strain of *Agrobacterium* used. For one study using the *Agrobacterium* stain C58C1 and the pGV2260 Ti plasmid, the variety Nantes proved to give high rates of transformation, with De Chantenay and 2027H somewhat lower, and 840217 even lower (Pawlicki et al. 1992). For a second study using carrot hypocotyls and the *Agrobacterium* strain LBA4404, the variety Denvers proved to give high rates of transformation, with Imperator 58 and Chantenay Red Cored somewhat lower, and Scarlet Nantes even lower (Thomas et al. 1989). Rates of transformation of recalcitrant varieties may be improved through the use of alternative *Agrobacterium* strains or with the use of plasmids containing multiple copies of the *virG* gene (Liu et al. 1992).

Acknowledgments. This work received support, in part, from the US Department of Agriculture NRICGP (95-37100-1618 and 96-35301-3144).

References

Ammirato PV (1986) Carrot. In: Evans DA, Sharp WR, Ammirato PV (eds) Handbook of plant cell culture vol 4. Macmillan, New York, pp 457–499

Ammirato PV (1987) Organizational events during somatic embryogenesis. In: Green CE, Somers DA, Hackett WP, Biesboer DD (eds) Plant tissue and cell culture. Plant biology vol 3. Alan R Liss, New York, pp 57–81

Ballas N, Zakai N, Loyter A (1987) Transient expression of the plasmid pCaMVCAT in plant protoplasts following transformation with polyethyleneglycol. Exp Cell Res 170:228–234

Banga O (1957) Origin of the European cultivated carrot. Euphytica 6:54–63

Banga O (1963) Origin and distribution of the western cultivated carrot. Genet Agrar 17:357–370

Bates GW, Carle SA, Piastuch WC (1990) Linear DNA introduced into carrot protoplasts by electroporation undergoes ligation and recircularization. Plant Mol Biol 14:899–908

Callis J, Fromm M, Walbot V (1987) Introns increase gene expression in cultured maize cells. Genes Dev 1:1183–1200

Cardarelli M, Spano L, De Paolis A, Mauro ML, Vitali G, Costantino P (1985) Identification of the genetic locus responsible for non-polar root induction by *Agrobacterium rhizogenes* 1855. Plant Mol Biol 5:385–391

Fromm M, Taylor LP, Walbot V (1985) Expression of genes transferred into monocot and dicot plant cells by electroporation. Proc Natl Acad Sci USA 82:5824–5828

Gallie DR (1991) The cap and poly(A) tail function synergistically to regulate mRNA translational efficiency. Genes Dev 5:2108–2116

Gallie DR (1993) PEG-mediated delivery of mRNA to plant protoplasts for transient expression. Plant Cell Rep 13:119–122

Gallie DR (1996) The role of post-transcriptional control in transgenic gene design. In: Owen MRL, Pen J (eds) Transgenic plants: a production system for industrial and pharmaceutical proteins. John Wiley, Chichester, pp 49–74

Gallie DR, Lucas WJ, Walbot V (1989) Visualizing mRNA expression in plant protoplasts: factors influencing efficient mRNA uptake and translation. Plant Cell 1:301–311

Gallie DR, Tanguay RL, Leathers V(1995) The tobacco etch viral 5′ leader and poly(A) tail are synergistic regulators of translation. Gene 165:233–238

Gilbert MO, Zhang YY, Punja ZK (1996) Introduction and expression of chitinase encoding genes in carrot following *Agrobacterium*-mediated transformation. In Vitro Cell Dev Biol Plant 32:171–178

Guivarc'h A, Caissard J-C, Brown S, Marie D, Dewitte W, Van Onckelen H, Chriqui D (1993) Localization of target cells and improvement of *Agrobacterium*-mediated transformation efficiency by direct acetosyringone pretreatment of carrot root discs. Protoplasma 174: 10–18

Jen GC, Chilton M-D (1986) Activity of T-DNA borders in plant cell transformation by mini-T plasmids. J Bacteriol 166:491–499

Kieth B, Chua N-H (1986) Monocot and dicot pre-mRNAs are processed with different efficiencies in transgenic tobacco. EMBO J 5:2419–2425

Last DI, Brettel RIS, Chamberlain DA, Chaudhury AM, Larkin PJ, Marsh EL, Peacock WJ, Dennis ES (1991) pEmu: an improved vector for gene expression in cereal cells. Theor Appl Genet 81:581–588

Leon P, Planckaert F, Walbot V (1991) Transient gene expression in protoplasts of *Phaseolus vulgaris* isolated from a cell suspension culture. Plant Physiol 95:968–972

Liu C-N, Li X-Q, Gelvin SB (1992) Multiple copies of *virG* enhance the transient transformation of celery, carrot and rice tissues by *Agrobacterium tumefaciens*. Plant Mol Biol 20: 1071–1087

Mackevic VI (1929) The carrot of Afghanistan. Bull Appl Bot Genet Plant Breed 20:517–557

Mattsson J, Borkird C, Engstrom P (1992) Spatial and temporal expression patterns directed by the *Agrobacterium tumefaciens* T-DNA gene 5 promoter during somatic embryogenesis in carrot. Plant Mol Biol 18:629–637

Murray EE, Buchholz WG, Bowen B (1990) Direct analysis of RNA transcripts in electroporated carrot protoplasts. Plant Cell Rep 9:129–132

Pawlicki N, Sangwan RS, Sangwan-Norreel BS (1992) Factors influencing the *Agrobacterium tumefaciens*-mediated transformation of carrot (*Daucus carota* L.). Plant Cell Tissue Organ Cult 31:129–139

Pitto L, Gallie DR, Walbot V (1992) Functional analysis of sequence for post-transcriptional regulation of the maize HSP70 gene in monocots and dicots. Plant Physiol 100:1827–1833

Rasmussen JO, Rasmussen OS (1993) PEG mediated DNA uptake and transient GUS expression in carrot, rapeseed and soybean protoplasts. Plant Sci 89:199–207

Rathus C, Bower R, Birch RG (1993) Effects of promoter, intron and enhancer elements on transient gene expression in sugar-cane and carrot protoplasts. Plant Mol Biol 23:613–618

Ryder MH, Tate ME, Kerr A (1985) Virulence properties of strains of *Agrobacterium* on the apical and basal surfaces of carrot root discs. Plant Physiol 77:215–221

Scott RJ, Draper J (1987) Transformation of carrot tissues derived from proembryogenic suspension cells: a useful model system for gene expression studies in plants. Plant Mol Biol 8:265–274

Steward FC, Mapes MO, Smith J (1958) Growth and organized development of cultured cells I. Growth and division of freely suspended cells. Am J Bot 45:693–703

Steward FC, Mapes MO, Kent AE, Holsten RD (1964) Growth and development of cultured plant cells. Science 143:20–27

Tepfer M (1984) Transformation of several species of higher plants by *Agrobacterium rhizogenes*: sexual transmission of the transformed genotype and phenotype. Cell 37:959–967

Thomas JC, Guiltinan MJ, Bustos S, Thomas T, Nessler C (1989) Carrot (*Daucus carota*) hypocotyl transformation using *Agrobacterium tumefaciens*. Plant Cell Rep 8:354–357

Van Sluys MA, Tempe J, Fedoroff N (1987) Studies on the introduction and mobility of the maize *Activator* element in *Arabidopsis thaliana* and *Daucus carota*. Embo J 6:3881–3889

Wurtele E, Bulka K (1989) A simple, efficient method for the *Agrobacterium*-mediated transformation of carrot callus cells. Plant Sci 61:253–262

12 Transgenic Strawberry (*Fragaria* Species)

H. Mathews[1] and R.K. Bestwick

1 Introduction

1.1 Distribution and Importance of Strawberry

The *Fragaria* species are found in a diverse array of climates, including temperate, grassland, mediterranean, and subtropical. They fall into four natural ploidy groups, each with a basic chromosome number of 7; diploids, tetraploids, hexaploids and octoploids. The cultivated strawberry, *Fragaria* x *ananassa* Duchesne, is a vegetatively propagated octoploid species ($2n = 8x = 56$) which originated in Europe around 1750 as a hybrid between the pistillate South American *F. chiloensis* Duch. and a North American *F. virginiana* Duch. (Martinelli 1992). *Fragaria* x *ananassa* Duch. is grown all over the arable world, while *F. virginiana* and *F. chiloensis* are native only to the New World. *Fragaria vesca* (wood strawberry) and *F. moschata* (musky strawberry) are also two species commercially propagated to a limited scale (Hancock and Luby 1993). Over the years a large number of different cultivars of strawberry, *Fragaria* x *ananassa* Duch. have evolved, each adapting to a specific region or environment.

The strawberry fruit is a rich source of vitamin C, sugar and ellagic acid, a putative inhibitor of chemically induced cancer (Maas et al. 1996). For many centuries, strawberries have been a favorite among the fruits of the temperate world. However, the worldwide consumption of the delectable strawberry fruit has increased steadily during the last 15 years.

The world production of strawberry is about 2.5 million tons. The major producing countries are the United States with 25%, Poland with 10% and Japan with 9%. The total US strawberry crop exceeds 0.81 million tons (USDA 1997) with a value of approximately $800 million. Of the US production, more than 70% is for the fresh market. The widespread marketability, enhanced by breeding efforts that select for post-harvest storage properties both for fresh market and processing industries has increased the value of this crop worldwide.

[1] Agritope Inc. 16160 SW Upper Boones Ferry Road Portland, Oregon 972 24-7744, USA

Biotechnology in Agriculture and Forestry, Vol. 47
Transgenic Crops II (ed. by Y.P.S. Bajaj)
© Springer-Verlag Berlin Heidelberg 2001

1.2 Need for Genetic Transformation

Conventional plant breeding and selection over the years have resulted in considerable improvement in yield, fruit size and quality traits of contemporary North American genotypes (Bringhurst and Voth 1984). However, the narrow genetic base of the cultivated strawberry combined with the polyploid nature and heterozygosity severely constrain traditional breeding methods in their ability to meet the growing needs of the strawberry industry (Sjulin and Dale 1987).

One of the greatest challenges facing breeders is to incorporate several different disease and pest resistances in commercially acceptable types (Galletta et al. 1989). Increasing pressure from consumers to minimize fungicide and pesticide input into agricultural systems without a negative impact on fruit quality is another area of major concern among strawberry growers. The highly perishable nature of strawberry fruit causes significant post harvest losses (28–42%) to the strawberry industry (Ceponis and Butterfield 1973; Wright and Billeter 1975; Kader 1991). The principal cause of post harvest losses appears to be gray mold (*Botrytis cinerea*) associated with softening, bruising, and leaking of the strawberries (El-Kazzaz et al. 1983). Processing markets also require berries with raised necks and reflexed calyxes to minimize cutting waste (Khanizadeh et al. 1991) and possibly fruit rot and gray mold susceptibility (Hondelman and Richter 1973; Gooding 1976; Barritt 1980; Popova et al. 1985; Testoni et al. 1989).

Strawberry growers are under some pressure to find alternatives to conventional breeding in order to meet the demands of the industry. Biotechnological approaches have several attractive attributes not shared by traditional breeding. Somaclonal variants with higher yield and plant vigor (Sansavini et al. 1980; Swartz et al. 1981; Marcotrigiano et al. 1984; Cameron et al. 1985, 1989; Cameron and Hancock 1986; Theiler-Hedtrich and Wolfensberger 1987; Nehra et al. 1992; Lopez-Aranda et al. 1994), resistance to fungus, *Fusarium oxysporum* f. sp. *fragaria* (Toyoda et al. 1991); and virus-free clones through meristem culture employed in commercial nurseries (Jungnickel 1988; Kondakova and Schuster 1991) are some of the positive outcomes of the new plant technologies.

Among the various biotechnological approaches, genetic engineering is by far the most desirable and efficient strategy implemented in crop improvement programs. The progress in plant genetic engineering has been spectacular since the recovery of the first transformed plants in the early 1980s. Plant genetic engineering is highly target specific and allows existing superior cultivars to be discretely altered for one or more traits. In this chapter we review the advances made in the area of genetic transformation of strawberry during the last decade.

2 Genetic Transformation

2.1 General Account

An efficient tissue culture regeneration system (Jones et al, 1988; Liu and Sanford 1988; Nehra et al. 1989, 1990c; Sorvari et al. 1993), combined with excellent micropropagation methods (Lee and de Foussard 1975; Damiano 1980; Waithaka et al. 1980) has facilitated transformation of strawberry cultivars. Genetic transformation of *Fragaria* species has primarily been through *Agrobacterium*-mediated gene transfer. Electroporation of protoplasts has resulted in transient expression of the *gus* gene in an octoploid breeding line of strawberry (Nyman and Wallin 1992). A recent report from the National Research Centre, Cairo, evaluated microprojectile bombardment parameters for successful stable integration of foreign genes into strawberry (Sawahel 1996). A summary of transformation studies in strawberry is given in Table 1.

The susceptibility of strawberry to wild species of *Agrobacterium tumefaciens* has been reported in *F. vesca* and *F.* x *ananassa* (Jelenkovic et al. 1986, 1991; Uratsu et al. 1991). The first successful transgenic plants in strawberry (*Fragaria* x *ananassa* Duch.) were developed by two independent research groups in Canada and the UK. Nehra et al. (1990a,b) obtained transformation of a North American strawberry cv. Redcoat using *Agrobacterium tumefaciens* carrying plasmid pBI121 containing genes for *npt*II *and gus*. During the same time period, cv. Rapella, an important European cultivar, was transformed with two different binary vectors, one with selectable marker genes, (nopaline synthase – *nos* and *npt*II) and the other with the *ipt* (isopentenyltransferase) gene by James et al. (1990). The leaf disk transformation system (Horsch et al. 1985) was used in the development of transgenic plants in both of these cultivars. In the initial studies, Nehra et al. (1990b) obtained transgenic callus at 3% frequency with recovery of < 1% transgenic plants. They further improved the transformation frequency to 7% by preculturing of leaf disks for 10 days prior to culture on selection medium (Nehra et al. 1990a). Transgenic plants of strawberry cv. Rapella with marker genes had normal phenotypes while the plants transformed with a vector carrying *npt*II and *ipt* genes exhibited abnormal phenotypes associated with overproduction of cytokinin by *ipt* gene (James et al. 1990). Southern hybridization and various biochemical assays were used to confirm stable integration of these genes into strawberry cultivars Redcoat and Rapella. As the cultivated strawberry is an octoploid there has been some interest in transforming diploid species such as *F. vesca* to develop it as a model system to study gene expression in *Fragaria* species (Haymes and Davis 1993; El Mansouri et al. 1996).

The development of a stable transformation system with marker genes was soon followed by insertion of genes of commercial importance to strawberry. There have been efforts to incorporate insect (James et al. 1992; Graham et al. 1995) and herbicide resistance to strawberry cultivars (Layton et al. 1996). Graham et al. (1995) reported very limited success with the leaf disc transformation method and they used stem explants for transforming straw-

Table 1. Summary of various studies conducted on transformation of Strawberry (*Fragaria* species)

Reference	Species; cv.	Explant	Method	Results Transgenes; Transformation freq.
Jelenkovic et al. (1986)	*Fragaria* x *ananassa*, cv?	Runner segments	*A. tumefaciens*	Transformed callus *Ti-plasmid*[a]
James et al. (1990)	*Fragaria* x *ananassa*, cv. Rapella	Leaf disc	*A. tumefaciens*	Transgenic plants; *NOS, NPTII, IPT (Isopentenyl transferase)*; 0.1–6.0%
Nehra et al. (1990a)	*Fragaria* x *ananassa*, cv. Redcoat	Leaf disc	*A. tumefaciens*	Transgenic plants; *GUS, NPTII*; <1%
Nehra et al. (1990b)	*Fragaria* x *ananassa*, cv. Redcoat	Leaf disc	*A. tumefaciens*	Transgenic plants; *GUS, NPTII*; ~7%
Jelenkovic et al. (1991)	*Fragaria* x *ananassa*, cv?	Runner segments	*A. tumefaciens*	Transgenic plants; *Ti-plasmid*[a]
Uratsu et al. (1991)	*Fragaria vesca*, Alpine clone	Runners (In planta)	*A. tumefaciens* and *A. rhizogenes*	Plant susceptibility to *Agrobacterium* Ti- and Ri-plasmid; tumor induction at 33–80%
Nyman and Wallin (1992)	*Fragaria* x *ananassa*, Breeding line 77101	Protoplasts	Electroporation	Transformed callus *GUS, HPT*; 0.01–0.05%
Haymes and Davis (1993)	*Fragaria vesca*; cv?	Leaf discs, petiole	*A. tumefaciens*	Transgenic plants *NPTII, GUS*[a]
Nyman (1993)	*Fragaria* x *ananassa*, cv?	Protoplasts	Electroporation	Transgenic plants *GUS, HPT*[a]
Cocci et al. (1994)	*Fragaria* x *ananassa*, cvs. Addie, Brio	Leaf segments?	*A. tumefaciens*	Towards pathogen control *NPTII, PGIP* (endo-polygalacturonase-inhibitory protein gene from *Phaseolus vulgaris*)[a]
Mathews et al. (1994a)	*Fragaria* x *ananassa*, cvs. Tristar, Totem	Leaf, petiole, shoot base	*A. tumefaciens*	Transgenic plants, control of ethylene biosynthesis *SAMase, NPTII, HPT, GUS*[a]

Reference	Species/cultivar	Explant	Method	Description
Mathews et al. (1994b)	*Fragaria* x *ananassa*, cvs. Tristar, Totem	Leaf, petiole, shoot base	*A. tumefaciens*	Transgenic plants, control of ethylene biosynthesis *SAMase, NPTII, HPT, GUS*[a]
Finstad and Martin (1995)	*Fragaria* x *ananassa*, cvs. Hood, Totem	Leaf segments	*A. tumefaciens*	Transgenic plants *NPTII, CP of SMYEPV* (coat protein gene of strawberry mild yellow edge virus)[a]
Graham et al. (1995)	*Fragaria* x *ananassa*, cvs. Melody, Rhapsody, Symphony.	Stem tissue	*A. tumefaciens*	Towards insect resistance *NPTII, CPTI* (cowpea trypsin inhibitor gene); 6%
Mathews et al. (1995)	*Fragaria* x *ananassa*, cvs. Tristar, Totem	Leaf, petiole, shoot base	*A. tumefaciens*	Transgenic plants, control of ethylene biosynthesis *SAMase, NPTII, HPT, GUS*; 12.5–58.8%
Nuutila et al. (1995)	*Fragaria* x *ananassa*, cv. Senga sengana	Leaves	*A. rhizogenes*	Hairy roots were used for in vitro growth of arbuscular mycorrhizal fungus, *Glomus fistulosum*. *Ri* – plasmid[a]
Layton et al. (1996)	*Fragaria* x *ananassa*, cvs. Redcoat, Pajaro	Leaf disc	*A. tumefaciens*	Transgenic plants *NptII*, glyphosate, *GUS*[a]
El Mansouri et al. (1996)	*Fragaria vesca*	Leaf disc	*A. tumefaciens*	Transgenic plants *NPTII, GUS*; 6.9%
Mathews (1996)	*Fragaria* x *ananassa*, cvs. Tristar, Totem	Leaf, petiole, shoot base	*A. tumefaciens*	Transgenic plants, control of ethylene biosynthesis *SAMase, NPTII, HPT, GUS*[a]
Sawahel (1996)	*Fragaria* x *ananassa*, cv?	Leaf disc	Microprojectile bombardment	Transgenic plants *GUS*?[a]
Asao et al. (1997)	*Fragaria* x *ananassa*, cv. Toyonoka	Leaf disc, petiole	*A. tumefaciens*	Transgenic plants Rice chitinase gene; 6.2–7.2%
Baker et al. (1998)	*Fragaria* x *ananassa*, cv?	Leaves, petioles, runners	*A. tumefaciens*	Transgenic plants *NPTII, GUS*; 20–150%
Mathews et al. (1998)	*Fragaria* x *ananassa*, cv. Totem	Leaf segments	*A. tumefaciens*	Transgenic plants, use of iterative culture in the elimination of chimera in transgenics. *NPTII, GUS*[a]

[a] Transformation frequency not reported.

berry cvs. Melody, Rhapsody and Symphony. The strategies proposed for the development of root weevil (*Otiorhynchus* species) resistance in strawberry include expression of the cowpea trypsin inhibitor gene (Graham et al. 1995) and the cysteine proteinase inhibitor gene (Vrain 1994) in transgenic plants. The expression of a proteinase inhibitor gene is also suggested as a possible control measure for *Botrytis* fruit rot in strawberry (Vrain 1994). Aphid-borne viruses viz. strawberry crinkle (SCV), mild yellow edge (SMYED), mottle (SMV) and veinbanding (SVBV) are a serious threat to strawberry plantations. The genetic engineering strategy using coat protein mediated resistance – CPMR – (Beachy et al. 1990) has been applied to cultivars Totem and Hood (Martin 1996).

In our laboratory we have transformed two Pacific Northwest cultivars of strawberry (*Fragaria* x *ananassa* Duch.), Tristar and Totem (Mathews et al. 1994a,b, 1995, 1998; Mathews 1996). Efficient methods of transformation were developed in cv. Tristar using marker genes, and cultivar Totem was transformed with a gene (SAMase, *S*-adenosylmethionine hydrolase) to control ethylene biosynthesis in plants with a view to increase the post-harvest shelf life of berries (Mathews et al. 1995). The SAMase gene, isolated from T3 bacteriophage (Studier and Movva 1976; Hughes et al. 1987a,b), catalyzes the conversion of SAM (*S*-adenosylmethionine) to methylthioadenosine (MTA). SAM is the metabolic precursor of 1-aminocyclopropane-1-carboxylic acid (ACC), the proximal precursor to ethylene ad therefore the expression of SAMase lowers the plant's ability to produce ethylene (Fig. 1). Since SAM (*S*-adenosylmethionine) is an important methylating agent in several biochemical reactions and is a substrate in polyamine biosynthesis, the tissue specific and timely expression of SAMase is critical (Good et al. 1994). Earlier, we demonstrated that incorporation of SAMase effectively reduced ethylene synthesis and extended the shelf life of tomatoes (Good et al. 1994, Kramer et al. 1996). We have applied the same strategy for testing the efficacy of this gene in strawberries for increasing the post-harvest life of berries. Details of our transformation methodology in strawberry and results are discussed in the following sections.

2.2 Methodoloy

2.2.1 Explants and Culture Conditions

In vitro maintained cultures are an ideal source of explants. In vitro shoots are generally maintained on the propagation medium containing Murashige and Skoog's basal medium (Murashige and Skoog, 1962), 1 mg/l indoleacetic acid (IAA), 1 mg/l benzylaminopurine (BA), 0.01 mg/l gibberellic acid (GA), pH5.8 gelled with 0.8% Sigma agar. Meristematic sections from the base of the proliferating cultures, longitudinal sections of young shoots, leaves and petioles can serve as explants for *Agrobacterium* treatment.

Petioles were cut into 4–6mm length, folded young leaves were cut longitudinally and cultured with their adaxial surfaces in contact with the

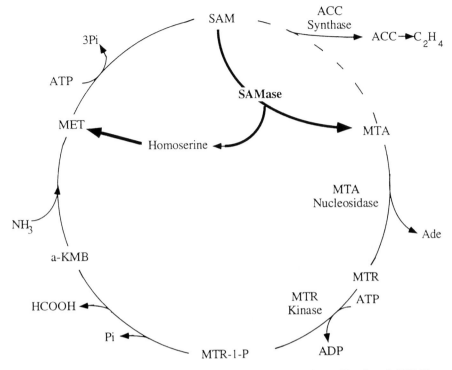

Fig. 1. Effect of SAMase expression on ethylene biosynthesis in plants. (Good et al. 1994, Plant Molecular Biology 26:781–790)

medium. Leaf, petiole and shoot base explants were cultured in Petri plates with 40 ml medium per plate while the shoots for proliferation or rooting were cultured in Phytatrays (Sigma) with 120 ml medium per tray. The Petri plates had 20–25 segments per plate and Phytatrays had 9–10 shoots per tray.

All media components were autoclaved at 120 °C at 1.1 kg.cm^{-2} except the antibiotics, acetosyringone and silver nitrate, which were filter sterilized before adding to the medium. All cultures were kept under 16-h photoperiods of 15–20 µmol·m^{-2}·s^{-1} provided by cool white fluorescent lamps. Observations were recorded every 3–4 weeks, followed by transfer to fresh medium.

2.2.2 Preparation of Agrobacterial Suspension

Single colonies of *Agrobacterium tumefaciens* (LBA4404, EHA101 or EHA105) inoculated into 30 ml of MG/L (Garfinkel and Nester 1980) liquid medium supplemented with 50 µM acetosyringone, pH5.6 were grown on a shaker at 200 rpm overnight (16–18h). The bacterial suspension at the start time of co-cultivation with plant tissues had an average count of 0.5 – 0.6 × 10^9 cells/ml.

2.2.3 Bacterial Strain and Binary Vectors Used

Disarmed strains of *Agrobacterium tumefaciens*, LBA4404, EHA101 and EHA105 were used for strawberry transformation. The binary vectors, pAG5110 and pAG5520, were pGA482-derived (An et al. 1985). Vector pAG5110 encoded DNA sequences for *gus* and *nptII* under the control of 35S promoter. Vector pAG5520 encoded the SAMase gene under the control of tomato E4 promoter (Cordes et al. 1989) and nptII under the 35S promoter (Fig. 2). The binary vectors pAG1552 and pAG1452 were constructed using the backbone of the pGPTV binary vector (Becker et al. 1994) and they contained the SAMase gene under the E4 promoter (Cordes et al. 1989) located near the right border and the marker gene nptII (pAG1552) or hpt (pAG1452) under the nos promoter located near the left border (Fig. 2).

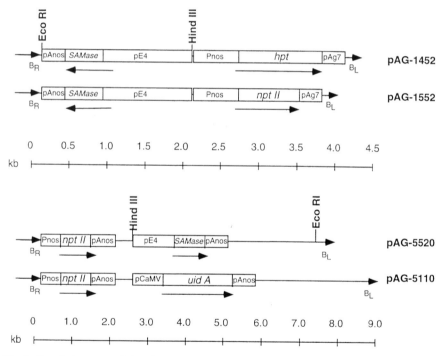

Fig. 2. Restriction maps of *Agrobacterium tumefaciens* binary vectors used in the present study. Vectors pAG1452 and 1552 contained SAMase gene with the E4 promoter (*pE4*) and were located near the *right border* and marker gene *nptII* (pAG1552) or *hpt* (pAG1452) with the nos promoter (*Pnos*) were located near the *left border*. Vector pAG5520 contained SAMase with the E4 (*pE4*) promoter located near the *left border* and *nptII* with the nos (*Pnos*) promoter located near the *right border*. Vector pAG5110 is similar to pAG5520 except the SAMase gene is replaced by *uidA* with the CaMV35S (*pCaMV*) promoter. Poly A addition signals were either from the *A. tumefaciens*, T-DNA nos gene (*pAnos*) or gene 7 (*pAg7*). (Mathews et al. 1995, In Vitro Cell Dev Biol 31:36–43)

2.2.4 Co-Cultivation with Agrobacterium tumefaciens

Soon after excision, isolated explants were soaked in a suspension of *A. tumefaciens* for 60–90 min followed by blotting on a sterile filter paper and cultured on MS (Murashige and Skoog, 1962) basal salts, B5 (Gamborg et al. 1968) vitamins, 3% sucrose, 2 mg/l BA, 0.5 mg/l IAA, 50–100 µM acetosyringone, pH 5.6 gelled with 0.25% phytagel. After 2 days of co-cultivation the explants were rinsed with a liquid medium of the above composition without acetosyringone and the explants were incubated with liquid medium of the same composition supplemented with 500–1000 mg/l cefotaxime/carbenicillin. The flasks were kept on a shaker at 100 rpm, and after 1 h the explants were blotted and placed on screening medium to be selected.

2.2.5 Screening Medium for Selection of Transformants

The screening medium contained MS (Murashige and Skoog 1962) salts, B5 (Gamborg et al. 1968) vitamins, 3% sucrose, 0.1–0.2 mg/l IBA, 5–10 mg/l BA, 500 mg/l carbenicillin and the selection agent kanamycin, geneticin or hygromycin at different levels. The initial selection levels of antibiotics used in transformation experiments were based on our studies on the tolerance of the non-transformed control tissues to antibiotics. From our earlier studies we determined that non-transformed control shoots do not withstand more than 25–50 mg/l kanamycin. (The dose response effect to a specific selection marker will need to be investigated for any particular cultivar under study. The tissue responses in strawberry are highly influenced by the age and culture conditions of the explant source.) Soon after co-cultivation, the segmented explants were exposed to 25–50 mg/l kanamycin, depending on the plasmid (pAG5110, pAG5520, pAG1552) and the cultivar. Selection levels were sequentially elevated during subcultures up to a maximum of 200 mg/l kanamycin in the shoot proliferation medium for the maintenance of transformants (flow chart, Fig. 3).

In the cultivar Tristar, meristematic segments co-cultivated with LBA4404 or EHA101 containing pAG5110 were cultured on regeneration medium with 0, 10 and 25 mg/l kanamycin. At the end of 3 weeks, the formation of shoot initials from the meristematic segments occurred in all three treatments. Explants from all treatments were transferred to medium with 50 mg/l kanamycin, followed by subsequent transfer to medium with 75 mg/l kanamycin. During subculture, completely bleached tissues were discarded and the fully or partially green tissues were maintained. Fully or partially green shoots were longitudinally segmented when transferred to fresh medium.

Totem was transformed with EHA101 or 105 containing the SAMase gene and the selectable marker genes *nptII* or *hpt* for resistance to kanamycin (geneticin) and hygromycin respectively. The regenerants were subjected to the iterative culture process with stepwise increase in selection pressure as shown in the flow chart (Fig. 3). After such screening, the regenerated shoots were multiplied on proliferation medium with 200 mg/l kanamycin or 70 mg/l

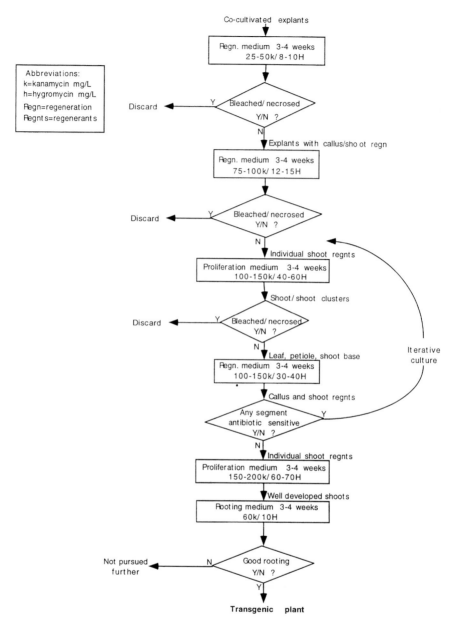

Fig. 3. Flow chart of transformation protocol for strawberry. (Mathews et al. 1995, In Vitro Cell Dev. Biol. 31:36–43)

hygromycin. Geneticin was started at 15 mg/l and taken stepwise to 40 mg/l. In the case of hygromycin, the initial level of selection was 10 mg/l and the final level used for maintenance of transformed shoots contained 70 mg/l. The effect of silver nitrate on the recovery of transformants in strawberry was also investigated in cv. Totem.

2.2.6 Histochemical Assay for gus Expression

Intact shoots (5–8 mm) and cut segments of the regenerants were subjected to 5-bromo-4-chloro-3-glucuronic acid (X-Gluc) treatment as per the protocols of Jefferson (1987).

2.2.7 Southern Hybridization

Genomic DNA was isolated from leaf tissue of transgenic plants and a non-transgenic control in culture as well as greenhouse established plants following the method of Doyle and Doyle (1990) except for the modification of adding polyvinylpolypyrrolydone (100 mg/g tissue) prior to CTAB isolation buffer. The DNA was digested with either *Eco*RI alone or in conjunction with *Hind*III. *Eco*RI cleaves once within the borders and produces junction fragments whereas *Eco*RI and *Hind*III cleave twice to generate an intra border 4.7 kb SAMase fragment. A probe for the strawberry alcohol dehydrogenase gene (ADH; Wolyn and Jelenkovic 1990) was used to confirm the complete digestion of the DNA and as a relative measure of the DNA content in each lane.

2.2.8 Induction of Rooting and Regeneration of Complete Plants

Individual shoots about 20- to 30-mm in size were isolated from multiple shoot clumps on proliferation medium followed by culture on half strength MS (Murashige and Skoog 1962) salts, B5 (Gamborg et al. 1968) vitamins, 1% sucrose, 100 mg/l carbenicillin and respective selection agents, depending on the plasmid used in transformation. In shoots transformed using plasmids, pAG5110, pAG5520 and pAG1552, 60 mg/l kanamycin or 15 mg/l geneticin was used in the rooting medium while for shoots transformed with pAG1452, 10 mg/l hygromycin was used. Our earlier studies had shown that kanamycin at 25 mg/l and hygromycin at < 5 mg/l completely inhibited root formation in control non-transformed shoots.

2.2.9 Transplantation

Well-rooted plants in Phytatrays were transferred to the greenhouse and the lids were left loosened. After about 2–4 days, the adherent media were rinsed off and the plants were potted in soil.

2.2.10 Analysis of Transgenic Berries

Expression of SAMase was examined at the mRNA transcript level using ribonuclease protection assays with an anti-sense SAMase riboprobe to detect SAMase mRNA from a total strawberry RNA population. Berries were picked at the ripe stage and stored frozen at −80 °C. Total RNA was isolated from the berries using the pine needle extraction procedure of Chang et al. (1993). RPA assays were performed using an Ambion's RPA II kit using 170–175 µg of total strawberry RNA, 1/90 dilution of RNAse A/Tl, and 40000 cpm of an 33P-UTP anti-sense SAMase probe per reaction. RPA reactions were electrophoresed on 8 M urea/5% polyacrylamide gels in 1 × TBE buffer. The gels were dried onto blotting paper and then exposed to X-ray film for 3–7 days at room temperature.

3 Results and Discussion

3.1 Elimination of Chimeras and Recovery of Transformed Clones

The protocol for strawberry transformation described above allows recovery of pure line transgenics at a significantly higher frequency than conventional methods. Elimination of chimeras is an integral part of strawberry transformation. Most of the primary regenerants were taken through at least one cycle of iterative culture in order to develop transgenic shoots until no parts showed sensitivity to selection (see flow chart – Fig. 3). In a separate set of experiments, we compared the effect of iterative versus non-iterative pathways in getting uniformly transformed plants (Mathews et al. 1998). Strawberry cv. Totem was transformed with *Agrobacterium tumefaciens* strain EHA 105 with pAG1501 containing genes for *nptII* and *gus* under 35S and nos promoters respectively. Random leaf samples of greenhouse-grown transgenics were analyzed for the presence of *gus* gene sequences by Southern hybridization as well as *gus* expression on leaf and petiole tissues by X-Gluc histological assay. All of the leaf samples from the transgenic events involving iterative culture protocol were positive for the *gus* insert, confirming the presence of transgenes and lack of chimeras. Leaf samples of the transgenic events from the non-iterative protocol were either positive or negative, indicating the chimeric nature of the transgenic plants (Table 2). The *gus* expression was highly variable irrespective of the iterative or non-iterative protocol used in transformation.

The occurrence of chimeric plants can easily be perceived due to the multicellular origin of organs (Stewart 1978; Poethig 1989; Irish 1991) and irregularity of cell lineage patterns in vitro combined with the phenomenon of cross protection of non-transformed cells by transformed cells. As the recovery of true transformants via seed is not a viable option in vegetatively propagated crops, all precautions to minimize this chimerism for transgenes are highly warranted.

Table 2. Comparison of iterative and non-iterative (conventional) pathways of transformation (Mathews et al. 1998)

Pathway	Event	Reaction with X-Gluc. + presence of blue tissue − no blue		Validation by Southern hybridization. Each event is represented by five random leaf samples. + and − indicate presence or absence of hybridization signal				
		leaflets	petiole	1	2	3	4	5
Non-iterative pathway (I)	2	−	−	−	−	−	−	−
	3	+	+	−	+	−	+	+
	4	−	−	−	−	−	−	−
	6	−	−	−	−	−	−	−
	8	−	−	−	−	−	−	−
	9	(not done)		+	+	−	−	+
Iterative pathway (II)	2	+	−	+	+	+	+	+
	4	+	−	+	+	+	+	+
	5	+	−	+	+	+	+	+
	9	−	−	+	+	+	+	+
	10	+	−	+	+	+	+	+
	11	+	+	+	+	+	+	+
Negative control		−	−	−	−	−	−	−

3.2 Cultivar Tristar

After 4 months of culture, the percent recovery of putative transformants which were consistently able to proliferate in the presence of selection showed a direct correlation to the level of selection in the initial screening medium. Exposure to a zero level of selection for the first 3 weeks resulted in recovery of no or few putative transformants. In explants treated with LBA4404, pAG5110, the frequency of transformed shoots was, 0.0%, 2.3%, 13.6% in treatments started with 0, 10 and 25 mg/l kanamycin respectively. This is in contrast to Nehra et al.'s (1990a) finding, that transformation frequency increased as a result of delayed selection pressure. They increased the transformation frequency of cv. Redcoat from <1% to 7% by pre-culture on a non-selection medium for 10 days. In explants treated with EHA101, pAG5110, the respective transformation frequencies were 2.5%, 11.6% and 16.7% in terms of shoot regeneration in the three treatments started with 0, 10 and 25 mg/l kanamycin respectively. A higher frequency of transformed shoots was observed in both cases when the explants were initially exposed to 25 mg/l kanamycin compared with an initial exposure of 10 mg/l kanamycin. Histochemical analysis of regenerants on selection medium showed a mixture of blue and non-blue regions on reaction with X-Gluc (Fig. 4a).

Fig. 4. a Shoot regenerant of cultivar Tristar transformed with pAG5110, showing blue and non-blue regions, on treatment with X-Gluc (H. Mathews, unpubl.). **b** Explanted segments of a primary shoot regenerant of cv. Totem on regeneration medium with 150 mg/l kanamycin. Segments capable of withstanding selection and capable of regeneration are indicated by *arrows*. (Mathews et al. 1995, In Vitro Cell Dev Biol 31:36–43). **c** Transgenic strawberry cv. Totem containing SAMase, in greenhouse. (Mathews et al. 1995, In Vitro Cell Dev Biol 31:36–43)

3.3 Cultivar Totem

Non-transformed regions of primary shoot regenerants were eliminated by iterative culture. Figure 4b shows one of the steps in the iterative process, the leaf segments resistant to antibiotic are green and undergoing callus formation and differentiation while the other segments from the same event are bleached or necrosed. The transformation frequency in cv. Totem ranged from 12.5% to 58.0% (Table 3). These frequencies are significantly higher than commonly reported so far for *Fragaria*. In experiments ST10, ST11, ST12, ST13 and ST14 equal numbers of explants were screened on regeneration medium containing silver nitrate. Addition of silver nitrate was not conducive for strawberry tissue as the explants produced yellow firable callus with very few

Table 3. Frequency of transformation in strawberry, cv. Totem (Mathews et al. 1995)

Exper. ID	*Agro* strain/binary vector	Selection	Explant,	no.	Trans. freq.[a] %	Trans. events[b] recovered	Events rooted
ST10-1A	EHA101/pAG5520	Kanamycin	Leaf	31	35.5	11	11
			Mer. Seg[c]	149	46.9	70	68
ST11-1A	EHA105/pAG5520	Kanamycin	Leaf	17	58.8	10	9
			Mer. Seg[c]	147	40.8	60	59
ST12-1	EHA101/pAG5520	Kanamycin	Leaf	40	32.5	13	13
ST13-1	EHA105/pAG5520	Kanamycin	Leaf	40	12.5	5	4
ST14-1	EHA101/pAG5520	Kanamycin	Young shoot segments	195	33.3	65	65
ST20	EHA105/pAG1542	Hygromycin	Young shoot segments	1222	15.6	191	182

[a] Trans. Freq. = transformation frequency.
[b] Trans. events = transgenic events.
[c] Mer. Seg. = meristematic segment.

Table 4. Effect of kanamycin vs. geneticin on recovery of transformants in cv. Totem (Mathews et al. 1995)

Exp. ID	*Agro* strain/binary vector	Selection	Explant,	no.	Trans. freq.[a] (%)	Trans. events[b] recovered
ST21-1	EHA105/pAG1552	Kanamycin	Leaf	329	15.5	51
			Petiole	131	16.0	21
ST21-2	EHA105/pAG5520	Geneticin	Leaf	293	1.3	4
			Petiole	132	2.3	3

[a] Trans. Freq. = transformation frequency.
[b] Trans. events = transgenic events.

shoot regenerants. Both strains of *Agrobacterium* EHA101 and 105 as well as the selection marker genes *nptII* and *hpt* were effective for generating transgenic plants in cultivar Totem although our data (Table 3) does not allow a strict comparison of the effectiveness of different *Agrobacterium* strains or the selection markers. Geneticin, a more potent aminoglycoside antibiotic than kanamycin, proved less favorable for recovering transformants in strawberry. The transformation frequencies with kanamycin were 7- to 12-fold higher than with geneticin in petiole and leaf explants (Table 4).

3.4 Rooting and Transplantation

Transgenic shoots of both cvs. Tristar and Totem were rooted in presence of antibiotic selection. In cv. Totem 95–100% of the transgenic shoots were successfully rooted on medium with 60 mg/l kanamycin or 10 mg/l hygromycin (Table 3). Well-rooted transgenic plants were quickly established in soil with almost 100% success (Fig. 4c). Nehra et al. (1990b) reported inadequate rooting ability of transgenics in the presence of 25 mg/l kanamycin and those which rooted had single long slender roots. They transferred transgenic shoots to medium without antibiotic selection to obtain well-rooted plants. In our experience such a step generated chimeric transgenics as shown in Table 2. The plants of the non-iterative pathway with reduced rooting in presence of selection agent were transferred to a medium without kanamycin for proper root development, following Nehra's protocol. The resultant plants failed to show hybridization signals in most of the leaf samples, confirming that the plants were chimeric for the transgene.

The total time taken from explant co-cultivation with *A. tumefaciens* to the transfer of transgenics to the soil was about 8–10 months.

3.5 Southern Hybridization

The Southern blot data (Fig. 5) showed a variety of transgene copy numbers and integration structures. The DNA from transgenic plant leaves (pAG5520) was digested with either *Eco*RI alone or in conjunction with *Hind*III. The *Eco*RI digest produces junction fragments that hybridize with the SAMase hybridization probe. The *Eco*RI and *Hind*III digested DNA produces a 4.8-kb fragment internal to the T-DNA borders that hybridizes to SAMase. The former allows determination of the number of individual integration events while the latter allows an estimation of the total transgene copy number for that event. The last panel in Fig. 5 is the *Eco*RI blot probed with the strawberry ADH gene which acts as an internal control for comparison of the relative amount of DNA in each lane. An example of how this system is used can be seen by comparing lanes 4 through 7 in each of the three panels in Fig. 5. First, the *Eco*RI blot indicates single integration events for lanes 4 and 6 and multiple events in lanes 5 and 7. The combined *Eco*RI and *Hind*III blot confirms the multiple gene copy number for lanes 5 and 7 and indicates an aberrant integration occurred in the event shown in lane 5 due to a smaller than expected fragment hybridizing to the SAMase probe. The signal strength in lanes 4 and 6 are identical to that seen in the *Eco*RI blot, confirming the single integration status for these two events. The differences in hybridization signal strength between these samples can not be attributed to a difference in the amount of DNA because the *Eco*RI blot probed with the strawberry ADH gene showed very similar signal intensities for lanes 4 through 7. Furthermore, the identical band pattern seen with the ADH probe clearly indicates a complete digestion occurred, validating the interpretation of multiple events for lanes 5 and 7. Because strawberry is octoploid, a single integration event can be expected to be eightfold lower than a native gene such as ADH. An approximate eightfold difference is seen in lanes 4 and 6 when comparing the band intensities from the SAMase probed blots and the ADH blot.

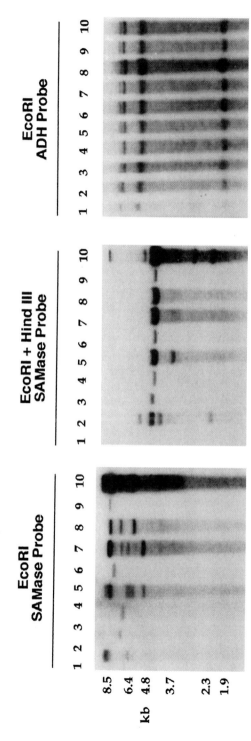

Fig. 5. Southern hybridization of transformed cv. Totem (pAG5520). *Lane 1* was from untransformed plant material and *lanes 2–10* were from nine independent transformation events. The *Eco*RI blot was stripped and reprobed with the strawberry alcohol dehydrogenase (ADH) gene. (Mathews et al. 1995. In Vitro Cell Dev Biol 31:36–43)

3.6 Evaluation of Transgenic Strawberry with SAMase

RPA assays were carried out with transgenic strawberries grown in the green-house. Presence of SAMase transcript was observed in 27 out of a total of 30 events tested (Fig. 6). Preliminary studies on the measurement of ethylene were not conclusive as the amount of ethylene produced was not consistent, probably due to the fact these were greenhouse-grown berries. All the trans-genics positive with the RPA assay, confirming the presence of the SAMase transcript, have recently been field planted (June 1997). About 600 plants from a total of 280 independent transgenic events are awaiting fruit evaluation both in terms of quality attributes and shelf life.

The role of ethylene in the ripening and post-harvest senescence of straw-berry fruit is not very clear (Perkins-Veazie 1991). However, there is increas-ing interest in the elucidation of biochemical and molecular events of fruit ripening in strawberry (Manning 1993). Nogata et al. (1993) reported the presence of a low level of exo- and endo-polygalacturonase in strawberry fruit (*Fragaria* x *ananassa*, Duch. cv. Toyonoka). Although strawberry is not con-sidered a typical climacteric fruit (Kader 1991), there is evidence that indicates that the removal of ethylene could play a role in reducing spoilage of fresh berries. El-Kazzaz et al. (1983) found that *B. cinerea* infection increased sub-stantially on fruits subjected to 20 ppm ethylene. Exogenous application of ACC (the immediate metabolic precursor to ethylene), to Earlyglow straw-berry at preclimacteric and climacteric stages induced higher ethylene production, while AVG (ethylene antagonist) application inhibited the biosynthesis of ethylene (Basiouny 1989). De la Plaza and Merodio (1989)

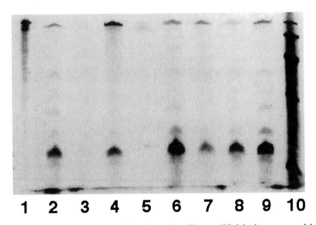

Fig. 6. RPA assay of SAMase transgenic berries of cv. Totem (H. Mathews, unpubl.). The ribonu-clease protection assay was performed with an anti-sense SAMase riboprobe for detection of SAMase mRNA from total strawberry RNA. For each RNA sample tested, 175 µg total RNA, 1/90 dilution of RNAse A/T1, and 40000 cpm of anti-sense SAMase riboprobe was used per assay reaction. Gels were exposed to film for 3–7 days. *Lane 1* SAMase antisense riboprobe; *lanes 2–9* eight independent transgenic events. Strong signals at 500 bp equal to size of SAMase were observed in all the events except for events AW1 and AL1. *Lane 10* RNA markers

reported that the treatment of strawberry cv. Chandler with an ethylene absorbent, gave increased firmness combined with a reduction in fungal attack from 26.3 to 10%.

In our laboratory we have analyzed ethylene evolution in picked strawberry fruit of non-transgenic control Totem at four stages of fruit development – turning, light red, near ripe and fully ripe – over a 3-day period. Ethylene production rates were undetectable in early stages of fruit ripening but rose to levels as high as 10 nl/g/h in fully ripe fruit. These results have led us to postulate that a strawberry genetically modified to have reduced ethylene synthesis will be less prone to senescence.

4 Summary and Conclusions

Since the first successful transgenic plants of strawberry with marker genes were developed in 1990 (James et al. 1990; Nehra et al. 1990a,b), there has been a steady increase in the generation of transgenics with various genes of interest such as insect resistance (James et al. 1992), fungal resistance (Cocci et al. 1994; Vrain 1994), herbicide resistance (Layton et al. 1996), virus resistance (Martin 1996) and enhancement of postharvest quality (Mathews et al. 1995). However, most of these transgenics are at various stages of field evaluation at the time of this report or have not yet resulted in the expected practical outcome, except for the report on virus resistant plants (see Table 5). Transgenic plants of cv. Totem containing the coat protein gene for SMYEPV (strawberry mild yellow edge potex virus) showed resistance to the strawberry mild yellow edge disease in the field (Martin 1996). Site-specific and developmentally regulated expression of transgenes is an important component for the success of many genetic engineering strategies. For example, the transgenic plants of strawberry cvs. Addie and Brio containing the PGIP gene (polygalacturonase inhibitory protein, from *Phaseolus vulgaris*) intended for resistance to *B. cinerea* and other fungal pathogens exhibited abnormal ripening characteristics, possibly due to a constitutive promoter driving PGIP (P. Rosati, pers. comm.).

Table 5. Field trails of strawberry transgenics

Target trait	Institution/s	Status as of 1998
Fruit ripening altered	Calgene/Monsanto, US	Under evaluation
Fruit ripening delayed	Agritope, US	Under evaluation
Insect (weevil) resistance	Scottish Research Institute, UK	Under evaluation
Fungal resistance	DNAP, US; Plant Sciences Research, US	Under evaluation
Product quality	DNAP, US	Under evaluation
Virus resistance SMYEP – strawberry mild yellow edge potex virus	Agriculture Canada/Research Station Vancouver, Canada	Transgenics showed delayed infection by virus

Two critical areas which would accelerate realization of the full potential of genetic engineering in strawberry would be (1) the identification and evaluation of useful genes (traits), and (2) regulation of foreign gene expression in transgenic plants. While the microbial-derived genes may be useful in improving some traits, the isolation and characterization of *Fragaria*-specific genes will become increasingly important not only for genetic engineering but also for studies of gene regulation of important processes such as fruit development and response to stress. Recent studies proposing an enzymatically controlled biosynthetic pathway in the formation of flavor compound DMHF (2,5-dimethyl-4-hydroxy-2H-furan-3-one) in strawberry (Zabetakis and Holden 1995, 1996) is indicative that genetic engineering may soon be able to address complex traits such as flavor and aroma.

The rapid progress displayed to date in the field of strawberry transformation demonstrates that the strawberry industry is on the verge of reaping benefits from biotechnology. As fruit production in general has an increasing impact on national and global economies, improvement of strawberry cultivars both in terms of desirable traits and year-round availability is a necessity. Genetic engineering may not be the entire answer but is an inevitable weapon in the arsenal of strawberry breeders. Genetic engineering programs need to be closely linked and integrated with broad-based, long-term breeding programs in order to fully affect the cultivar development program in strawberry.

Acknowledgments. We thank Jana Dieter for her excellent assistance in the graphics, proofreading and preparation of the manuscript.

References

An G, Watson BD, Stachel S, Gordon MP, Nester EW (1985) New cloning vehicles for transformation of higher plants. EMBO J 4:277–284

Asao H, Nishizawa Y, Arai S, Sato T, Hirai M, Yoshida K, Shinmyo A, Hibi T (1997) Enhanced resistance against a fungal pathogen *Sphaerotheca humuli* in transgenic strawberry expressing rice chitinase gene. Plant Biotechnol 14:145–149

Baker C, Lea M, Engler D, Morgan A (1998) Efficient transformation of commercial strawberry cultivars. Congr on In Vitro Biology, May 30 – June 3, 1998, Las Vegas, NV, USA. Abstract # P-1060

Barritt BH (1980) Resistance of strawberry clones to *Botrytis* fruit rot. J Am Soc Hortic Sci 105:160–164

Basiouny FM (1989) Ethylene evolution in strawberry (*Fragaria* x *ananassa* Duch) during fruit development. Acta Hortic 265:363–368

Beachy R, Loesch-Fries S, Turner NE (1990) Coat protein-mediated resistance against virus infection. Annu Rev Phytopathol 28:451–474

Becker D, Kemper E, Schell J, Masterson R (1994) New plant binary vectors with selectable markers located proximal to the left T-DNA border. Plant Mol Biol 20:1195–1197

Bringhurst RS, Voth V (1984) Breeding octoploid strawberries. Iowa State J Res 58:371–381

Cameron JS, Hancock JF (1986) Enhanced vigor in vegetative progeny of micropropagated strawberry plants. HortScience 21:1225–1226

Cameron JS, Hancock JF, Nourse TM (1985) The field performance of strawberry nursery stock produced originally from runners or micropropagation. Strawberry Prod 4:56–58

Cameron JS, Hancock JF, Flore JA (1989) The influence of micropropagation on yield components, dry matter partitioning and gas exchange characteristics of strawberry. Sci Hortic 38:61–67

Ceponis MJ, Butterfield JE (1973) The nature and extent of retail and consumer losses in apples, oranges, lettuce, peaches, strawberries and potatoes marketed in greater New York. USDA, USDA Marketing Res Rep 996, Washington, DC

Chang C, Kwok SF, Bleecker AB, Meyerowitz EM (1993) *Arabidopsis* ethylene-response gene ETR1: similarity of product to two-component regulators. Science 262:539–544

Cocci C, Mezetti B, Rosati P (1994) Regeneration and transformation of strawberry. In VIIIth Int Congr of Plant Tissue and Cell Culture, Firenze, Italy, 152 pp

Cordes S, Deikman J, Margossian LJ, Fischer RL (1989) Interaction of a developmentally regulated DNA-binding factor with sites flanking two different fruit-ripening genes from tomato. Plant Cell 1:1025–1034

Damiano C (1980) Strawberry micropropagation. In: Conf on Nursery Production of Fruit Plants through Tissue Culture April 21–23, Beltsville, Maryland, pp 11–22

De la Plaza JL, Merodio C (1989) Effect of ethylene chemisorption on refrigerated strawberry fruit. Acta Hortic 265:427–433

Doyle JJ, Doyle JL (1990) Isolation of plant DNA from fresh tissue. Focus 12:13–15

El-Kazzaz MK, Sommer NF, Fortlage RJ (1983) Effect of different atmospheres on postharvest decay and quality of fresh strawberries. Phytopathology 72:282–285

El Mansouri I, Mercado JA, Valpuesta V, Lopez-Aranda JM, Pliego-Alfaro F, Quesada MA (1996) Shoot regeneration and *Agrobacterium*-mediated transformation of *Fragaria vesca* L. Plant Cell Rep 15:642–646

Finstad K, Martin RR (1995) Transformation of strawberry for virus resistance. Acta Hortic 385:86–90

Galletta GJ, Draper AD, Maas JL (1989) Combining disease resistance, plant adaptation and fruit quality in breeding short day and day-neutral strawberries. Acta Hortic 265

Gamborg OL, Miller RA, Ojima K (1968) Nutrient requirement of suspension cultures of soybean rootcultures. Exp Cell Res 50:151–158

Garfinkel DJ, Nester EW (1980) *Agrobacterium tumefaciens* mutants affected in crown gall tumorigenesis and octopine catabolism. J Bacteriol 144:732–743

Good X, Kellogg JA, Wagoner W, Langhoff D, Matsumura W, Bestwick RK (1994) Reduced ethylene synthesis by transgenic tomatoes expressing S-adenosylmethionine hydrolase. Plant Mol Biol 26:781–790

Gooding HJ (1976) Resistance to mechanical injury and assessment of shelf-life in fruits of strawberry (*Fragaria* x *ananassa*). Hortic Res 16:71–82

Graham J, McNichol RJ, Grieg K (1995) Towards genetic based insect resistance in strawberry using the cowpea trypsin inhibitor gene. Ann Appl Biol 127:163–173

Hancock JF, Luby JJ (1993) Genetic resources at our doorstep: the wild strawberries. BioScience 43:141–147

Haymes KM, Davis TM (1993) *Agrobacterium*-mediated transformation and regeneration of the diploid strawberry. Acta Hortic 348:440

Hondelman W, Richter E (1973) Susceptibility of strawberry clones to *Botrytis cinerea* Pers in relation to pectin quantity and quality of fruits. Gartenbauwissenschaft 38:311–314

Horsch RB, Fry SE, Hoffman NL, Eicholtz D, Rogers SG, Fraley RT (1985) A simple and general method for transferring genes into plants. Science 227:1229–1231

Hughes JA, Brown LR, Ferro AJ (1987a) Expression of cloned coliphage T3 S-adenosylmethionine hydrolase gene inhibits DNA methylation and polyamine biosynthesis in *E. coli*. J Bacteriol 169:3625–3632

Hughes JA, Brown LR, Ferro AJ (1987b) Nucleotide sequence analysis of the coliphage T3 S-adenosylmethionine hydrolase gene and its surrounding ribonuclease III processing sites. Nucleic Acids Res 15:717–729

Irsh VF (1991) Cell lineage in plant development. Curr Opin Genet Dev 1:169–173

James DJ, Passey A, Barbara DJ (1990) *Agrobacterium*-mediated transformation of the cultivated strawberry (*Fragaria* x *ananassa* Duch.) using disarmed binary vectors. Plant Sci 69:79–94

James DJ, Passey AJ, Eastebrook MA, Solomon MG, Barbara DJ (1992) Progress in the introduction of transgenes for pest resistance in apples and strawberries. Phytoparasitica 20:83–87

Jefferson RA (1987) Assaying chimeric genes in plants. Plant Mol Biol 5:387–405

Jelenkovic G, Chin CK, Billings S (1986) Transformation studies of *Fragaria* x *ananassa* Duch. by Ti plasmid of *Agrobacterium tumefaciens*. HortScience 21:695

Jelenkovic G, Chin C, Billings S, Eberhardt J (1991) Transformation studies in cultivated strawberry, *Fragaria* x *ananassa* Duch. In: Dale A, Luby JJ (eds) The Third North American Strawberry Conference. Houston, Texas. Timber Press, Portland, OR

Jones OP, Waller BJ, Beech MG (1988) The production of strawberry plants from callus cultures. Plant Cell Tissue Organ Cult 12:235–241

Jungnickel F (1988) Strawberries (*Fragaria* spp. and hybrids. In: Bajaj YPS (ed) Biotechnology in agriculture and forestry. Springer, Berlin Heidelberg New York

Kader AA (1991) Quality and its maintenance in relation to the postharvest physiology of strawberry. In: Lyby DA, Lyby JJ (eds) The strawberry into the 21st century. Timber Press, Portland, OR, pp 145–152

Khanizadeh S, Buszard D, Lareau MJ, Pelletier R (1991) Evaluation of strawberry cultivars with different degrees of resistance to red stele. Fruit Var J 45:12–17

Kondakova V, Schuster G (1991) Elimination of strawberry mottle virus and strawberry crinkle virus from isolated apices of three strawberry varieties by addition of 2,4-dioxohexahydro-1,3,5-triazine (5-azadihydrouracil) to the nutrient medium. J Phytopathol 132:84–86

Kramer MG, Kellogg JA, Wagoner W, Matsumura W, Good X, Peters ST, Clough G, Bestwick RK (1996) Reduced ethylene synthesis and ripening control in tomatoes expressing S-adenosyl-methionine hydrolase. In: Kanellis AG (ed) Biology and biotechnology of the plant hormone ethylene. Crete, Greece. Kluwer, Dordrecht, pp 307–319

Layton JG, Dhir S, Broyles D, Morrish F, Wilkinson J, DeBrecht G, Kaniewska M, Wu G, DeLaquil P, Hinchee M (1996) Transformation of strawberry using *Agrobacterium tumefaciens*. In: World Congress on In Vitro Biology, Hot Topics, June 22–27 San Francisco, California

Lee ECM, de Foussard RA (1975) Regeneration of strawberry plants from tissue cultures. Comb Proc Int Plant Prop Soc 25:277–285

Liu ZR, Sanford JC (1988) Plant regeneration by organogenesis from strawberry leaf and runner tissue. HortScience 23:1057–1059

Lopez-Aranda JM, Pliego-Alfaro F, Lopez-Navidad I, Barcelo-Munoz M (1994) Micropropagation of strawberry (*Fragaria* x *ananassa* Duch.). Effect of mineral salts, benzyladenine levels and number of subcultures on the in vitro and field behavior of the obtained microplants and the fruiting capacity of their progeny. J Hortic Sci 69:625–637

Maas JL, Wang SY, Galletta GJ (1996) Health enhancing properties of strawberry fruit. In: North American Strawberry Growers Association Annual Meeting, Feb 11–14, Newport Beach, California, pp 11–18

Manning K (1993) Soft fruit. In: Seymour GB, Taylor JE, Tucker GA (eds) Biochemistry of fruit ripening. Chapman & Hall, London, pp 347–377

Marcotrigiano M, Swartz HJ, Gray SE, Tokarcik D, Popenoe J (1984) The effect of benzyl amino purine on the in vitro multiplication rate and subsequent field performance of tissue-culture propagated strawberry plants. Adv Strawberry Prod 3:23–25

Martin RR (1996) Strawberry viruses and their control: what's new? North American Strawberry Growers Association Annual Meeting, Feb 11–14, Newport Beach, California, pp 24–28

Martinelli A (1992) Micropropagation of strawberry (*Fragaria* spp.). In: Bajaj YPS (ed) Biotechnology in agriculture and forestry. Springer, Berlin Heidelberg New York

Mathews H (1996) Invited feature presentation, "Control of ethylene synthesis in small fruits by genetic engineering" at the Annu Conf of Northwest Centre for Small Fruit Conference, Red Lion Hotel Columbia River, Portland, Oregon. Dec 4–5, 1996

Mathews H, Wagoner W, Kellogg J, Bestwick RK (1994a) Strawberry transgenics: genetic transformation for control of ethylene biosynthesis. Int Congr of Plant Tissue and Cell Culture. Firenze, Italy, June 12–17, 1994. Abstract # S7–26

Mathews H, Cohen C, Wagoner W, Kellogg J, Dewey V, Wanek D, Lupulesa E, Schuster D, Bestwick R (1994b) Transformation of strawberry and raspberry for control of ethylene biosynthesis. 4th Int Congr of Plant Molecular Biology, Amsterdam, 19–24 June 1994, Abstr 2041

Mathews H, Wagoner W, Kellogg J, Bestwick R (1995) Genetic transformation of strawberry: stable integration of a gene to control biosynthesis of ethylene. In Vitro Cell Dev Biol 31:36–43

Mathews H, Dewey V, Wagoner W, Bestwick RK (1998) Molecular and cellular evidence of chimeric tissues in primary transgenics and elimination of chimerism through improved selection protocols. Transgenic Research 7:123–131

Murashige T, Skoog FA (1962) A revised medium for rapid growth and bioassays with tobacco tissue cultures. Physiol Plant 15:473–497

Nehra NS, Stushnoff C, Kartha KK (1989) Direct shoot regeneration from strawberry leaf disks. J Am Soc Hortic Sci 114:1014–1018

Nehra NS, Chibbar RN, Kartha KK, Datla RSS, Crobsy WL, Stushnoff C (1990a) *Agrobacterium*-mediated transformation of strawberry calli and recovery of transgenic plants. Plant Cell Rep 9:10–13

Nehra NS, Chibbar RN, Kartha KK, Datla RSS, Crobsy WL, Stushnoff C (1990b) Genetic transformation of strawberry by *Agrobacterium tumefaciens* using leaf disk regeneration. Plant Cell Rep 9:293–298

Nehra NS, Stushnoff C, Kartha KK (1990c) Regeneration of plants from immature leaf-derived callus of strawberry (*Fragaria* x *ananassa*). Plant Sci 66:119–126

Nehra NS, Kartha KK, Stushnoff C (1992) Plant biotechnology and strawberry improvement. Adv Strawberry Res 11:1–11

Nogata Y, Ohta H, Voragen AGJ (1993) Polygalacturonase in strawberry fruit. Phytochemistry 34:617–620

Nuutila AM, Vestberg M, Kauppinen V (1995) Infection of hairy roots of strawberry (*Fragaria* x *ananassa* Duch.) with arbuscular mycorrhizal fungus. Plant Cell Rep 14:505–509

Nyman M (1993) Protoplast technology in strawberries. Doctoral Thesis at Uppsala University 1993. Acta Universitatis Upsaliensis. Comprehensive Summarien of Uppsala Dissertations from the Faculty of Science 451, Uppsala, pp 44

Nyman M, Wallin A (1992) Transient gene expression in strawberry (*Fragaria* x *ananassa* Duch.) protoplasts and the recovery of transgenic plants. Plant Cell Rep 11:105–108

Perkins-Veazie P (1991) Clues in the mystery of strawberry fruit ripening. In: Dale A, Luby JJ (eds) The strawberry into the 21st century. Timber Press, Portland pp 172–173

Poethig S (1989) Genetic mosaics and cell lineage analysis in plants. Trends Genet 5:273–277

Popova IV, Konstantinova AE, Zekalashvili A, Zhananov BK (1985) Features of breeding strawberries for resistance to berry molds. Sov Agric Sci 3:29–33

Sansavini S, Rosati P, Gaggioli D, Toschi MF (1980) Inheritance and stability of somaclonal variations in micropropagated strawberry. Acta Hortic 280:375–384

Sawahel WA (1996) Transgenic strawberry. Rice Biotechnol Q 26:33

Sjulin TM, Dale A (1987) Gametic diversity of North American strawberry cultivars. J Am Soc Hortic Sci 112:375–385

Sorvari S, Ulvinen S, Hietaranta T, Hiirsalmi H (1993) Preculture medium promotes direct shoot regeneration from micropropagated strawberry leaf disks. HortScience 28:55–57

Stewart RN (1978) Ontogeny of the primary body in chimeral forms of higher plants. In: Subtenly S, Sussex IM (eds) The clonal basis of development. Academic Press New York, pp 131–160

Studier FW, Movva NR (1976) SAMase gene of bacteriophage T3 is responsible for overcoming host restriction. J Virol 19:136–145

Swartz HJ, Galletta GJ, Zimmerman RH (1981) Field performance and phenotypic stability of tissue culture-propagated strawberries. J Am Soc Hortic Sci 106:667–673

Testoni A, Lovatti L, Faedi W (1989) Shelf life and fruit quality of strawberry varieties and selections after storage. Acta Hortic 265:435–442

Theiler-Hedtrich R, Wolfensberger H (1987) Comparison of plant and yield characteristics of in vitro and normal propagated strawberry plants. Acta Hortic 212:445–458

Toyoda H, Horikoshi K, Yamano Y, Ouchi S (1991) Selection for *Fusarium* wilt disease resistance from regenerants derived from leaf callus of strawberry. Plant Cell Rep 10:167–170

Uratsu SL, Ahmadi H, Bringhurst RS et al. (1991) Relative virulence of *Agrobacterium* strains on strawberry. HortScience 26:196–199

USDA (1997) Fruit and tree nuts: situation and outlook report: Economic Research Service, Fruits and Tree Nuts/FTS-279/March-Washington DC

Vrain TC (1994) Engineering genetic resistance to *Botrytis* fruit rot in strawberry and blueberry. In: 3rd Annual Conference, Portland, Oregon, 30 Nov – 1 Dec 1994. Northwest Center for Small Fruit Research, 15 pp

Waithaka K, Hildebrandt AC, Dana MN (1980) Hormonal control of strawberry axillary bud development in vitro. J Am Soc Hortic Sci 105:428–430

Wolyn DJ, Jelenkovic G (1990) Nucleotide sequence of an alcohol dehydrogenase gene in octoploid strawberry (*Fragaria* x *ananassa* Duch.). Plant Mol Biol 14:855–857

Wright WR, Billeter BA (1975) Marketing losses of selected fruits and vegetables at wholesale, retail and consumer levels in the Chicago area. USDA Marketing Res Rep 1017, Washington, DC

Zabetakis I, Holden MA (1995) A study of strawberry flavour biosynthesis. In: Etiviant P, Schrier P (eds) Bioflavour 95: Analysis-Precursor studies-Biotechnology. INRA Paris, pp 211–216

Zabetakis I, Holden MA (1996) The effect of 6-deoxy-D-fructose on flavour bioformation from strawberry (*Fragaria* x *ananassa*, cv. Elsanta) callus cultures. Plant Cell Tissue Organ Cult 45:25–29

13 Transgenic Sweet Potato (*Ipomoea batatas* L. Lam.)

M. OTANI and T. SHIMADA

1 Introduction

The sweet potato, *Ipomoea batatas* (L.) Lam. (2n = 6X = 90), belongs to the section Batatas genus *Ipomoea*, as do 11 wild species, none of which produces edible storage roots. The cultivation area of this species is localized mainly at low latitudes of South America, South Asia and South Africa, and goes northwards as far as some regions of Europe and the USA. Sweet potato ranks seventh among food crops in annual production in the world (Jabsson and Raman 1991). However, conventional breeding, based on sexual hybridization of sweet potato is not well developed because of its sterility and cross incompatibility. To overcome these limitations, novel approaches such as somatic hybridization and genetic transformation must be incorporated into sweet potato breeding.

On the other hand, sweet potato is an attractive plant species for a target for "molecular farming", because of its high production of biomass. Recent developments in genetic engineering enable the production of various biomolecules such as carbohydrates, fatty acids, high-value pharmaceutical polypeptides, industrial enzymes and biodegradable plastics in transgenic plants (Goddijin and Pen 1995). Transgenic plants may become attractive and cost effective alternatives to microbial and animal systems for the production of biomolecules.

In sweet potato, a few reports on the transgenic plants produced by using *Agrobacterium rhizogenes* and *A. tumefaciens*-mediated methods and the direct gene delivery into protoplasts has been published (Dodds et al. 1991; Otani et al. 1993; Murata et al. 1995; Newell et al. 1995; Okada et al. 1995; Gama et al. 1996), (Table 1). However, there is severe limitation of genotypes for transformation and efficiency is still low.

We have established an efficient and variety-independent method for embryogenic callus production from meristem tissues of sweet potato using altered plant growth regulators, picloram, dicamba or 4FA in the medium (Otani and Shimada 1996). These embryogenic calli have been utilized as suitable target materials for *Agrobacterium*-mediated transformation of sweet potato (Otani et al. 1998).

Research Institute of Agricultural Resources, Ishikawa Agricultural College, Nonoichi-machi, Ishikawa 921-8836, Japan

Biotechnology in Agriculture and Forestry, Vol. 47
Transgenic Crops II (ed. by Y.P.S. Bajaj)
© Springer-Verlag Berlin Heidelberg 2001

Table 1. Summary of various studies conducted on transformation of sweet potato

Reference	Explant/culture used	Cultivar	Vector/method used	Exogenous DNA	Observations/remarks
Dodds et al. (1991)	In vitro whole plants	Not stated	*A. rhizogenes*		Transgenic plants
Prakash and Varadarajan (1992)	Leaves and petioles	Jewel, TIS-70357	Particle bombardment	*gusA, nptII*	Transformed calli
Otani et al. (1993)	Leaves	Five cultivars (Chugoku 25, etc.)	*A. rhizogenes*	*gusA, nptII*	Transgenic plants
Newell et al. (1995)	Storage roots	Jewel	*A. tumefaciens*	*gusA, nptII*, cowpea trypsin inhibitor, snowdrop lectin	Transgenic plants
Murata et al. (1995)	Mesophyll protoplasts	Chugoku 25, Chikei 682-11	Electroporation	*hpt*, SPFMV-S coat protein gene	Transgenic plants
Gama et al. (1996)	Embryogenic callus	White star	*A. tumefaciens*	*gusA, nptII*	Transgenic plants
Otani et al. (1998)	Embryogenic callus	Kokei 14	*A. tumefaciens*	*gusA, hpt*	Transgenic plants

This chapter describes a simple, efficient and reproducible method for the production of transgenic sweet potato plants by *Agrobacterium tumefaciens*- and *A. rhizogenes*-mediated transformation. We also describe the morphology of transgenic plants and stability of integrated genes.

2 Transformation Protocol

2.1 *A. tumefaciens*-Mediated Transformation

Embryogenic Callus Induction and Tissue Culture Media. Sweet potato [*Ipomoea batatas* (L.) Lam.] cultivar Kokei 14 and Beniazuma were grown in media for tissue culture and transformation as shown in Table 2. Embryogenic calli were induced from shoot meristems on LS medium (Linsmaier and Skoog 1965) containing 1 mg/l 4-fluorophenoxyacetic acid (4FA) or picloram according to the methods of Otani and Shimada (1996). These embryogenic calli were maintained at 26 °C in the dark and were proliferated by subculture on the same fresh medium.

Bacterial Strain and Plasmid. The sweet potato embryogenic callus was infected using *Agrobacterium tumefaciens* strain EHA101 harboring the binary vector plasmid pIG121-Hm, which carries the chimeric neomycin phosphotransferase (*nptII*), β-glucuronidase (*gus*A) and hygromycin phosphotransferase (*hpt*) genes (Akama et al. 1992). *A. tumefaciens* strain LBA4404/pTOK233 (Hiei et al. 1994) and *A. tumefaciens* strain R1000/pBI121 (Otani et al. 1996) were also examined for transient GUS expression in embryogenic callus.

Table 2. Media used for tissue culture and transformation of sweet potato. (Otani et al. 1998)

Stage	Medium composition
Embryogenic callus induction and proliferation	LS medium, 1 mg/l 4FA, 30 g/l sucrose, 3.2 g/l gellan gum, pH 5.8
Bacterial infection	LS medium, 1 mg/l 4FA, 10 mg/l acetosyringone, 30 g/l sucrose, pH 5.8
Co-culture	LS medium, 1 mg/l 4FA, 10 mg/l acetosyringone, 30 g/l sucrose, 3.2 g/l gellan gum, pH 5.8
Selection	LS medium, 1 mg/l 4FA, 25 mg/l hygromycin, 500 mg/l carbenicillin, 30 g/l sucrose, 3.2 g/l gellan gum, pH 5.8
Somatic embryo formation	LS medium, 4 mg/l ABA, 1 mg/l GA_3, 25 mg/l hygromycin, 500 mg/l carbenicillin, 30 g/l sucrose, 3.2 g/l gellan gum, pH 5.8
Plant formation	LS medium, 0.05 mg/l ABA, 25 mg/l hygromycin, 500 mg/l carbenicillin, 30 g/l sucrose, 3.2 g/l gellan gum, pH 5.8

Transformation. The transformation procedure was performed according to the method of Rashid et al. (1996) with some modification. The *Agrobacterium* was grown for 2 days at 27°C on LB medium supplemented with 50 mg/l kanamycin, 50 mg/l hygromycin and 1.5% (w/v) agar. The colony of bacteria was transferred to liquid LS medium supplemented with 10 mg/l acetosyringone and 1 mg/l 4FA, and shaken at 100 rev./min for 1 h in the dark at 27°C. The embryogenic calli were soaked in a bacterial suspension for 2 min, and then blotted dry with sterile filter paper to remove excess bacteria. The calli were then transferred onto co-culture medium (Table 2). After 3 days of co-cultivation, the infected calli were washed three times with sterile distilled water supplemented with 500 mg/l carbenicillin and then transferred onto selection medium (Table 2). The cultures were kept at 26°C in the dark.

Selection and Regeneration of Transgenic Plants. After selection for 2 weeks, the calli were washed three times with sterile distilled water supplemented with 500 mg/l carbenicillin and then transferred to the fresh selection medium. The calli were subcultured onto the fresh medium every 2 weeks. After 60 days of culture on the selection medium, the calli were transferred onto the somatic embryo formation medium (Table 2). After 21 days of culture on the somatic embryo formation medium, somatic embryos formed from hygromycin-resistant calli were transferred onto the plant formation medium (Table 2) for germination. Regenerated plants derived from somatic embryos were cultured on the plant growth medium (Table 2). The cultures were maintained at 26°C under a 16-h photoperiod at $38 \mu mol \cdot m^{-2} \cdot s^{-1}$ using daylight fluorescent tubes.

DNA Isolation and Southern Hybridization. Total DNA was isolated from leaves of in vitro plants by the sodium dodecyl sulfate (SDS) extraction method according to Honda and Hirai (1990). After complete digestion with *Hind*III, DNA fragments were separated by 1% agarose gel electrophoreses, and transferred to Amersham's Hybond-N nylon membrane as described by Southern (1975). The filter was hybridized with a GUS DNA fragment labeled with $[\alpha-^{32}P]dCTP$ using a Megaprime DNA labeling system (Amersham International, England).

2.2 *A. rhizogenes*-Mediated Transformation

Plant Materials. In vitro plants of 14 cultivars of sweet potato [*Ipomoea batatas* (L.) Lam.], Bise, Chugoku 25, Chugoku 35, Hi-starch, Kanto 18, Kanto 94, Kokei 14, Kyukei 17-3043, Naeshirazu, Norin 2, Okinawa 100, Shinya, Yamakawamurasaki and W51 were used (Otani et al. 1987). In vitro plants were grown at 26°C under continuous illumination at $38 \mu mol \cdot m^{-2} \cdot s^{-1}$ from daylight fluorescent tubes.

Bacterial Strains and Vectors. *Agrobacterium rhizogenes*, one agropine type strain (15834) and seven mikimopine (Isogai et al. 1988) type strains A5, A13,

H4, C8, D6, NIAES1724 and NIAES1725 (Daimon et al. 1990) were grown for 16 h at 27 °C in liquid YEB medium (Vervliet et al. 1974). The binary vector plasmid pBI121 (Jefferson et al. 1987) was mobilized from *Escherichia coli* C600 to *A. rhizogenes* C8, harboring a wild mikimopine-type Ri plasmid, by triparental mating using pRK2013 as a helper plasmid (Ditta et al. 1980), and cultured under the same conditions.

Transformation and Establishment of Hairy Root Cultures. The apical third to fifth of fully expanded leaves of in vitro plants were inoculated with *A. rhizogenes*, according to the methods of Noda et al. (1987). The inoculated leaf disks were placed on sterilized moist paper in a glass Petri dish and incubated at 26 °C in the dark. After 3–5 days of incubation, leaf disks were transferred to a 1% (w/v) agar or 0.32% (w/v) gellan gum-solidified LS medium supplemented with 500 µg/ml Claforan (Hoechst) and incubated under the same conditions. Bacteria-free root lines were obtained after excision of single roots and propagation on LS medium supplemented with 400 µg/ml Claforan and 0.32% (w/v) gellan gum during three subcultures.

Plant Regeneration from Hairy Roots. The hair roots (30–40 mm in length) of sweet potato were transferred onto 0.32% (w/v) gellan gum-solidified LS medium lacking both antibiotics and plant growth regulators (PGR), and cultured at 26 °C under continuous illumination at 38 µmol · m^{-2} · s^{-1} using daylight fluorescent tubes.

The percentage of hairy roots with shoot formation was calculated by the number of hairy roots with shoot formation per number of hairy roots transferred to the regeneration medium.

Detection of Opines (Agropine, Mannopine and Mikimopine). The opines in both hairy roots and leaves of regenerated plants from hairy roots were detected by silver-staining for agropine and mannopine (Petit et al. 1983), and Pauly reagent-staining for mikimopine after paper electrophoresis according to the method of Petit et al. (1986).

Southern Blot Hybridization. Total DNA isolated from leaves of in vitro plants digested with *Eco*RI was subjected to electrophoresis and transferred to Amersham's Hybond-N nylon membrane. Southern analysis was carried out using the TL-DNA probe, pLJ1, with digoxigenin (DIG) labelling and an AMPPDR (Tropix, Inc.) detection system (Boehringer Mannheim) according to the supplier's instructions. The DNA of hairy root-derived plants transformed by the mikimopine-type strain of *A. rhizogenes* were also analyzed by the method of Handa (1992).

Polymerase Chain Reaction (PCR) Analysis of Introduced Foreign Genes. For analysis of the plants regenerated from the hairy roots by PCR, the sequence of two genes, NosP-*nptII*-NosT and 35SP-*gus*A-NosT, was amplified with the combination of primers, primer PNosP (5'AAATGCTCCACTGA CGTTCC3') located in the NosP region, 81 bp 5' of the translation initiation site (ATG), primer PNosT (5'CGCAAGACCGGCAACAGGAT3') in the 80 bp 3' of the Nos translational stop signal (TAA) and primer P35SP (5'GATGTGATATCTCCACTGAC3') located in the CaMV35S promoter region.

3 Results

3.1 *A. tumefaciens*-Mediated Transformation

3.1.1 *Factors Affecting Transient Expression of GUS*

We examined three factors; strains of *Agrobacterium*, inclusion of acetosyringone and culture period of embryogenic calli before bacterial inoculation, for efficient transformation in sweet potato. Transient expression of GUS was examined for 1.5 g fresh weight of embryogenic calli per treatment soon after 3 days of co-cultivation. The transient GUS expression varied among the bacterial strains. *A. tumefaciens* strain EHA101/pIG121-Hm gave the highest number of GUS spots, while other bacterial strains gave no (for LBA4404/pTOK233) or very few spots (average six spots per g fresh weight for R1000/pBI121).

The effect of acetosyringone on transient GUS expression was tested using 14-day-old calli of cultivar Kokei 14 and *A. tumefaciens* strain EHA101/pIG121-Hm. As shown in Table 3, addition of acetosyringone to both infection and co-cultivation media clearly promoted the transient expression of GUS (Fig. 1a).

Embryogenic calli were infected by *A. tumefaciens* strain EHA101/pIG121-Hm at 3, 6, 10, 14 and 21 days after the beginning of subculture. The calli cultured for a short period (3 and 6 days) gave few GUS spots, while 14-day-old calli gave the most (Table 3). These findings clearly showed a culture period before bacterial infection of 14 days was needed to obtain efficient expression of *gus*A gene in embryogenic callus of Kokei 14.

Table 3. Effect of acetosyringone and pre-culture period on transient GUS expression in embryogenic callus of sweet potato. (Otani et al. 1998)

Culture period (days)	No. of GUS spots/g fresh weight of embryogenic calli	
	Acetosyringone (+)[a]	Acetosyringone (−)[a]
3	514.7 ± 21.5	−
6	840.9 ± 242.2	−
10	1054.4 ± 255.6	−
14	1526.6 ± 266.7	382.2 ± 115.4
21	922.8 ± 132.4	−

[a] + and − denote the presence and absence of acetosyringone in the medium, respectively.

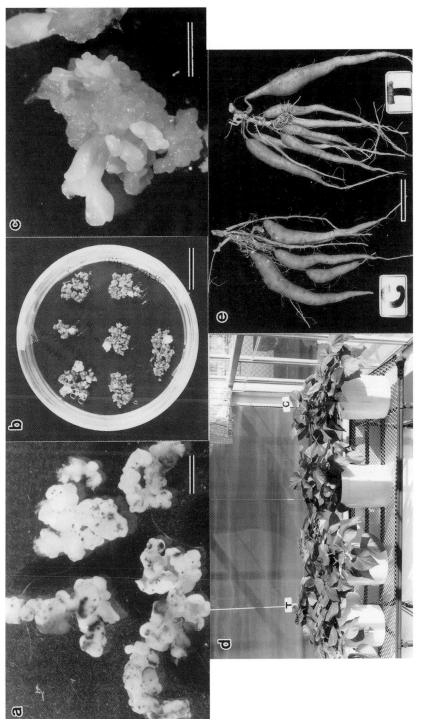

Fig. 1 a–e. Production and histochemical GUS assay of transgenic plants of sweet potato cv. Kokei 14 by *Agrobacterium tumefaciens* strain EHA101/pIG121-Hm. **a** Transient GUS expression in embryogenic callus of sweet potato after 3 days of co-cultivation with *Agrobacterium*: *bar* = 15 mm. **b** Hygromycin-resistant cell clusters developing on the selection medium containing 25 mg/l hygromycin: *bar* = 20 mm. **c** Numerous somatic embryos produced from hygromycin-resistant calli: *bar* = 10 mm. **d** Transgenic plants (*left*, six pots) and an untransformed plant (*right*, two pots) established in soil and grown for 2 months in a greenhouse. **e** Storage roots formed on untransformed (*left*) and transgenic plants (*right*): *bar* = 50 mm. (Otani et al. 1998)

3.1.2 Plant Regeneration from Hygromycin-Resistant Calli

The embryogenic calli infected with *A. tumefaciens* strain EHA101/pIG121-Hm were cultured on hygromycin-containing media for 60 days, and these embryogenic calli could produce several hygromycin-resistant calli, while uninfected embryogenic calli failed to form them on the same media. The average number of hygromycin-resistant calli produced was 10.7 per g fresh weight of infected embryogenic calli (Fig. 1b). These hygromycin-resistant calli produced numerous somatic embryos on the somatic embryo formation medium containing hygromycin (Fig. 1c). The hygromycin-resistant plantlets were developed from these somatic embryos on the plant formation medium. An average of 53.1% of hygromycin-resistant calli regenerated plantlets (Table 4). All of the regenerated plants grew further and rooted on the LS plant growth regulator-free medium supplemented with 25 mg/l hygromycin. Regenerated plants were transferred to pots containing a vermiculite and perlite mixture (3:1) and maintained at 26°C under a 16h photoperiod in a growth chamber for 14 days. Then these regenerated plants were grown in a greenhouse. These transgenic plants grew normally and formed storage roots after 3 months (Fig. 1d,e). Five transgenic plants regenerated from independent hygromycin-resistant calli were analyzed to determine various phenotypic characteristics, such as apical immature leaf color, mature leaf color, mature leaf shape, stem color, number of storage roots per plant, fresh weight of a storage, skin color of storage roots and flesh color of storage roots. No morphological differences were observed between untransformed plants and the transgenic plants.

3.1.3 Histochemical GUS Assay of Transgenic Plants and Their Vegetatively Propagated Progenies

Histochemical analysis of GUS activity was carried out on fully expanded leaves and storage roots of the regenerated plant from hygromycin-resistant calli. The tissue of regenerated plants was stained blue, indicating the expression of the *gus*A gene, but the control plants were not stained (Fig. 2a). GUS expression was also observed in storage roots harvested after 3 months of cul-

Table 4. Transformation efficiency of sweet potato using *Agrobacterium tumefaciens* strain EHA101/pIG121-Hm. (Otani et al. 1998)

Experiments	No. of hyg[r] calli obtained[a]	No. of hyg[r] calli produced plants (%)	No. of hyg[r] plants regenerated
1	20	9 (45.0)	26
2	12	8 (66.7)	14

[a] Number of hygromycin-resistant cell clusters obtained from 1.5 g fresh weight of embryogenic calli.

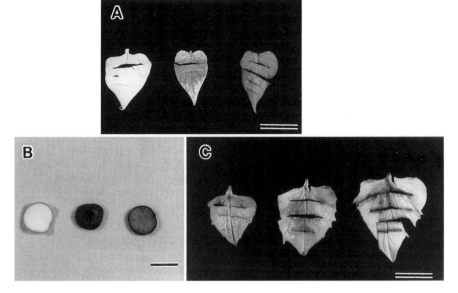

Fig. 2 A–C. Histochemical GUS assay of transgenic plants. Leaves (**A**) and storage roots (**B**) of two independent transgenic plants (*center* and *right*) and untransformed plant (*left*). **C** Leaves of freshly sprouting shoots from harvested storage roots of two independent transgenic plants (*center* and *right*) and untrasformed plant (*left*). *Bar* = 15 mm. (Otani et al. 1998)

tivation in pots, while roots of untransformed control plants did not show any GUS activity (Fig. 2b).

Leaves of freshly sprouting shoots from harvested storage roots of transgenic plants also showed GUS activity (Fig. 2c), suggesting that the *gus*A gene was transmitted to their vegetatively propagated progenies. We observed GUS activity in the fourth generation vegetatively propagated progenies. Since sweet potato is commonly propagated using storage roots, this suggests the stability of the introduced gene and the usefulness of genetically engineered sweet potato for the practical breeding of sweet potato.

3.1.4 Integration of Foreign DNA in the Genome of Transgenic Plants

To measure the copy number of the integrated T-DNA directly, genomic blot analysis was performed for transgenic plants regenerated from five independent hygromycin-resistant callus lines. A *Sac*I/*Xba*I fragment of the *gus*A gene was used as a hybridization probe (Fig. 3a). Since there is only one *Hin*dIII site in the T-DNA region, hybridizing fragment(s) of different lengths indicated that the T-DNA was integrated at different location(s) in the sweet potato plant genome. The copy number was determined by the number of hybridizing fragments. Figure 3b shows that the number and size of hybridiz-

a

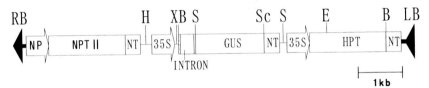

B

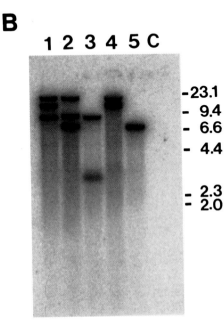

Fig. 3 a,b. Transformation vector and Southern blot analysis. **a** Schematic diagram of a part of the T-DNA region of transformation vector pIG121-Hm. *RB* right border, *LB* left border, *NP* nopaline synthase promoter, *NT* nopaline synthase terminator, *35S* 35S promoter of cauliflower mosaic virus, *INTRON* the first intron of catalase gene of castor bean, *NPTII* gene for neomycin phosphotransferase, *GUS*, gene for β-glucuronidase, *HPT* gene for hygromycin phosphotransferase. Cutting sites of restriction enzymes are indicated; *Bam*HI (*B*), *Eco*RI (*E*), *Hind*III (*H*), *Sal*I (*S*), *Sac*I (*Sc*), *Xba*I (*X*). **b** Southern blot analysis of five independent transgenic plants. DNA was digested with HindIII and allowed to hybridize to the gus probe. *Lanes* 1–5, transgenic plants (Nos. 1, 5, 6, 7, and 11 shown in Table 4), which were regenerated from independent hygromycin-resistant calli; *lane C* untransformed control plant. (Otani et al. 1998)

ing fragments varied among the different transgenic plant lines. Transgenic plant lines possessed one to three copies of T-DNA.

3.1.5 Inheritance of Introduced Genes

One of transgenic Kokei 14 plants, line 11, and a non-transformed Kokei 14 plant were crossed with non-transformed sweet potato cv. Chikei 682-11 plants. The F_1 seeds were germinated and the seedlings were investigated in relation to GUS expression and hygromycin resistance.

The F_1 plants showed the segregation ratio of 1 GUS positive and hygromycin resistant to 1 GUS negative and hygromycin sensitive (Table 5), indicating that T-DNA was inserted at single locus.

Table 5. Characterization of independent transgenic plant lines

Characters	Control	Transgenic plant lines					
		1[a]	5	6	7	11	
Copy number of T-DNA	0	3	3	2	2	1	
Apical immature leaf color	Green	Green	Green	Green	Green	Green	
Mature leaf color	Green	Green	Green	Green	Green	Green	
Mature leaf shape	Heart shape with small lobe	Heart shape with small lobe	Heart shape with small lobe	Heart shape with small lobe	Heart shape with small lobe	Heart shape with small lobe	
Stem color	Green	Green	Green	Green	Green	Green	
Number of storage roots per plant	5	4	7	5	5	5	
Fresh weight of a storage root (g) $\pm$ SE	26.1 ± 8.8	13.4 ± 2.9	24.2 ± 4.7	21.7 ± 7.7	25.7 ± 7.3	34.3 ± 6.8	
Skin color of storage roots	Red	Red	Red	Red	Red	Red	
Flesh color of storage roots	Light yellow	Light yellow	Light yellow	Light yellow	Light yellow	Light yellow	

[a] Plant line 1 was grown for 2 months in the greenhouse.

3.2 A. *rhizogenes*-Mediated Transformation

3.2.1 Pathogenicity of A. rhizogenes Strains and Plant Regeneration from Hairy Root

Hairy roots were first observed within 7 days of inoculation (Fig. 4a). They were produced from wounded sites, for example, from the cut surface of the excised petiole. Mikimopine type strains of *A. rhizogenes* gave similar pathogenicity to the agropine type strain 15834; the percentage of leaf disks producing hairy roots in cv. Chugoku 25 ranged from 20 to 40% in both strains. On the other hand, genotypic differences in hairy root formation frequency were observed among the cultivars tested when leaf disks were inoculated with *A. rhizogenes* strain A13 (Table 6). Chugoku 25, Kanto 18, Kokei 14 and Shinya produced hairy roots at a low frequency (23.5–27.8%), while the other cultivars produced hairy roots at a higher (51.4–88.6%) frequency. Adventitious shoot formation was observed when hairy roots were cultured on 0.32% gellan gum-solidified LS medium lacking both antibiotics and PGR, under continuous illumination (Fig. 4b). Regeneration of complete plants was obtained in five cultivars (Chugoku 25, Chugoku 35, Hi-starch, Kyukei 17-3043 and Yamakawamurasaki) among the ten cultivars tested. The percentage of hairy roots with shoot formation was 50% for Chugoku 25, 100% for Chugoku 35, 80% for High-starch, 40.9% for Kyukei 17-3043 and 41.7% for Yamakawamurasaki. These findings suggest that there are varietal differences in shoot regeneration from the hairy roots. Shoot formation was not improved by the addition of BA (0.5 and 2 mg/l) to the regeneration medium (data not shown).

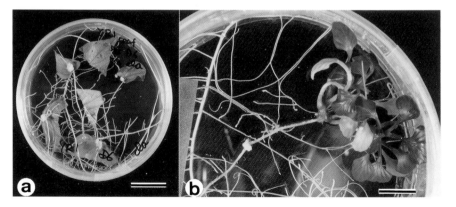

Fig. 4 a,b. Plant regeneration from hairy root of sweet potato transformed by *Agrobacterium rhizogenes*. **a** Hairy root formation from leaf disk inoculated by *A. rhizogenes*: *bar* = 20 mm. **b** Shoot formation from hairy root transformed by *A. rhizogenes*: *bar* = 10 mm. (Otani et al. 1993)

Table 6. Varietal differences in hairy root formation induced by
A. rhizogenes A13. (Otani et al. 1993)

Cultivars	No. leaf disks inoculated	No. leaf disks forming hairy roots[a]
Bise	31	25 (80.6)
Chugoku 25	73	18 (24.7)
Chugoku 35	30	13 (43.3)
Hi-starch	37	25 (67.6)
Kanto 18	34	9 (26.5)
Kanto 94	33	19 (57.6)
Kokei 14	36	10 (27.8)
Kyukei 17-3043	78	52 (66.7)
Naeshirazu	35	18 (51.4)
Norin 2	38	25 (65.8)
Okinawa 100	31	18 (58.1)
Shinya	34	8 (23.5)
Yamakawamurasaki	35	31 (88.6)
W 51	40	30 (75.0)

[a] Percentage in parentneses.

3.2.2 Analysis of Regenerants via Opine Synthesis by Hairy Roots and Hairy Root-Derived Plants

Both agropine and mannopine were detected in the extracts of the hairy roots and leaves of the plants regenerated from the hairy roots obtained by infection with the agropine-type strain of *A. rhizogenes*. Mikimopine was also detected in the hairy roots and leaves of the plants regenerated from the hairy roots obtained by infection with mikimopine-type bacterial strains (Fig. 5). Since the opine synthesis of *A. rhizogenes*-inoculated plants is encoded by T-DNA of the Ri plasmid (Chilton et al. 1982), the present findings indicate stable maintenance and expression of T-DNA in these hairy roots and the plants regenerated from them.

3.2.3 Morphology of Plants Regenerated from Hairy Roots

The adventitious shoots formed on the hairy roots grew and rooted on the 0.8% (w/v) agar-solidified LS medium. These plantlets were habituated and transplanted to the experimental field of Ishikawa Agricultural College. All of these plants survived.

All plants regenerated from the hairy roots showed Ri T-DNA-induced morphological changes. Aerial parts showed decreased apical dominance and shortened internodes (Fig. 6a). The average stem length was shortened to about one-fifth that of normal plants. A difference in stem length was observed among the transgenic plants regenerated from the independent hairy root lines. The leaves were wrinkled and smaller than normal (Fig. 6b). The shape of the flower was changed dramatically; hairy root-derived plants had small and star-shaped flowers (Fig. 6c). The pollen fertility of hairy root-derived

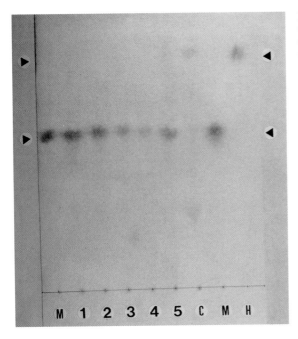

Fig. 5. Opine assay of extracts of five regenerated plants derived from independent hairy root clones of cv. Yamakawamurasaki. Ten microliters of extract corresponding to approximately 50 mg of leaves (fresh weight) were blotted on Advantec No. 2 filter paper (Toyo Roshi, Inc.) and subjected to electrophoresis at 20 V/cm. *Lane M* standard mikimopine, *lanes 1, 2, 3, 4* and *5* five independent regenerated plants, *lane H* standard histidine. *Upper arrow* indicates histidine and *lower arrow* indicates mikimopine. Mikimopine was not detected in untransformed plants (*lane C*), but was observed in hairy root-derived plants (*lanes 1–5*). (Otani et al. 1993)

plants was not altered compared with normal sweet potato plants, since more than 90% of pollen grains of both the hairy root-derived plants and untransformed plants were stained by 0.5% acetocarmine solution. Subterranean parts showed abundant roots with extensive branching and smaller storage roots (Fig. 6d).

Interestingly, the numbers of storage roots in the plants regenerated from the hairy root was equal to that of normal plants. The average fresh weight of the storage roots from the plant regenerated from the hairy roots was 3.6 g, while that of a normal plant was 39.4 g.

The storage roots of both hairy root-derived plants and normal plants were incubated at 28 °C to allow adventitious shoots to sprout from them. In the storage roots of hairy root-derived plants few adventitious shoots formed, and the shoots grew more slowly than those produced from the storage roots of normal plants, while the storage roots of the former showed excessive rooting with intensive branching (Fig. 6e). These vegetative progenies also exhibited similar morphological alterations.

3.2.4 T-DNA Analysis

One of the regenerated plants, SE3-5, which was confirmed to contain both agropine and mannopine, was further analyzed to confirm the transformation of Ri-TL-DNA (Fig. 7). The DNA of a regenerated plant hybridized to the TL-DNA probe and contained four internal fragments comigrating with

Fig. 6 a-e. Morphological abnormalities of regenerated plants transformed by *A. rhizogenes*. **a** Stem of plant regenerated from hairy root transformed by *A. rhizogenes* A13 (*lower*) and untransformed plant (*upper*) of sweet potato cv. Chugoku 25. Morphological abnormalities such as smaller leaves, shortened internodes and reduced apical dominance were observed in the transformed plant: *bar* = 30 mm. **b** Wrinkled leaves observed in transformed plant: *bar* = 30 mm. **c** Flowers of transformed plant (*left*) and untransformed plant (*right*). Bar = 15 mm. **d** Aerial parts (*upper*) and subterranean parts (*lower*) of transformed plant (*left, t*) and untransformed plant (*right, c*): *bar* = 20 cm. **e** Adventitious shoot formation from the storage root of untransformed (*left*) and transformed plant (*right*): *bar* = 20 cm. (Otani et al. 1993)

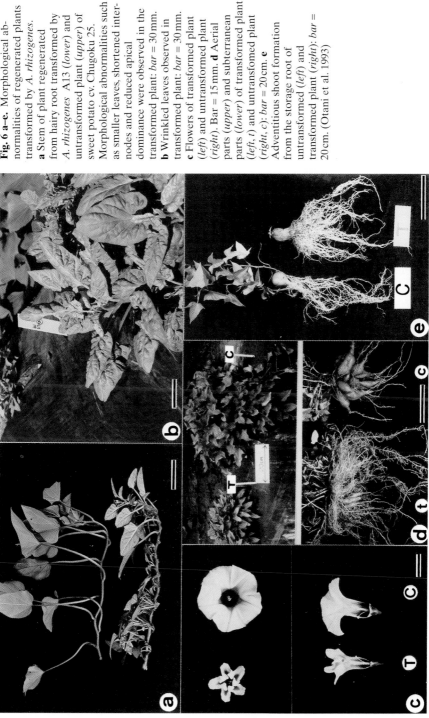

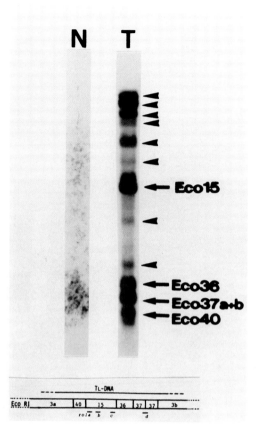

Fig. 7. Southern blot analysis of EcoRI digested DNA from a regenerated plant transformed by *A. rhizogenes* 15834. Digested DNAs were subjected to electrophoresis in a 0.8% agarose gel, blotted onto a nylon filter, and then hybridized against the pLJ1 probe which was labeled with non-radioactive digoxigenin-dUTP using a DNA Labeling and Detection Kit (Boehringer Mannheim). *N* untransformed plant, *T* transformed plant. *Arrows* indicate possible border fragments. A physical map of the TL-DNA digested with *Eco*RI is illustrated *below*. Intact T-DNAs are expected to contain Eco15 (4.3 kb), 36 (1.8 kb), 37 (1.6 kb) and 40 (1.4 kb) fragments present in pLJ1. Eight bands observed in the transformant are border fragments which are junctions between T-DNA (Eco 3a or 3b) and the adjacent plant DNA. (Otani et al. 1993)

*Eco*RI fragments of the probe (*Eco*RI-40, 15, 36, and 37a+b). This regenerated plant also had eight border fragments corresponding to junctions between TL-DNA (*Eco*RI-3a and 3b) and plant DNA. This suggested that at least four copies of TL-DNA were present in the regenerated plant, SE3-5.

DNA from hairy root-derived plants transformed by the mikimopine-type strain of *A. rhizogenes* hybridized to the T-DNA probe, a 7.5 kbp *Eco*RI fragment, including the core T-DNA region (Handa 1992).

3.2.5 Transgenic Sweet Potato Plants Possessing nptII and gusA Genes

Six plants were regenerated from different hairy root clones of cv. Yamakawa-murasaki induced by *A. rhizogenes* C8 possessing a binary vector, pBI121. All regenerated plants produced mikimopine. Shoots of these regenerated plants were transferred to LS medium containing 100 mg/l kanamycin. Three of the regenerated plants produced many roots 3 days after transfer, while the other plants did not form any roots. In our preliminary study using hairy roots transformed by wild type *A. rhizogenes*, no root formation or root growth was observed on the medium containing 100 mg/l kanamycin. Thus, these three plants which formed many roots on kanamycin-containing medium were defined as being resistant to kanamycin. Root tips and leaf disks of the kanamycin-resistant plants were stained with 5-bromo-4-chloro-3-indolyl-glucuronide (X-gluc) for 16 h to show the expression of the *gus* gene. Integration of *nptII* and *gus* genes was also confirmed by PCR (Fig. 8).

In the present study, 50% (3/6) of hairy roots obtained were doubly transformed by Ri plasmid T-DNA and pBI121 T-DNA without any selection pressure.

3.2.6 Inheritance of the Ri-Transformed Phenotype

Two transgenic Chugoku 25 plants (one was transformed by *A. rhizogenes* strain A13, and the other by strain 15834) and a non-transformed Chugoku 25 plant were crossed with non-transformed Yamakawamurasaki plants.

The germination rate was 84.6–100%. There were no clear differences between the transformed and non-transformed F_1 seeds at germination. The F_1 seedlings derived from the cross between the transgenic plants and non-transformed Yamakawamurasaki segregated Ri-transformed *and non-*

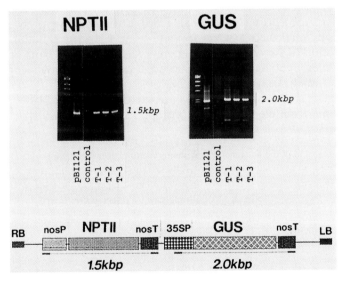

Fig. 8. PCR experiment on DNA samples from three kanamycin-resistant plants showing GUS activity. PCR amplification of the *nptII* gene and *gus*A gene. EtBr staining pattern after agarose gel electrophoresis of PCR amplified samples. *Lane 1* λ*Hind*III DNA size standard (23.1, 9.4, 6.5, 4.3, 2.3, 2.0 and 0.5 kbp fragments from *top* to *bottom*), *lane 2* plasmid pBI121 showing the 1.5 kbp expected band for the *nptII* gene or the 2 kbp expected band for the *gus*A gene, *lane 3* untransformed plant, *lanes 4, 5 and 6* three independent transformed plants. (Otani et al. 1993)

transformed types. About 70% of these progeny demonstrated typical hairy root traits such as shortened internodes, reduced hypocotyl length and wrinkled leaf. The segregation data were consistent with the integration of two independent, dominant Ri T-DNA loci (Table 7).

4 Disussion

4.1 *A. tumefaciens*-Mediated Transformation

In the present study we succeeded in the transformation of sweet potato mediated by *A. tumefaciens.* There are various factors which affect transient GUS expression in embryogenic callus of sweet potato cv. Kokei 14 and Beniazuma. We examined three factors; inclusion of acetosyringone, culture period of embryogenic calli before bacterial inoculation and the differences between *Agrobacterium* strains, and confirmed that all of these factors affected the transformation efficiency of sweet potato cells. The most interesting finding was the differences in transformation efficiency among the bacterial strains. The transient GUS expression varied with the bacterial strain. *A. tumefaciens* strain LBA4404, harboring a "super-binary" vector pTOK233, has been reported to be a more efficient strain than EHA101/pIG121-Hm in rice transformation (Hiei et al. 1994) and *A. tumefaciens* strain R1000, harboring pRiA4 and pBI121, has been reported to be an efficient strain for transformation of *Ipomoea trichocarpa* (Otani et al. 1996), a wild relative of sweet potato. However, *A. tumefaciens* "super-virulent" strain EHA101 is better in genetic transformation of sweet potato using embryogenic callus. Gama et al. (1996) also succeeded in obtaining transgenic sweet potato plants by using the same bacterial strain, EHA101.

Murata et al. (1995) and Okada et al. (1995) obtained transformed sweet potato plants from electroporated protoplasts and biolistically transformed

Table 7. Segregation of the Ri transformed phenotype in progeny of crosses between two transformed Chugoku 25 plants and the cultivar Yamakawamurasaki

Cross	No. of seeds	No. germinating (%)	Phenotype Ri type[a]	Normal	χ^2 value[b]
Untransformed plants × Yamakawamurasaki	13	11 (84.6)	0	11	–
A13-1[c] × Yamakawamurasaki	9	9 (100)	7	2	0.04_{ns}
15834-1[d] × Yamakawamurasaki	26	25 (96.2)	17	8	0.65_{ns}

[a] Ri type: Ri plasmid-transformed type.
[b] Segregation ratio tested was 3:1. ns: not significant.
[c] Transgenic plant A13-1 was transformed by *A. rhizogenes* strain A13.
[d] Transgenic plant 15834-1 was transformed by *A. rhizogenes* strain 15834.

suspension cultures, respectively. However, some problems such as genotypic differences in transformation efficiency and inefficient selection of transformed cells still remain. Moreover, transformation by direct gene transfer methods such as electroporation and particle bombardment often leads to complex integration of multiple copies of the introduced genes (Klein et al. 1989; Register et al. 1994). The production of transgenic plants having a low copy number (one to three) of integrated genes by the *A. tumefaciens*-mediated transformation method presented here was an advantage of this method.

In contrast to the direct gene transfer method, Newell et al. (1995) obtained seven transgenic plants from 140 disks of storage roots of sweet potato cv. Jewel by *Agrobacterium*-mediated transformation. The storage root may not be a good material for obtaining regenerated plants in sweet potato, because the culture responses vary among the genotypes and cultivars with the ability to regenerate plants from storage root disks are rare (Yamaguchi 1978). Using the present transformation method for sweet potato, based on the culture system of embryogenic callus induction from meristem tissue at high frequency using the medium containing 4FA or picloram, more than 50% of meristem tissues formed embryogenic callus in all 11 cultivars tested (Otani and Shimada 1996). Therefore, the method of *A. tumefaciens*-mediated transformation using embryogenic callus might overcome the genotypic differences in genetic transformation of sweet potato.

Although it is difficult to compare our result with those of Gama et al. (1996) in which they selected transformed embryogenic cells by using kanamycin, in this study the transformation efficiency has been considerably improved by using hygromycin for the selection of transformed cells.

We have obtained transgenic Kokei 14 plants possessing the coat protein gene of sweet potato feathery mottle virus (Mori et al. 1995), a severe pathogen of Kokei 14 in Japan. Furthermore, we also produced transgenic Kokei 14 and Beniazuma plants possessing ω-3 fatty acid desaturase gene from tobacco. Further investigation of these transgenic plants is now in progress. The *A. tumefaciens*-mediated gene transfer system using embryogenic callus may be useful as a routine method for the genetic modification of sweet potato.

4.2 *A. rhizogenes*-Mediated Transformation

We reported here not only the successful transformation of Ri-T-DNA into sweet potato, but also transformation of foreign genes mediated by *A. rhizogenes* C8 harboring a binary vector, pBI121. Since whole plants can be regenerated at a high frequency from hairy roots transformed by *A. rhizogenes*, a binary vector system based on the Ri plasmid could be used efficiently for the genetic transformation of some genotypes of sweet potato.

On the other hand, *A. rhizogenes* was chosen to allow the possibility of using morphological alterations produced by the Ri-T-DNA for sweet potato breeding programs. In fact, the aerial parts of transformed plants showed shortened internodes. Since dwarfness is an important characteristic in crop breeding, shortening of internodes in transformed plants is a desirable change. However, the subterranean parts of transformed plants showed abundant roots with extensive branching and smaller storage roots. The transgenic tomato plant transformed with only the *rol B* gene of Ri-T-DNA was characterized by a reduction in both internode length and apical dominance, while the root system of this transgenic plant was similar to that of untransformed plants (van Altvorst et al. 1992). The subterranean parts of the transgenic sweetpotato plant with the *rol B* gene or *rol C* gene should be compared with those of transgenic plants with the intact Ri plasmid T-DNA introduced.

5 Summary

A simple, efficient and reproducible method for the transformation of commercial varieties of sweet potato mediated by *Agrobacterium tumefaciens* was described. We used the embryogenic calli as target tissues for transformation. *A. tumefaciens* strain EHA101/pIG121-Hm used in the present study contained a binary vector with genes for β-glucuronidase (*gus*A) and hygromycin resistance (*hpt*). Around ten hygromycin-resistant cell clusters were produced from 1 g fresh weight of the infected embryogenic calli. The hygromycin-resistant plantlets were regenerated from 53.1% of the hygromycin-resistant calli. Histochemical GUS assay and Southern hybridization analysis indicated that these plants were stably transformed with a copy number of introduced genes of one to three. Transgenic plants grew normally and formed storage roots after 3 months of cultivation in a greenhouse.

Transformation was also accomplished by infection by *Agrobacterium rhizogenes*. Transgenic sweet potato possessing not only Ri-T-DNA but also *nptII* and *gus*A genes were obtained from the hairy roots infected by *A. rhizogenes* containing the binary vector pBI121 in addition to the wild type Ri-plasmid. Leaf disks of in vitro plants were inoculated with different *A. rhizogenes* strains. Numerous hairy roots were induced on leaf disks by both agropine-type and mikimopine-type strains. Whole plants transformed with Ri-T-DNA were regenerated from the hairy roots in five cultivars. These plants had wrinkled leaves, altered flower shape, reduced apical dominance, shortened internodes, small storage roots and abundant, frequently branching roots that showed reduced geotropism.

Acknowledgements. The authors thank Drs. K. Nakamura and K. Toriyama for the gift of the binary vector plasmid, pIG121-Hm, and helpful advice. We are also grateful to Dr. M. Mii and Mr. K. Komaki for useful advice and Dr. T. Komari and Dr. T. Daimon for the gift of *A. tumefaciens* LBA4404/pTOK233 and *A. rhizogenes* strains, respectively.

References

Akama K, Shiraishi H, Ohta S, Nakamura K, Okada K, Shimura Y (1992) Efficient transformation of *Arabidopsis thaliana*: comparison of the efficiencies with various organs, plant ecotype and *Agrobacterium* strains. Plant Cell Rep 12:7–11

Chilton MD, Tepfer DA, Petit A, David C, Casse-Delbart F, Tempe J (1982) *Agrobacterium rhizogenes* inserts T-DNA into the genome of the host plant root cells. Nature 295:432–435

Daimon H, Fukami M, Mii M (1990) Hairy root formation in peanut by the wild type strains of *Agrobacterium rhizogenes*. Plant Tissue Cult Lett 7:31–34

Ditta, G, Stanfield S, Corbin D, Helinski DR (1980) Broad host-range DNA cloning system for Gram-negative bacteria – construction of a gene bank of *Rhizobium melilotis*. Proc Natl Acad Sci USA 77:7347–7351

Dodds JH, Merzdorf C, Zambrano V, Siguenas C, Jaynes J (1991) Potential use of *Agrobacterium*-mediated gene transfer to confer insect resistance in sweet potato. In: Jansson RK, Raman KV (eds) Sweet potato pest management: a global perspective. Westview, Boulder, pp 203–219

Gama MICS, Leite RP Jr, Cordeiro AR, Cantliffe DJ (1996) Transgenic sweet potato plants obtained by *Agrobacterium tumefaciens*-mediated transformation. Plant Cell Tissue Organ Cult 46:237–244

Goddijin OJM, Pen J (1995) Plants as bioreactors. Trend Biotechnol 13:379–387

Handa T (1992) Genetic transformation of *Antirrhinum majus* L. and inheritance of altered phenotype introduced by Ri T-DNA. Plant Sci 81:199–206

Hiei Y, Ohta S, Komari T, Kumashiro T (1994) Efficient transformation of rice (*Oryza sativa* L.) mediated by *Agrobacterium* and sequence analysis of the boundaries of the T-DNA. Plant J 6:271–282

Honda H, Hirai A (1990) A simple and efficient method for identification of hybrids using non-radioactive rDNA as probe. Jpn J Breed 40:339–348

Isogai A, Fukuchi N, Hayashi M, Kamada M, Harada H, Suzuki A (1988) Structure of a new opine, mikimopine, in hairy root induced by *Agrobacterium rhizogenes*. Agric Biol Chem 52:3235–3237

Jabsson RK, Raman KV (1991) Sweet potato pest management: a global overview. In: Jabsson RK, Raman KV (eds) Sweet potato pest management: a global perspective. Westview, Boulder, pp 1–12

Jefferson RA, Kavanagh TA, Bevan MW (1987) GUS fusions: β-glucuronidase as a sensitive and versatile gene fusion marker in higher plants. EMBO J 6:3901–3907

Klein TM, Kornstein L, Sanford JC, Fromm ME (1989) Genetic transformation of maize cells by particle bombardment. Plant Physiol 91:440–444

Linsmaier EM, Skoog F (1965) Organic growth factor requirement of tobacco tissue culture. Physiol Plant 18:100–127

Mori M, Sakai J, Kimura T, Usugi T, Hayashi T, Hanada K, Nishiguchi M (1995) Nucleotide sequence analysis of two nuclear inclusion body and coat protein genes of a sweet potato feathery mottle virus severe strain (SPFMV-S) genomic RNA. Arch Virol 140:1473–1482

Murata T, Okada Y, Fukuoka H, Saito A, Kimura T, Mori M, Nishiguchi M (1995) Genetic transformation of sweet potato, *Ipomoea batatas* (L.) Lam. In: Liu QC, Kokubu T (eds) Proc 1st Chinese-Japanese Symp on sweetpotato and potato. Beijin Agricultural University Press, Beijin, pp 369–374

Newell CA, Lowe JM, Merryweather A, Rooke LM, Hamilton WDO (1995) Transformation of sweet potato [*Ipomoea batatas* (L.) Lam.] with *Agrobacterium tumefaciens* and regeneration of plants expressing cowpea trypsin inhibitor and snowdrop lectin. Plant Sci 107:215–227

Noda T, Tanaka N, Mano Y, Nabeshima S, Ohkawa H, Matsui C (1987) Regeneration of horeseradish hairy roots incited by *Agrobacterium rhizogenes* infection. Plant Cell Rep 6:283–286

Okada Y, Murata T, Saito A, Kimura T, Mori M, Nishiguchi M, Fukuoka H (1995) Production of transgenic sweet potato plants by particle bombardment. Breed Sci 46 (Suppl 1): 256

Otani M, Shimada T (1996) Efficient embryogenic callus formation in sweet potato [*Ipomoea batatas* (L.) Lam.]. Breed Sci 46:257–260

Otani M, Shimada T, Niizeki H (1987) Mesophyll protoplast culture of sweet potato [*Ipomoea batatas* L.]. Plant Sci 53:157–160

Otani M, Mii M, Handa T, Kamada H, Shimada T (1993) Transformation of sweet potato [*Ipomoea batatas* (L.) Lam.] by *Agrobacterium rhizogenes*. Plant Sci 94:151–159

Otani M, Shimada T, Kamada H, Teruya H, Mii M (1996) Fertile transgenic plants of *Ipomoea trichocarpa* Ell. induced by different strains of *Agrobacterium rhizogenes*. Plant Sci 116:169–175

Otani M, Shimada T, Kimura T, Saito A (1998) Transgenic plant production from embryogenic callus of sweet potato [*Ipomoea batatas* (L.) Lam.] using *Agrobacterium tumefaciens*. Plant Biotechnol 15:11–16

Petit A, David C, Dahl GA, Ellis JG, Guyon P, Casse-Delbart F, Tempe J (1983) Further extension of the opine concept: plasmids in *Agrobacterium rhizogenes* cooperate for opine degradation. Mol Gen Genet 190:204–214

Petit A, Berkaloff A, Tempe J (1986) Multiple transformation of plant cells by *Agrobacterium* may be responsible for the complex organization of T-DNA in crown gall and hairy root. Mol Gen Genet 202:388–394

Prakash CS, Varadarajan U (1992) Genetic transformation of sweet potato by particle bombardment. Plant Cell Rep 11:53–57

Rashid H, Yokoi S, Toriyama K, Hinata K (1996) Transgenic plant production mediated by *Agrobacterium* in indica rice. Plant Cell Rep 15:727–730

Register JC III, Peterson DJ, Bell PJ, Bullock WP, Evans IJ, Frame B, Greenland AJ, Higgs NS, Jepson I, Jiao S, Lewnau JL, Sillick JM, Wilson HM (1994) Structure and function of selectable and non-selectable transgenes in maize after introduction by particle bombardment. Plant Mol Biol 25:951–961

Southern EM (1975) Detection of specific sequences among DNA fragments separated by gel electrophoresis. J Mol Biol 98:503–517

van Altvorst AC, Bino RJ, van Dijk AJ, Lamers AM, Lindhour WH, van der Mark F, Dons JJM (1992) Effect of the introduction of *Agrobacterium rhizogenes* rol genes on tomato plant and flower development. Plant Sci 83:77–85

Vervliet G, Holsters M, Teuclay H, Montagu V, Schell J (1974) Characterization of different plaque forming and defective temperate phages in *Agrobacterium* strains. J Gen Virol 26:33–48

Yamaguchi T (1978) Hormonal regulation of organ formation in cultured tissue derived from root tuber of sweet potato. Bull Univ Osaka Pref Ser B 30:54–88

14 Genetic Transformation in *Luffa* (*L. cylindrica* L. Roem)

L. Spanò

1 General Account

The genus *Luffa* (Cucurbitaceae family) comprises seven species, distributed in both the Old and the New World (Heiser and Schilling 1990). The genus is well known because of the fibrous, spongelike nature of its fruits and the wide cultivation of two of its species, *Luffa acutangola* and, especially, *Luffa aegyptiaca*. In English-speaking countries the common names loofah, sponge gourd, rag gourd and dishrag gourd are used for *Luffa aegyptiaca* Mill, the most extensively cultivated species.

Three varieties have been recognized in *Luffa acutangola*: one of which, var. *acutangola*, is a large-fruited, cultivated form. *L. aegyptiaca* comprises a domesticated (var. *aegyptiaca*) and a wild variety; the domesticated differs from the wild form in its more deeply furrowed, less bitter and larger fruits, that sometimes reach lengths of 50 cm (see Fig. 1A). The name to be used for this species is still controversial; *L. cylindrica* (L.) Roem. is in wide usage (Jeffrey 1962), but the name *L. aegyptiaca* has been adopted by the United States Department of Agriculture (Terrell et al. 1986).

The young fruits of both domesticated varieties are cooked and used as vegetables, particularly in India and south-west Asia. The older fruits become too fibrous and bitter to be edible. However, the persistent fibrovascular bundles have given the fruits a large number of other uses, mostly in the replacement of plastic and rubber tools, such as shopping bags, bath-mats and sponges. This is partculary true for *L. aegyptiaca*, which has the stronger and more durable vascular network. Probably the greatest use is for washing and scrubbing utensils. Luffas, most of which are imported, have become quite popular as cosmetic sponges in western countries in recent years. Dried fruits are also utilized as starting material in the production of sound absorbing panels, engine filters, helmet padding and slipper stuffing. The abundant residual biomass (leaves and stems) left in the fields after the fruits have been harvested can be utilized in the production of livestock feeding and/or organic manure.

Luffa acutangola, *L. aegyptiaca* and *L. echinata* are used as medicines in several parts of tropical Africa and Asia (Chopra et al. 1965; Burkill 1985);

Dipartimento di Biologia di Base ed Applicata, Università degli Studi dell'Aquila 67010 Coppito l'Aquila, Italy

Biotechnology in Agriculture and Forestry, Vol. 47
Transgenic Crops II (ed. by Y.P.S. Bajaj)
© Springer-Verlag Berlin Heidelberg 2001

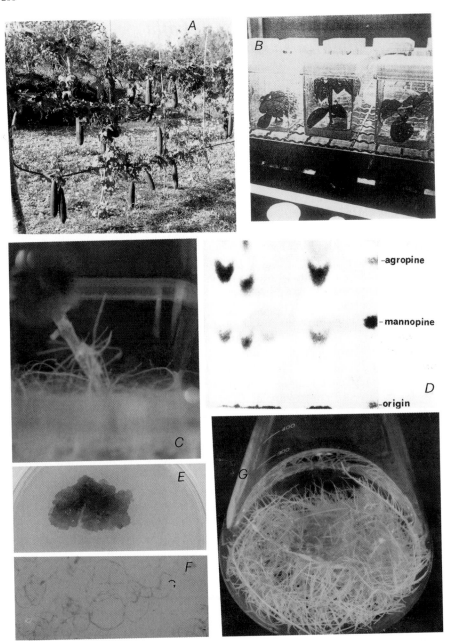

Fig. 1. Soil grown (**A**) and in vitro cultured (**B**) plants of *Luffa cylindrica* L. (Roem). **C** Infected plantlet, 4 weeks after inoculation with virulent *Agrobacterium rhizongenes* 1855, showing hairy root development. **D** Opine detection in acidic extracts from different hairy root clones. Callus (**E**) and cell suspension (**F**) derived from transformed roots. **G** Suspension culture of *Luffa* hairy roots, 2 weeks after last subculture, growing on hormone-free B5 liquid medium

most frequently cited uses are as laxative, purgative and diuretic as well as for the treatment of skin ailments.

In 1983 Kishida and co-workers purified a protein of about 26kDa – denominated luffin – with strong inhibitory activity on protein synthesis from the seeds of *L. cylindrica* L. (Roem). Further work by Kamenosono et al. (1988) and Ramakrishnan et al. (1989) led to the identification of two isoforms, luffin a and luffin b, having similar amino acid composition and the biochemical and enzymatic characteristics of the type 1 RIPs (ribosome-inactivating proteins).

Ribosome-inactivating proteins (RIPs) are enzymes of exclusive plant origin which de-adenylate both eukaryotic (28S) and prokaryotic (23S) ribosomal RNA at a specific and conserved site, thus arresting protein synthesis (Barbieri et al. 1993). RIPs are classified into two major groups, depending upon their subunit composition: type-2 RIPs consist of two (sometimes four) polypeptide chains bound together by disulphide bridges. One of the subunits (the B chain) binds to a galactose-containing receptor, while the other subunit (the A chain) catalytically inactivates ribosomes by means of a specific de-adenylation of 28S rRNA. Type-1 RIPs are single chain proteins enzymatically equivalent to the A chains of type-2. Due to the lack of a galactose-binding subunit, type-1 RIPs have very low aspecific toxicity for intact eukaryotic cells; moreover, they are easier to handle and more stable than type-2 RIPs or type-2 A-chains. For these reasons, the use of type-1 RIPs has recently attracted considerable interest as the active moiety in the preparation of cytotoxic immunoconjugates (Barbieri et al. 1993).

Interest in ribosome-inactivating proteins is growing due to their potential use both in agriculture, as antiviral or antifungal factors to be utilized for plant protection, and in clinical medicine, as components of specific cytotoxic conjugates (immunotoxins). The expression of recombinant proteins in heterologous systems, such as yeast or bacteria, to achieve large-scale production, has often been impaired by the toxicity of the new synthesized enzymes on the protein synthesis apparatus of the host cells. An alternative approach to obtain large quantities of RIPs could be the development and optimization of in vitro culture systems derived from the same plant of origin and capable of producing active enzymes without detrimental effect.

In this context, one of the major drawbacks for industrial exploitation is the genetic instability shown by most cell cultures, which is usually correlated with rapid losses in biosynthetic potential and great variability in productivity (Charlwood and Rhodes 1990). Somaclonal variation, described as the result of genetic instability of cultured cells (Bajaj 1990), has often been correlated with the undifferentiated state of the cultures; a possible solution could therefore be to use cultures of differentiated tissues such as shoot or roots. Because of their structural simplicity, root cultures would appear to be easier to handle and more suitable than shoot cultures for large-scale industrial production in liquid medium.

Transformation of plants by infection with *Agrobacterium rhizogenes* results the transfer to the host plant cells of part of the Ri (root-inducing) plasmid, and in the proliferation of adventitious "hairy roots" (Spanò et al.

1982). Roots transformed by *A. rhizogenes* have an altered phenotype that allows them to grow readily in culture. Moreover, they are genetically stable, easy to manipulate, and can be engineered with foreign genes.

Over the past 10 years the hairy root culture approach has been used by numerous laboratories worldwide for the transformation of a wide range of species and the production of a large variety of phytochemicals (Flores and Medina-Bolivar 1995). In addition to low molecular weight compounds, roots are also known to produce a wide range of macromolecules, including proteins and mucilages. Results reported by Savary and Flores (1994) have shown that hairy root cultures of *Thricosanthes kirilowii* are capable of expressing a species-specific pattern of constitutive and inducible bioactive proteins, including defense-related proteins such as chitinases and the type 1 RIP thricosanthin.

2 Hairy Root Transformation

Transformed roots were obtained by pinching the stems of in vitro grown plantlets of *L. cylindrica* germinated on B5 basal medium (Gamborg 1970) and infected with *Agrobacterium rhizogenes* NCPPB 1855 after 1 month. Hairy root cultures were maintained in the light (ca. $65\,mE\,m^{-2}s^{-1}$) on hormone-free B5 liquid medium at 25 °C on a rotatory shaker (100 rpm) and subcultured every 30 days.

To induce callus formation, explants from young leaves, stems and roots of in vitro grown plantlets were cultured on B5 solid medium supplemented with 2,4-D ($1\,mg\,l^{-1}$) NAA ($1\,mg\,l^{-1}$) and BAP ($0.5\,mg\,l^{-1}$). In some experiments kinetin was substituted for BAP without modifying the results. Callus cultures were maintained on this medium in a growth chamber (18-h illumination period ca. $65\,mE\,m^{-2}s^{-1}$, 25 °C) and subcultured every 25 days. Cell suspensions were initiated by inoculating about 1.5 g of callus in 250 ml flasks containing 50 ml of B5 liquid medium supplemented with NAA ($1\,mg\,l^{-1}$) and BAP ($0.5\,mg\,l^{-1}$). The suspension was grown at 25 °C in the light (ca. $65\,mE\,m^{-2}s^{-1}$) on a rotatory shaker (120 rpm) and subcultured every 25 days.

Young plants, 4- to 5-weeks-old, grown in vitro under sterile conditions (see Fig. 1B) were used for the transformation experiments. Explants of different organ origin were obtained from the same plantlets as for the initiation of callus cultures (see Fig. 1E).

About 80% of the inoculated plantlets formed vigorous hairy roots within 4 weeks of infection with the bacteria; an average of 8–15 roots per inoculated stem could be obtained (see Fig. 1C). If multiple infection sites were present on the same stem, a single responsive site was usually observed; occasionally (less than 5% of the tested plantlets) multiple rooting sites could be obtained on the same plant.

Excised roots were grown successfully on hormone-free medium. These hairy roots appeared very thick and highly branched (Fig. 1F) compared with control non-transformed roots obtained from seedlings grown in culture under the same conditions.

Not all the transformed root clones showed the presence of agropine in their acidic extracts (see Fig. 1D); this finding is consistent with the localization of agropine synthesizing functions on a different transforming element (so-called TR-DNA) that is physically separated from the TL-DNA and is transferred independently from it.

Optimal growth conditions included light, but no greening was ever observed in either hairy or control roots. In some plant species, such as *Convolvolus*, *Catharantus* and *Lotus*, hairy roots cultured under these conditions were reported to spontaneously regenerate shoots on medium lacking phytohormones (Tepfer 1990). In the case of *Luffa* hairy roots, however, no spontaneous shoot regeneration was recorded, even after long periods in culture. The addition to the culture medium of different combinations of growth regulators did not succeed in promoting organogenesis from *Luffa* hairy roots.

Hairy as well as normal roots were also grown in suspension in liquid B5 medium; under these culture conditions transformed roots grew very fast (roughly ten times faster than untransformed controls), showing a sigmoidal curve and reaching the stationary phase within 3 weeks from inoculation; this result extends to *Luffa* the potential of hairy root cultures for biomass production and raises the possibility of using the transformed root clones for large-scale purification of luffin. Experiments at the laboratory scale by Sanità et al. (1996) have demonstrated that *Luffa* hairy roots can express and accumulate the ribosome-inactivating protein luffin without any evident detrimental effect either on culture viability or on the rate of growth (see Table 1).

Callus cultures were easily obtained from both normal and *A. rhizogenes*-transformed tissues of *L. cylindrica* in the presence of 1 mgl^{-1} 2,4-D; for the maintenance of these callus in culture, however, 2,4-D was omitted from the medium composition and NAA was substituted at the same concentration (Fig. 1E). No major difference could be observed between callus from normal and transformed tissues in respect to texture, growth rate, capability to regen-

Table 1. Translational inhibitory activity in extracts from *Luffa cylindrica* L. (Roem) tissues cultured in vitro. One unit is defined as the amount of protein necessary to inhibit protein synthesis by 50% in a rabbit reticulocyte lysate system. (Sanita et al. 1996)

Tissue	Specific activity (U mg protein^{-1})	Total activity (U g tissue^{-1})
Callus	Inactive	–
Cell suspension	Inactive	–
Hairy roots (1st week)	1 600	11 300
Hairy roots (2nd week)	2 300	28 000
Hairy roots (3rd week)	13 000	162 700
Seeds	5 400	73 000

erate shoots and luffin biosynthesis. In particular, none of these calli were able either to produce detectable amounts of the ribosome-inactivating protein luffin (Table 1), or to regenerate shoots.

From calli, suspension cultures were also initiated; they showed maximal growth when the medium was supplemented with $30\,g\,l^{-1}$ sucrose, $1\,mg\,l^{-1}$ NAA and $0.5\,mg\,l^{-1}$ BAP. Other hormone combinations had detrimental effects on cell growth and viability. In all cases, non-translational inhibitory activity was detectable either in cell extracts or in the culture medium. The same lack of RIP activity is found in hairy root suspension culture media.

The translational inhibitory activity found in extracts from hairy root cultures is the highest that has been found in various tissues of *Luffa*, including seeds. The identity of the form of luffins produced by hairy roots has not been yet clarified and could be different from the naturally occurring RIPs of *L. cylindrica* seeds. Results by Savary and Flores (1994) have in fact indicated that, in the case of *Thricosanthes*, hairy root cultures synthesize and secrete a protein of higher molecular weight, which may represent a precursor form of thricosanthin.

3 Summary and Conclusions

Transformed lines of *Luffa cylindrica* L. Roem, obtained by inoculation of in vitro grown plantlets with wild type *Agrobacterium rhizogenes*, were tested for the production of ribosome-inactivating proteins. Crude extracts from transformed roots showed inhibitory activity increasing during the culture period and reaching a maximum value in the stationary phase; no activity was present in callus and/or cell suspensions nor in the culture medium. Results confirm that hairy root cultures can be successfully utilized for the production of RIPs.

The potential of in vitro cultures for the mass production of plant metabolites has been successfully extended to plant defense proteins, a broadly distributed class of commercially important plant proteins. Of particular interest in the class of plant defense enzymes known as ribosome-inactivating proteins (RIPs); potent inhibitors of ribosomal function with potential applications in agriculture, as pesticides and as resistance factors in transgenic plants, and in medicine, as components of immunotoxins in experimental cancer and AIDS therapy.

In vitro production of RIPs from plant cell suspensions and hairy root cultures has been reported only for a few plant species, but it appears a very promising field for the discovery of novel RIPs, the study of their biosynthesis, processing and targeting and for the possibility of expressing recombinant proteins at a high level with no or reduced toxic effects on the host cells.

Acknowledgments. Seeds of *Luffa cylindrica* L. (Roem) were kindly provided by Prof. F. Tammaro, University of L'Aquila, Italy. Agropine type *Agrobacterium rhizogenes* strain NCPPB

1855 and purified opines (mannopine and agropine) were a generous gift of Dr. J. Tempè, Gif-sur-Yvette, France.

References

Bajaj YPS (1990) Somaclonal variation – origin, induction, cryopreservation and implications in plant breeding. In: Bajaj YPS (ed) Biotechnology in agriculture and forestry, vol 11. Somaclonal variation in crop improvement I. Springer, Berlin Heidelberg New York, pp 3–48

Barbieri L, Battelli MG, Stirpe F (1993) Ribosome-inactivating proteins from plants. Biochim Biophys Acta 1154:237–282

Burkill HM (1985) The useful plants of Western Tropical Africa, 2nd edn, vol 1. Royal Botanic Gardens, Kew

Charlwood BV, Rhodes MJ (1990) Secondary products from plant tissue cultures. Clarendon Press, Oxford

Chopra RN, Badhwar R, Ghosh S (1965) Poisonous plants of India, 2nd edn, vol 1. Indian Council of Agric Res, New Delhi

Flores HE, Medina-Bolivar F (1995) Root culture and plant natural products: "unearthing" the hidden half of plant metabolism. Plant Tissue Cult Biotechnol 1:59–74

Gamborg O (1970) The effects of amino acids and ammonium on the growth of plant cells in suspension cultures. Plant Physiol 45:372–375

Heiser CB, Schilling EE (1990) The genus *Luffa*: a problem in phytogeography. In: Bates DM, Robinson RW, Jeffrey C (eds) Biology and utilization of Cucurbitaceae. Comstock Publishing Associates, Cornell University Press, Ithaca, pp 120–133

Jeffrey C (1962) Notes on Cucurbitaceae, including a proposed new classification of the family. Kew Bull 15:337–371

Kamenosono M, Nishida H, Funatsu G (1988) Isolation and characterization of two luffins, protein-biosynthesis inhibitory proteins from the seeds of *Luffa cylindrica*. Agric Biol Chem 52:1223–1227

Kishida K, Masuho Y, Hara T (1983) Protein-synthesis inhibitory protein from seeds of *Luffa cylindrica roem*. FEBS Lett 153:209–212

Ramakrishnan S, Enghlid JJ, Bryant HL, Xu FJ (1989) Characterization of a translation inhibitory protein from *Luffa aegyptiaca*. Biochem Biophys Res Commun 160:509–516

Sanità di Toppi L, Gorini P, Properzi G, Barbieri L, Spanò L (1996) Production of ribosome-inactivating protein from hairy root cultures of *Luffa cylindrica* L. (Roem). Plant Cell Rep 15:910–913

Savary BJ, Flores HE (1994) Biosynthesis of defense-related proteins in transformed root cultures of *Thricosanthes kirilowii* Maxim. var. japonicum (Kitam.). Plant Physiol 140:1195–1204

Spanò L, Pomponi M, Costantino P, van Slogteren GMS, Tempè J (1982) Identification of T-DNA in the root-inducing plasmid of the agropine type *Agrobacterium rhizogenes* 1855. Plant Mol Biol 1:291–300

Tepfer D (1990) Genetic transformation using *Agrobacterium rhizogenes*. Physiol Plant 79:140–146

Terrell E, Hill S, Wiersema J, Rice W (1986) A checklist of names of 3000 vascular plants of economic importance. USDA, Agricultural Handbook No 505, Washington, DC

15 Transgenic Tomato (*Lycopersicon esculentum*)

R. Barg, S. Shabtai, and Y. Salts

1 Introduction

1.1 Distribution and Importance of the Plant

The cultivated tomato, *Lycopersicon esculentum*, is one of the most important vegetable crops worldwide. It is the number one vegetable crop for fresh consumption and approximately one third of the total world yield is consumed in processed form (FAO 1995; Tomato News 1998). The present wide geographical distribution of this species, which originated in South America, was achieved by breeding for adaptation to diverse growth conditions, and for resistance to old and new plant diseases. Development of sophisticated agricultural techniques, mainly closely controlled greenhouse practices, has required breeding for highly reproductive elite cultivars specifically adapted to such growth regimes. Breeding for resistance to plant pathogens has gained extreme importance with increasing public awareness of the negative environmental effects of various pesticides and fungicides, and interest in "health food" has promoted breeding for a high content of certain substances, such as anti-oxidants and vitamins. The development of the processing industry has called for breeding of a completely different array of agronomic traits, including a vegetative growth habit and fruit qualities suitable for mechanical harvest and processing. Moreover, today the processing industry demands different fruit properties from varieties intended for diverse processed products ranging from juice, ketchup and pastes to canned peeled tomatoes and dried raisin tomatoes.

1.2 Need for Genetic Transformation

Tomato had been the subject of genetic breeding using classical methods for over 200 years (see review by Stevens and Rick 1986). The two main factors which limit the progress of breeding efforts are the availability of sources for traits of interest in sexually related plants, and the duration of the

Department of Plant Genetics, Institute of Field and Garden Crops, Agricultural Research Organization, The Volcani Center, Bet Dagan 50 250, Israel

Biotechnology in Agriculture and Forestry, Vol. 47
Transgenic Crops II (ed. by Y.P.S. Bajaj)
© Springer-Verlag Berlin Heidelberg 2001

reproductive life cycle, namely the number of back-cross generations per time unit.

The wild relatives of the cultivated tomato are the natural sources of genes both for pathogen resistance and for required physiological characteristics such as drought resistance, salt resistance, cold tolerance, sugar content etc.

The wild species *L. pimpinellifolium*, *L. cheesmanii*, *L. parviflorum*, *L. chmielewski* and *L. hirsutum*, cross-hybridize with *L. esculentum* rather easily, at least uni-directionally, and indeed have been exploited in many breeding programmes. However, there are serious barriers to interspecific hybridization with *L. peruvianum* which is considered the richest source of resistance genes as well as genes for other agronomically important traits (reviewed by Taylor 1986, and references therein). Moreover, there are traits required in modern tomato cultivars for which there are no available sources among the wild relatives of the domesticated tomato, or where the available sources are governed by complex genetic systems, making them much less attractive for the breeders.

In tomato breeding programmes it is hard to progress beyond three generations per year, and that only if the trait of interest is tightly linked to a defined molecular marker. In the absence of such a marker, the selection for the requested phenotype is frequently restricted to defined growth conditions, which slows down the rate of introgression programmes. On the one hand, classical breeding relies on crosses with remote genetic sources and back-crossing into an elite cultivar, while on the other hand the "life span" of modern cultivars is progressively shortened; hence genetic engineering, which enables us to shorten the time invested in breeding new varieties, is of utmost agronomic and economic importance.

Actually, it has been acknowledged for over a decade that the impressive development of genetic engineering techniques together with the development of transformation protocols suitable for tomato, opened a new era in tomato breeding (Fobes 1980; Nevins 1987). In this review we present a brief account of the development of transformation protocols for tomato; and of the impact of genetic transformation, both on the study of the molecular basis of various physiological phenomena, and on the development of new and improved transgenic tomato cultivars. An updated protocol for tomato transformation is also elaborated.

2 Genetic Transformation in Tomato

2.1 Brief Review of the Development of Tomato *Agrobacterium*-Mediated Transformation Methodology

The bacterium *Agrobacterium tumefaciens* was found to infect tomato cells readily, and proved to be an effective vehicle for delivering foreign DNA into plant cells (e.g. Zambryski et al. 1983). Therefore, *Agrobacterium* was embraced by the tomato biologists and was developed as the first and most favorable tool for the introduction of foreign genes into the tomato genome. Genetic transformation of tomato via *Agrobacterium* was first reported in 1985 by Horsch et al. from Monsanto who regenerated transgenic L2 tomato plants as well as transgenic tobacco and petunia plants. This was achieved by inoculation and co-cultivation of leaf disks (generated by punching leaves with a 6-mm-diameter paper punch), with *A. tumefaciens* containing a disarmed co-integrate Ti plasmid. During co-cultivation the leaf disks were incubated on nurse culture plates for 2–3 days and transferred to selective plates without the nursing feeder cells.

During the years 1986–1987 a wealth of papers reporting successful transformation of tomato were published by several research groups. In the early tomato transformation protocols as well as those published later on, the co-cultivation technique of Horsch et al. (1985) was adopted. The early transformation experiments proved that both disarmed co-integrate Ti vectors and binary Ti vectors can be utilized for tomato transformations. The Monsanto laboratory (McCormick et al. 1986) utilized both kinds of Ti vectors and reported the production of over 300 transgenic tomato plants from eight cultivars. Still, binary Ti vectors were used in most of the early protocols (e.g. An et al. 1986; Shahin et al. 1986), and subsequently binary vectors were used almost exclusively. This was due to the convenience of handling these small plasmids compared with the large co-intergate Ti plasmids.

Many different factors influencing the efficiency of tomato transformation have been investigated, and most of the early transformation reports elaborated on the optimal conditions for efficient transformation. Over the years, the following factors were found to substantially influence the efficiency of transformation by co-cultivation: (1) the genotype of the transformed tomato line; (2) the type of explant; (3) the use of nursing feeder cells or acetosyringone; and (4) the bacterial strain and type of helper plasmid (summarized in Table 1).

Effect of the Tomato Genotype. Many tomato (*L. esculentum*) cultivars exhibit a low efficiency of transformation and are recalcitrant to regeneration following transformation. The genotype effect on the frequency of transformation was recorded in the earliest reports on tomato transformation (e.g. McCormick et al. 1986). This finding prompted Koornneef et al. (1986, 1987) and Chyi and Phillips (1987) to prepare and utilize *Lycopersicon* hybrid lines which exhibit a high regeneration potential. Koornneef et al. (1986, 1987) used the tomato line MsK93 which contained 25% *L. peruvianum* in its ancestry. The efficient shoot regeneration from roots and calli of line MsK93 was later attributed to the *L. peruvianum* dominant allele *Rg-1* which was mapped to chromosome 3 (Koornneef et al. 1993). Chyi and Phillips (1987) used the highly regenerative interspecific tomato hybrid *L. esculentum* × *L. pennellii*. It should be noted that the enhanced transformation efficiency of tomato line MP-1 (Barg et al. 1997) might also be attributed to genes donated by *L. peruvianum*, which is an ancestor of this line. However, this hypothesis has not yet been tested.

Table 1. Summary of various studies conducted on regeneration of transgenic tomato plants via *Agrobacterium tumefaciens* transformation[a]

Reference	Explant/tissue used	Vector/method used	Remarks
Horsch et al. (1985)	Leaf discs	Strain GV3Ti11SE, co-integrate and binary vectors, CC[b] on feeder layer	Tested on several cultivars
McCormick et al. (1986)	Leaf discs	Strain C58C1, co-integrate and binary vectors, CC on feeder layer	Tested on 16 cultivars
An et al. (1986)	Leaf slices and stem sections	Binary vectors with two helper plasmids, CC	Tested on seven cultivars
Shahin et al. (1986)	Punctured or inverted cotyledons	LBA4404 or *A. rhizogenes* A4 with binary vector pARC8, bacterial suspension smeared on cut or puncture	Tested on five cultivars
Koornneef et al. (1986)	Protoplasts and leaf discs	C58(pGV3850:1103) co-integrate vector and direct DNA transfer	Utilized line MsK93 (*L. esculentum* X *L. peruvianum*). Many tetraploid plants were obtained
Koornneef et al. (1987)	Stem and leaf	C58(pGV3850:1103) co-integrate and LBA4404(pAL4404, pAGS112) binary vector	Line MsK93 (*L. esculentum* X *L. peruvianum*)
Chyi and Phillips (1987)	Stem segments (1 cm)	Strain C58C1(pGV3850), co-integrate vector, a scoop of agar-grown bacteria smeared on stem's apical surface	Interspecific hybrid VF36 X LA716
Fillatti et al. (1987a)	Cotyledons	LBA4404(pPMG85/587) binary vector CC on feeder layer	Studied glyphosate resistance
McCormick (1991)	Cotyledons	Binary and co-integrate vectors, CC on feeder layer or with acetosyringone	Calibrated for VF36, less suitable for VFNT cherry
Hamza and Chupeau (1993)	Cotyledon sections	Binary vector p35GUSINT, CC on feeder layer	Optimized conditions for cv. UC82 Three other culivars and *L. peruvianum* were studied
Lipp Joao and Brown (1993)	Cotyledons, hypocotyls	C58C1(pGSFR1161) co-integrate vector, CC	Acetosyringone effects tested
van Roekel et al. (1993)	Cotyledons	Strain C58 harboring binary vector with different helper plasmids. CC on feeder layer	Calibrated for Moneymaker
Frary and Earle (1996)	Cotyledons, hypocotyls	Strain LBA4404, binary vector, CC on feeder layer	Calibrated for Moneymaker
Barg et al. (1997)	Cotyledons	Strain EHA105 and PGV3101, binary vectors, with acetosyringone	Calibrated for line MP-1, suitable also for cv. UC82
Pfitzner (1998)	Cotyledons	Strain LBA4404, binary vectors, with acetosyringone	Calibrated for cv. Craigella, suitable also for Moneymaker

[a] Kanamycin was used as the selective agent in all the cited reports.

[b] In all the reports, except Chyi and Phillips (1987), transformation was performed by co-cultivation (CC) of the explant with bacterial suspension.

Effect of Type of Explant. The type of explant used for transformation was reported to have a profound effect on tomato transformation efficiency. McCormick et al. (1986) reported the effect of the leaf section size. They found that 2 × 2-cm leaf sections were superior to the small 6-mm leaf disks originally used by Horsch et al. (1985). The use of stem sections was reported by An et al. (1986) and Chyi and Philips (1987). Cotyledons and hypocotyl sections from young seedlings were reported to be superior to leaf sections (e.g. Shahin et al. 1986; Fillatti et al. 1987a,b). In all the current protocols, cotyledons are the recommended explant for transformation (e.g. McCormick 1991; van Roekel et al. 1993; Frary and Earle 1996; Barg et al. 1997; Pfitzner 1998).

Effect of Nurse Culture Cells (Feeder Layer). The use of nurse culture cells in feeder plates for preconditioning of the plant sections and/or during co-cultivation, was reported by many groups to improve tomato transformation, and has been routinely used in many protocols (e.g. Horsch et al. 1985; McCormick et al. 1986; Fillatti et al. 1987a,b; Koornneef et al. 1987; McCormick 1991; Hamza and Chupeau 1993; van Roekel et al. 1993). The positive effect is mediated, most probably, through signal molecules secreted by the feeder cells which, in turn, affect the bacterial virulence. An example of such a molecule is acetosyringone which was found to increase transformation frequency in many species, including tomato. Incorporating acetosyringone into the bacterial cultures greatly increased transformation efficiency (e.g. Davis et al. 1991; Lipp Joao and Brown 1993) and is recommended in many current transformation protocols.

Effects of the Transforming Vector, the Helper Plasmid and the Bacterial Strain. The effect of type of bacterial strain and Ti plasmid on the virulence and efficiency of transformation can be highly significant (e.g. Hood et al. 1986; Torisky et al. 1997). Therefore, a careful assessment of the effect of the origin of the *vir* region on transformation efficiency was performed by Hood et al. (1993) and by van Roekel et al. (1993). They compared the effect of helper plasmids originating from octopine, nopaline, and L,L-succinamopine plasmids, and found profound effects on the transformation frequency of several plant species. They reported that the L,L-succinamopine plasmid is the most efficient for tomato transformations and the bacterial strain used influenced transformation frequency as well.

It has been known for quite some time that the right border of the T-DNA is the leading part of T-DNA integration into the plant genome (see Zambryski 1988). Thus, it became obvious that developing T-DNA vectors in which the selective marker resides near the left border will ensure that the selected transgenic plants contain the transgene of interest, and such vectors were constructed and offered for plant genetic engineering (e.g. McBride and Summerfelt 1990; Becker et al. 1992), as well as the set of binary vectors developed by PCAMBIA (www.cambia.org.au).

The recent development of binary-BAC vectors (Hamilton et al. 1999) enables the transformation of large genomic DNA sequences (ca. 100 kb) via

Agrobacterium. Hanson et al. (1999) incorporated a lethal gene in binary vectors, outside the T-DNA borders, in order to prevent regeneration of transgenic plants containing sequences flanking the T-DNA.

Other Factors Affecting Transformation Efficiency. The effects of many factors on transformation efficiency have been reported by several groups. The tested factors included: explant size, explant orientation and age, type and concentration of hormones in the regeneration media, the antibiotic resistance genes used to select the transgenic plants, the antibiotics used to select against the bacteria (e.g. Ling et al. 1998), the type of gelling agent, sealant type, bacterial age and density, etc. A thorough examination of many of these factors was performed by Frary and Earle (1996), who developed an improved transformation protocol for the tomato cultivar Moneymaker. These researchers found that explant size (i.e., a whole cotyledon with its tip and base removed vs. cutting the cotyledon into three pieces), explant orientation, gelling agent and plate sealant, all affected transformation frequency. The type of explant (hypocotyl or cotyledon) and the frequency of explant transfer to fresh regeneration medium, did not affect the transformation efficiency of Moneymaker.

Dillen et al. (1997) examined the effect of temperature during co-cultivation on the efficiency of T-DNA transfer from a binary plasmid into the plant. They showed that transient expression of β-glucuronidase (GUS) was greatly affected by the temperature during co-cultivation. The extent of the temperature effect varied for the different helper plasmids utilized. The optimal temperature for co-cultivation of both infiltrated tobacco leaves and *Phaseolus* callus was 22 °C. Incubation at 19 or 25 °C reduced GUS activity by two to sixfold, and incubations at higher or lower temperatures reduced GUS expression considerably. Dillen et al. concluded that "the efficiency of many published transformation protocols can be improved by reconsidering the factor of temperature."

In all the reports cited above, and most transformation protocols utilized to date, the bacterial *nptII* gene, which encodes for neomycin phosphotransferase, and consequently confers kanamycin resistance, was used as the selective marker to isolate transgenic plants. However, binary vectors carrying other selective genes are also available (e.g. Becker et al. 1992, and PCAMBIA vectors).

In summary, many factors influence tomato transformation efficiency, but it is difficult to compare their relative contributions since the rate of regeneration of transgenic plants is calculated differently in different reports. We have included the factors we think are the most important in the protocol detailed below.

2.2 Utilization of Transgenic Tomato Plants for Scientific Purposes

Since plant transformation became feasible, this technology has become a major tool in the study of all aspects of plant biology, from the advancement of knowledge concerning basic biological phenomena to the development of plants with new and improved agronomic traits.

There are countless articles in the literature in which transgenic tomato plants were used for a wide array of research activities. Here, we shall mention a few of the areas in which basic biological problems were approached using transgenic tomatoes.

Promoter Analysis. Basically, the analysis relies on splicing DNA sequences – residing in a putative promoter region – to reporter genes such as β-glucuronidase (Jefferson et al. 1987), luciferase (e.g. Schneider et al. 1990), and the green fluorescent protein from jellyfish (Chalfie et al. 1994). Transgenic plants harboring such DNA constructs are regenerated and the pattern of expression of the reporter genes is analyzed. Examples of tomato promoters analyzed by this method include genes expressed in fruit (e.g. Van Haaren and Houck 1991; Montgomery et al. 1993; Santino et al. 1997), in pollen (e.g. Eyal et al. 1995; Bate et al. 1996), genes induced by wounding (e.g. Mohan et al. 1993) or circadian signals (Piechulla et al. 1998), and the *RUBISCO* small subunit gene family (e.g. Ueda et al. 1989; Meier et al. 1995). These studies identified various promoter sequences that regulate the level of transcription, translation, and the mode of temporal and spatial expression.

Study of Basic Developmental Processes. Transgenic tomato plants have been exploited for the study of developmental phenomena such as the control of compound leaf architecture, which was analyzed most elegantly by Hareven et al. (1996). Many other developmental processes, such as ethylene perception (Tieman and Klee 1999) or biosynthesis (mentioned below), and systemin production (Dombrowski et al. 1999) have also been studied using transgenic tomatoes.

Gene Identification by Insertional Mutagenesis. Transgenic tomatoes harboring the maize transposable elements *Ac* and *Ds* have been constructed (e.g. Knapp et al. 1994; Cooley et al. 1996; Meissner et al. 1997). Transposition into genes or their promoters may induce mutated phenotypes which can be utilized for the isolation of the interrupted mutated gene. Examples of important genes cloned by this method include the interesting *dwarf* gene (Bishop et al. 1996), and the fungus resistance gene *Cf-9* (Jones et al. 1994).

Functional Analysis of Cloned Genes. Transgenic plants have been used to determine the function of different genes either by overexpression of the gene by its fusion to a constitutive promoter, such as the 35S-CaMV viral promoter (e.g. the *AGAMOUS* homeotic gene by Pnueli et al. 1994), or by turning off the expression of genes by antisense or sense suppression (e.g. Hamilton et al. 1990, 1998; Picton et al. 1993; Theologis et al. 1993).

2.3 Transformation of Genes of Agronomic Importance

Many traits of agricultural importance have been investigated in, or intro-gressed to, tomato using genetic transformation. The first two reports on trans-formation of genes of economic value into tomato appeared in 1987, and in both cases the plants were transformed with bacterial DNA. Fillatti et al. (1987a,b) reported the introduction of a mutated *aroA* bacterial gene confer-ring tolerance to the herbicide glyphosate. Fischhoff et al. (1987) reported transformation of the native bacterial gene encoding for a Bt-endotoxin, which conferred insect resistance. In later transformation experiments, the native Bt gene was replaced by a synthetic gene in which the codon usage was optimized for Bt-protein production by the plant translation machinery (Kuiper and Noteborn 1996). Another important trait that has been engineered in tomato is male sterility. This trait allows the production of hybrid seeds without the tedious manual removal of the anthers in the maternal parent (Mariani et al. 1992).

The first commercial product of recombinant DNA technology was the Flavr Savr tomato characterized by prolonged shelf life. This was achieved by introducing an antisense polygalacturonase gene (Redenbaugh et al. 1995). Several other genes and strategies were used to prolong tomato shelf life by engineering for delayed ripening. Following one strategy, ethylene evolution in the ripening fruits was reduced through antisense and sense suppression of the genes encoding for two enzymes, ACC synthase and the ethylene-forming enzyme, which are involved in the ethylene biosynthetic pathway (Hamilton et al. 1990, 1995; Theologis 1992; Picton et al. 1993; Theologis et al. 1993). Another strategy for inhibition of ethylene biosynthesis relied on catabolizing the ethylene precursor ACC, which was done by expressing the bacterial gene ACC deaminase (Klee et al. 1991; Reed et al. 1996). Other transgenic tomato lines were generated with the aim of improving additional tomato fruit attrib-utes. Among the genes transformed were those involved in starch biosynthe-sis (e.g. Stark et al. 1996), carotenoid synthesis (Schuch et al. 1996), pectin metabolism (Thakur et al. 1996a,b), sugar metabolism (Klann et al. 1996), volatile production (Speirs et al. 1998), and parthenocarpic competence (Carmi et al. 1997; Ficcadenti et al. 1999).

One of the most fruitful contributions of genetic engineering to agricul-ture has been the introduction of viral genes to plants, including tomato, as a means of conferring resistance to viral diseases (tomato transformation for virus resistance was reviewed by Toyoda 1993). Coat protein, satellite RNA, defective polymerase, antisense RNA and ribozymes designed to cleave viral RNAs, have all been successfully utilized to generate resistant plants, and tomato is one of the most important crops to benefit from these technologies (e.g. Nelson et al. 1988; Kim et al. 1994; Fuchs et al. 1996; Gielen et al. 1996; Bendahmane and Gronenborn 1997; Gal-On et al. 1998).

During recent years, dozens of pathogen-resistance genes, derived from several plant species, have been identified and cloned. These genes were shown to confer resistance to various pathogens, including viruses, bacteria, fungi and nematodes, following their transformation to pathogen-sensitive plants.

Among the isolated resistance genes, the tomato is well represented: six different resistance genes against bacteria and fungi have already been cloned from tomato (see reviews by De Wit 1997; Hutcheson 1998; Ronald 1998). Several tomato genes have been isolated by positional cloning, e.g. the *Pseudomonas syringae* resistance gene *Pto* (Martin et al. 1993), the *Cladosporium fulvum Cf-2* resistance gene (Dixon et al. 1996), and the *Fusarium oxysporum I2* resistance gene (Ori et al. 1997; Simons et al. 1998). Other resistance genes, e.g. *Cladosporium fulvum Cf-9* resistance gene (Jones et al. 1994), have been cloned by transposon tagging. All the isolated resistance genes can be transformed into sensitive plants belonging to the same species, turning them into resistant plants. Some of the genes were shown to confer the resistance trait when transformed across the species boundaries. One such example is the tobacco *N* gene, which confers resistance against the TMV virus. Upon transformation of this gene to a TMV-sensitive tomato line, the transgenic plants acquired resistance to TMV (Whitham et al. 1996).

3 Methodology

The following detailed protocol for transformation of tomato is based essentially on that previously published by McCormick (1991). (Permission granted by S. McCormick to publish a modified version of her protocol is deeply appreciated). The protocol was modified for the indeterminate tomato line MP-1, which we found to be highly amenable for transformation (Barg et al. 1997). The protocol is designed for transformation with binary vectors containing kanamycin resistance as the selective marker, and does not include feeder cells.

3.1 Protocol

Seed Sterilization. It is convenient to prepare batches of approximately 100 dry seeds in small "bags" made of Miracloth clipped with staples (see note no. 1 below, for treatment of freshly harvested seeds). A bag is then inserted into a capped 50-ml plastic tube filled with a 30% household bleach solution (i.e. commercial solution of sodium hypochloride, containing 30g/1 of active C1) + 0.1% Tween-20, or a drop of any household detergent, and incubated for 20min. It is important to rotate or shake the tube during the sterilization. Then the bags are sequentially transferred into three containers (e.g. sterile magenta boxes) containing large volumes of sterile distilled water. The bags are soaked in each water container for 15min, and shaken occasionally.

Seed Germination. The sterilized seeds are placed almost confluently on top of MSG medium (approximately 100 seeds in a magenta box), and are maintained at 26°C under 16h light/8h dark conditions for 9–12 days.

Cotyledon Pre-incubation. Cotyledons of line MP-1 are most amenable to transformation 10–12 days post "sowing" when there are no or only minute true leaves visible. Groups of approximately 20 seedlings are gently pulled up with a forceps and cut at the upper third of the hypocotyls, directly onto liquid MSO medium (in Petri dishes 90mm in diameter). The cotyledons are cut at the proximal (wide) side of the blade, approximately 1–3mm from the pedicle (to discard pre-existing meristem). To avoid wilting, the dissection is performed while the cotyledon is submerged in the medium. The cotyledons are placed upside down in Petri dishes (90mm diameter) containing ca. 20ml D1 medium. The cotyledons (75–100) are placed very close together (sides touching) and are incubated in the culture room (16h light/8h dark, 26 °C) for 2 days.

Co-cultivation with Agrobacterium. *Agrobacterium* is cultured in 5ml LB medium containing the appropriate selective antibiotics (at 28 °C, 250rpm, usually for 2 days). On the day of transformation, the 2-day-old culture is diluted (approximately 1:100) into 5ml LB medium without antibiotics and grown for another 2–3h (to attain 0.3–0.5 OD at 600nm). The bacterial culture is centrifuged (ca. 1700g) for 10min, and the pellet is resuspended in 5ml MSO medium (vortex for thorough resuspension) and diluted (1:5) in MSO medium. Acetosyringone (3,5-dimethoxy-4-hydroxy-acetophene) is added to the diluted bacterial culture to a final concentration of 375µM. A concentrated solution of acetosyringone (14.8mM, i.e., 2.9mg/ml ethanol) is freshly made for each experiment and kept in the dark until added to the culture (add 25µl to 1ml culture), because it is susceptible to light. Between 5 and 7ml of the diluted bacterial culture is poured over the cotyledons which had been pre-incubated for 2 days, the plates are covered with aluminum foil and incubated for 1–2h. Ensure that the cotyledons are dipped but not "swimming" in the bacterial culture. Then the bacterial culture is sucked off, and if necessary the cotyledons are re-organized (upside down!) so as to confluently cover the Petri dish. The Petri dishes are sealed (either by Time tape or parafilm) and incubated in the dark (e.g. covered with aluminum foil) in a culture room for 2 days.

Induction of Regeneration. After 2 days of co-cultivation, the cotyledons are transferred to Petri dishes (90mm) containing ca. 20ml of D1 medium + 100mg/l kanamycin (or another appropriate agent to select transgenic plant cells) + 500mg/l Claforan (cefotaxime, to select against the bacteria), hereafter designated "antibiotics". Alternatively, 500mg/l carbenicillin can be added as an anti-*Agrobacterium* antibiotic. The plates are kept in the culture room under 16h light/8h dark, 26 °C, and the cotyledons are transferred to fresh D1 + antibiotics plates at intervals of 7–10 days. Usually the cotyledons are kept on D1 medium for two or three successive growth cycles (i.e. 21–30 days). In line MP-1 small regenerated shoots are detectable after 10–14 days of incubation. During the transfers on D1 medium, while the shoots are still small, the entire cotyledon is transferred; however, if a substantial mass of callus develops together with the shoot, it is preferable to remove the excess callus before transfer to fresh medium. If more than one shoot is selected per cotyledon, these should be considered "identical twins" unless Southern analysis points to independent transformation events. On the other hand, approximately 30% of the regenerated transgenic plants of line MP-1 were found to contain two or more independent insertions of the transgene. In these cases an additional generation might be required in order to separate the independent transformation loci (Barg et al. 1997).

Shoot Development. Shoots containing two or three well-developed leaves and an apical meristem are transferred to small glass or plastic containers, for example "Gerber" baby food jars on which a transparent plastic lid is placed, or a disposable plastic container (e.g. 120 mm diameter × 70 mm height) containing D2 + antibiotics medium. The transferred shoots should contain only a minimal mass of callus at their base. We found that addition of 0.04 mg/l IAA improves shoot development in D2 medium; however, if the transformed cultivar shows a high tendency for callus development, auxin should be omitted from the D2 and probably also from the D1 medium. Usually two successive growth cycles on D2 + antibiotics medium, at a 10-day interval, are sufficient for development of well-organized shoots 2–3 cm in height and with four to five leaves with a defined apical meristem. These shoots are ready for rooting. In some cases an additional growth cycle on D2 is required for sufficient shoot development. However, if three growth cycles do not promote acceptable shoot development, usually the aberrant or arrested shoot will not recover and should be discarded.

Well-organized shoots should be transferred to D2 and later to MSR medium as soon as they reach the appropriate stage of organogenesis. An extra transfer on D1 or D2 medium does not necessarily improve them, and frequently they will start to develop extra callus at the base of the shoots, while development of the shoot itself is retarded.

Rooting. Well-organized shoots (any developed callus should be removed) are transplanted to containers with MSR medium + antibiotics, where transformed shoots will develop a branched root system. Keeping the containers on shelves which are not illuminated from below improves the rooting. Once a branched root system develops, the plant is released from the medium, the agar is gently but thoroughly washed off the roots, and the plant is potted either in preimbibed peat pellets (e.g. Jiffy-7, Jiffy Products, Ltd., Norway) or in a very wet soil mixture. The plants are kept in a loosely closed container, under culture room light regime for 1–3 weeks until the plant undergoes hardening (the cover should be gradually removed to avoid wilting) and develops several new leaves. Then the plant is transplanted into a large pot (10 l) and grown further in a greenhouse. Developed shoots that for some unknown reason do not root in MSR, or shoots that were contaminated before developing a root system, should be transplanted to peat pellets or soil mixture after dipping the bottom of their stem in "rooting powder" (e.g. Hormex, Brooker Chemicals, CA, USA, which contains 0.1% indole butyric acid). We frequently found those were transformed shoots, despite a lack of rooting in the presence of the selective antibiotic.

Scoring of R_1 Transgenic Plants. The simplest means of scoring transgenic progenies is by testing for kanamycin resistance. Sterilized R_1 seeds are germinated in a magenta box on MSG medium + 100 mg/l kanamycin. Transgenic progenies will develop a branched root system, whereas non-transgenic progenies will develop an unbranched primary root only, frequently with a "kink" of over 60°. The hypocotyl of the sensitive seedlings may have enhanced anthocyanin pigmentation (but some cultivars are naturally characterized by anthocyanin pigmented hypocotyls). Unlike seedlings of VF36 (McCormick 1991), the wild-type seedlings of MP-1 are frequently as tall and their cotyledons as developed, as the transgenic seedlings on this medium, and only the root phenotype is unequivocally indicative of transgeneity.

One drawback of this test is that the seedlings germinated on this medium in the culture room are much more etiolated, fragile and spindly than seeds germinated in a tray containing a soil mixture and kept in a greenhouse; as a result, not all recover from the hardening shock when transplanted to the greenhouse. A test for kanamycin-resistance using the spray method (Weide et al. 1989) was not informative on young MP-1 seedlings (at the third true leaf stage) germinated in a tray, since the wild-type plants did not manifest distinct symptoms. It became informative only when applied to older potted plants in the greenhouse, making it impractical for screening large populations of transgenic MP-1 tomato plants (D. Granot and N. Dai, Israel, pers. comm.). However, PCR analysis of very young R_1 seedlings, germinated under optimal growth conditions, has become practical due to rapid methods for DNA extraction from cotyledons and very young leaves (e.g. Edwards et al. 1991; Fulton et al. 1995).

3.2 Notes and Troubleshooting

1. When collecting seeds either for transformation or for scoring transgenic progenies, it is recommended to incubate the freshly harvested seeds in the fruit juice for 24 h, then add to the incubated ('fermented') seeds a similar volume of 2.2% sulfuric acid (H_2SO_4), incubate for another 4 h, and then wash the seeds thoroughly and dry. This treatment improves the germination rate and reduces the pathogen populations on the seeds. For long storage, the seeds should be kept at low temperatures (4–6 °C).
2. Contamination is the major cause of loss of transgenic shoots. It is therefore recommended not to include more than 200–250 cotyledons per experiment, and not to transplant all the cotyledons to fresh medium on the same day. It is also recommended to leave the plates for 48 h at room temperature following pouring of the media into them. This enables one to spot contamination before use, or before storage of the plates at 4 °C for later use (plates can be stored at 4 °C for 3–4 weeks).
3. Development of a branched root system in the presence of 100 mg/l kanamycin is a strong indication of regeneration of a transgenic plant. Therefore, it is recommended that all the regeneration steps (following the 2 days of co-cultivation) are performed in the presence of 100 mg/l kanamycin.

3.3 Media

Hormones and kanamycin are filter-sterilized and added to cooled (approximately 45 °C) autoclaved media, immediately before dispensing the media into plates. It is preferable to dispense the stock of kanamycin in small aliquots before storing (at −20 °C) because it tends to lose activity following repeated cycles of freezing and thawing.

MSG Medium (for Seed Germination). MS (2.3 g/l; Murashige and Skoog 1962) macro- and micro-elements (e.g. Duchefa or Gibco). Vitamins content: Gamborg's B5 vitamins, namely: myo-inositol 100.0 mg/l (0.56 mM), nicotinic acid 1.0 mg/l (8.12 μM), pyridoxin HCl 1.0 mg/l (4.86 μM), thiamine HCl 10.0 mg/l (29.65 μM), with the addition of 2 mg/l glycine (26.64 μM). 1.5% sucrose, 0.8% agar, pH = 5.8 (adjust with 1 M KOH).

MSO Medium (Liquid Medium for Cotyledon Cutting). MS (4.3 g/l) macro- and micro-elements (e.g. Duchefa or Gibco). Vitamins: as in MSG medium. 1.5% sucrose, pH = 5.8.

D1 Medium (for Initiation of Regeneration). Salts and vitamins as in MSO medium, hormones: 1 mg/l zeatin, 0.1 mg/l IAA, 3% glucose, 0.8% agar, pH = 5.8.

D2 Medium (for Shoot Development). Salts and vitamins as in MSO medium, hormones: 0.1 mg/l zeatin, 0.04 mg/l IAA (omit the latter if the tomato line shows high tendency for callus formation at the base of the regenerating shoots), 3% glucose, 0.8% agar, pH = 5.8.

MSR Medium (for Rooting). Salts and vitamins as in MSO medium, hormones: 2 mg/l IBA (indole butyric acid), 1.5% sucrose, 0.8% agar, pH = 5.8.

LB (Luria-Bertani) Medium. Bacto-tryptone (10 g/l), 5 g/l bacto-yeast extract, 10 g/l NaCl, adjust pH to 7.5 with 1 M NaOH.

4 Results and Discussion

4.1 Modification of the Transformation Protocol for Tomato Line MP-1

The protocol detailed above is based essentially on that previously published by McCormick (1991). It was modified for the indeterminate tomato line MP-1 which we found to be highly amenable for transformation compared with other cultivars such as VF36 frequently used for tomato transformation (Barg et al. 1997). Transformation of MP-1 with 'benign' transgenes usually results in the appearance of distinct shoots on over 50% of the cotyledons within 10–14 days (Fig. 1), and this occurs with almost no prior callus formation.

Fig. 1. Shoot regeneration on cotyledons of line MP-1 transformed with a binary vector carrying the chimeric transgene *TPRP-F1:iaaH*. The plate containing D1 medium + kanamycin (100 mg/l) was photographed 14 days post co-cultivation with transformed *Agrobacterium tumefaciens* (strain EHA 105). Note that the cotyledons are positioned upside down

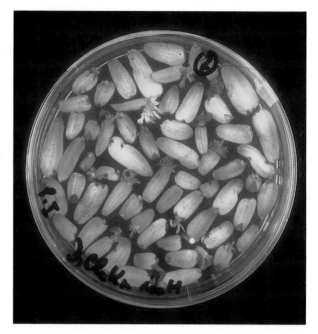

However, the rate of regeneration of fully developed transgenic plants is significantly lower. The fully developed transgenic plants regenerated from this line are usually of normal phenotype, as there is a very low tendency for somaclonal variation among plants regenerated from this tomato line. Even plants transgenic for less benign genes such as the *rolB* gene (conferring hypersensitivity to auxin) expressed under a young fruit specific promoter (Carmi et al. 1997) were found to have normal vegetative development (Fig. 2) while bearing parthenocarpic fruits, as expected.

Based on our experience, the less callus formed on the surface of the cotyledon's cut prior to shoot regeneration, the greater the chance of obtaining well-organized, fast-growing transgenic shoots. Hence, it is more beneficial to perform several successive transformation experiments rather than to wait for shoots to appear following repeated growth cycles on regeneration-inducing medium. Moreover, there is a greater chance to select for 'escapees' among shoots developed from a large callus mass at late growth cycles on this medium.

Seeds of line MP-1, which is not available commercially, can be obtained from our laboratory. An important characteristic of this indeterminate line is tolerance to the devastating viral disease TYLCV.

4.2 Factors Affecting Uniformity Among Transformation Experiments

Variation among transformation experiments is a known problem, yet not all the factors contributing to this problem are known. One factor which con-

Fig. 2. Regenerated tomato plant of line MP-1 carrying the transgene *TPRP-F1::rolB*. The vegetative development of the R_0 plant was normal since the transgene is expressed under a fruit-specific promoter (Carmi et al. 1997). The plant was photographed 16 weeks after transfer to the greenhouse

tributs to this variation is the physiological status of the seedlings used for transformation. We found that including only 1.5% sucrose in the MSG medium accelerates seed germination and improves its uniformity. Another factor is the explant itself: like McCormick (1991), we found that cotyledons are the best organ for transformation. It is not recommended to wait until the true leaves expand in order to have more biological material available for transformation, especially since the aging cotyledons rapidly lose their enhanced transformation tendency. Moreover, the differences in regeneration rate among cotyledons increase with aging. It is also not recommended to include the hypocotyls as explants, because they tend to produce profuse callus which is frequently associated with regeneration of aberrant shoots.

The status of the infecting bacteria also contributes to variation among experiments. There are differences among *Agrobacterium* strains with regard to efficiency of tomato transformation. The strain EHA105 (Hood et al. 1993) gives very satisfying results. Growing the 2-day bacterial cultures in the presence of antibiotics ensures maintenance of the binary plasmid, and we found that diluting the culture at the day of transformation and growing it on for a couple of hours to 0.3–0.5 OD, improves the rate of transformation and reduces variation among experiments.

Pre-incubation of the cotyledons on a feeder layer as a measure to improve infection also increases the variation among experiments, mainly due to variation in the physiological status of the cell suspension used as the nursing cells. Instead, application of acetosyringone to 2-day pre-incubated cotyledons is technically simple and the reproducibility among experiments is quite high.

4.3 Modification of the Protocol for Other Cultivars

The protocol will most probably have to be modified if another selective marker gene is included in the binary vector, or if applied to other cultivars. It should be noted that the rate of regeneration from non-transformed cotyledons does not predict the transformation competence; in a preliminary test we found that varieties with very poor transformation competence regenerate profusely from cut cotyledons on D1 medium.

Currently, advanced lines based on MP-1 with somewhat larger fruit size and resistance to TMV, besides the tolerance to TYLCV, are being evaluated for their transformation competence. This protocol was used by us to recover transgenic plants from a parental line of advanced greenhouse cultivars. In this line regeneration on D1 was preceded by callus formation and the rate of regeneration of fully developed transformants was substantially lower than from line MP-1. On the other hand, successful transformation into such parental lines is of substantial breeding importance.

4.4 Analysis of Transgenic Plants

In the regenerated plants at R_0, genomic transformation can only be confirmed by Southern analysis and not by PCR analysis, since carryover of the bacteria within the mature plant cannot be ruled out. Determination of the transgene copy number in R_1 (and R_2) plants should be based not only on the ratio of kanamycin-resistant progenies but also on Southern analysis. Bias from expected ratios may reflect partial transgene silencing, and in line MP-1, insertion of more than one copy into the same locus (i.e., in tandem) has been documented (Barg et al. 1997).

5 Summary

During the last decade a vast effort has been invested into the improvement of transformation protocols for tomato, which is still more recalcitrant to transformation than some other solanaceous species. Because of the agricultural

importance of this vegetable crop, it is widely appreciated that any biotechnological improvement made in tomato bears substantial agronomic and economic benefits. The accelerating impact of genetic engineering on tomato
breeding can be appreciated from the fact that among the 30 transgenic cultivars that have been exempted already from regulation by the USDA, five
transgenic tomato cultivars characterized by modified fruit ripening characteristics, are included (BSS Biotechnology Update, 1998). Unquestionably, this
is just the tip of the iceberg, since many more characteristics of tomato have
been successfully modified following transgenic manipulation.

Since in tomato the fruit is the organ consumed, it is not surprising that a
considerable effort has been invested in manipulating genes expressed during
the stages of reproductive development. A transgene for male sterility facilitated the production of hybrid tomato seeds (Mariani et al. 1992), while
expressing the *rolB* gene in the young fruit (Carmi et al. 1997) induced
parthenocarpy. Characteristics related to the quality of the fruit have also
been genetically engineered. Stark et al. (1996) enhanced the content of total
soluble solids (TSS) in the ripening fruit by introducing a gene altering starch
accumulation in young fruits, whereas Klann et al. (1996) improved the TSS
by silencing the gene for acid invertase. Fruit color has been modified by
manipulating carotenoid biosynthesis (Schuch et al. 1996). The shelf life of
tomato fruit has been significantly prolonged following transformation with
several transgenes that suppressed ethylene evolution in the maturing fruit
(Hamilton et al. 1990, 1995; Klee et al. 1991; Picton et al. 1993; Theologis et al.
1993; Reed et al. 1996). Transformation with genes affecting cell wall degradation also attenuated fruit softening (Thakur et al. 1996a,b).

The scarcity of cloned plant genes for resistance to various pathogens has
driven the recruitment of transgenes from other organisms. This includes
transgenes for pest resistance such as the bacterial Bt-toxins genes (e.g.
Fischhoff et al. 1987; Kuiper and Noteborn 1996), or viral genes conferring
resistance to viral diseases such as: TMV (Nelson et al. 1988), spotted wilt
virus (Kim et al. 1994), CMV (Fuchs et al. 1996; Gielen et al. 1996), or TYLCV
(Bendahmane and Gronenborn 1997). In addition, TMV resistance was conferred to tomato by transformation of the tobacco-derived *N* gene (Whitham
et al. 1996).

The present short review points to the immense potential inherent in the
utilization of transgenes for tomato breeding. Nevertheless, utilization of
transgenes from non-plant organisms still provokes serious public concern,
while crops expressing plant-derived transgenes either in sense or antisense
orientation are broadly acceptable. Until recently, the rate of cloning of
plant genes of known function has been relatively slow. However, the
novel methods developed for rapid cloning of plant genes of known functions combined with the ongoing EST tomato project (www.tigr.org) will
rapidly expand the array of defined plant genes available for transformation
(e.g. Bouchez and Hofte 1998). That, together with the development of
improved transformation techniques that enable direct introgression of
unique genes into parental lines of elite tomato cultivars, paves the way to the
routine exploitation of transgenes in advanced breeding programmes.

References

An G, Watson BD, Chiang CC (1986) Transformation of tobacco, tomato, potato, and *Arabidopsis thaliana* using a binary Ti vector system. Plant Physiol 81:301–305

Barg R, Pilowsky M, Shabtai S, Carmi N, Szechtman AD, Dedicova B, Salts Y (1997) The TYLCV-tolerant tomato line MP-1 is characterized by superior transformation competence. J Exp Bot 48:1919–1923

Bate N, Spun C, Foster DG, Twell D (1996) Maturation specific translation enhancement mediated by the 5′ UTR of a late pollen transcript. Plant J 10:613–623

Becker D, Kemper E, Schell J, Masterson R (1992) New plant binary vectors with selectable markers located proximal to the left border. Plant Mol Biol 20:1195–1197

Bendahmane M, Gronenborn B (1997) Engineering resistance against tomato yellow leaf curl virus (TYLCV) using antisense RNA. Plant Mol Biol 33:351–357

Bishop GJ, Harrison K, Jones JDG (1996) The tomato dwarf gene isolated by heterologous transposon tagging encodes the first member of a new cytochrome P450 family. Plant Cell 8:959–969

Bouchez D, Hofte H (1998) Functional genomics in plants. Plant Physiol 118:725–732

BSS Biotechnology Update (1998) Field testing of new agricultural products continues. <http://www.aphis.usda.gov/biotech/newsletter.html>, February 1998

Carmi N, Salts Y, Shabtai S, Pilowsky M, Dedicova B, Barg R (1997) Transgenic parthenocarpy due to specific over-sensitization of the ovary to auxin. In: Altman A, Ziv M (eds) Hort Biotech In Vitro Cult and Breeding. Acta Hortic 447:579–581, ISAS

Chalfie M, Tu Y, Euskirchen G, Ward WW, Prasher DC (1994) Green fluorescent protein as a marker for gene expression. Science 263:802–805

Chyi YS, Phillips GC (1987) High efficiency *Agrobacterium*-mediated transformation of *Lycopersicon* based on conditions favorable for regeneration. Plant Cell Rep 6:105–108

Cooley MB, Goldsbrough AP, Still DW, Yoder JI (1996) Site-selected insertional mutagenesis of tomato with maize Ac and Ds elements. Mol Gen Genet 252:184–194

Davis ME, Miller AR, Lineberger RD (1991) Temporal competence for transformation of *Lycopersicon esculentum* (L. Mill.) cotyledons by *Agrobacterium tumefaciens*: relation to wound-healing and soluble plant factors. J Exp Bot 42:359–364

De Wit PJGM (1997) Pathogen avirulence and plant resistance: a key role for recognition. Trends Plant Sci 12:452–458

Dillen W, De Clercq J, Kapita J, Zamber M, Van Montagu M, Angenon G (1997) The effect of temperature on *Agrobacterium tumefaciens*-mediated gene transfer to plats. Plant J 12:1459–1463

Dixon MS, Jones DA, Keddie JS, Thomas CM, Harrison K, Jones JDG (1996) The tomato *Cf-2* disease resistance locus comprises two functional genes encoding leucine-rich repeat proteins. Cell 84:451–458

Dombrowski JE, Pearce G, Ryan CA (1999) Proteinase inhibitor-inducing activity of the prohormone prosystemin resides exclusively in the C-terminal systemin domain. Proc Natl Acad Sci USA 96:12947–12952

Edwards KC, Johnstone C, Thumpson C (1991) A simple and rapid method for the preparation of plant genomic DNA for PCR analysis. Nucleic Acids Res 19:1349

Eyal Y, Durie C, McCormick S (1995) Pollen specificity elements reside in 30 bp of the proximal promoters of two pollen expressed genes. Plant Cell 71:373–384

FAO (1995) Q Bull Statist 8 (1/2):59

Ficcadenti N, Sestili S, Pandolfini T, Cirillo C, Rotion GL, Spena A (1999) Genetic engineering of parthenocarpic fruit development in tomato. Mol Breed 5:463–470

Fillatti JJ, Kiser J, Rose B, Comai L (1987a) Efficient transformation of tomato and the introduction and expression of a gene for herbicide tolerance. In: Nevins DJ, Jones RA (eds) Tomato biotechnology. Alan R Liss, New York, pp 199–210

Fillatti JJ, Kiser J, Rose R, Comai L (1987b) Efficient transfer of a glyphosate tolerance gene into tomato using a binary *Agrobacterium tumefaciens* vector. Bio/Technology 5:726–730

Fischhoff DA, Bowdish KS, Perlak FJ, Marrone PG, McCormick SM, Niedermeyer JG, Dean DA, Kusano-Kretzmer K, Mayer EJ, Rochester DE, Rogers SG, Fraley RT (1987) Insect tolerant transgenic tomato plants. Bio/Technology 5:807–813

Fobes JF (1980) The tomato as a model system for the molecular biologist. PMB Newsl 1:64–67

Frary A, Earle ED (1996) An examination of factors affecting the efficiency of *Agrobacterium*-mediated transformation of tomato. Plant Cell Rep 16:235–240

Fuchs M, Provvidenti R, Slightom JL, Gonsalves D (1996) Evaluation of transgenic tomato plants expressing the coat protein gene of cucumber mosaic virus strain WL under field conditions. Plant Dis 80:270–275

Fulton TM, Chunwongse J, Tanksley SD (1995) Microprep protocol for extraction of DNA from tomato and other herbaceous plants. Plant Mol Biol Rep 13:207–209

Gal-On A, Wolf D, Wang Y-Z, Faure JE, Pilowsky M, Zelcer A, Wang YZ (1998) Transgenic resistance to cucumber mosaic virus in tomato: blocking of long-distance movement of the virus in lines harboring a defective viral replicase gene. Phytopathology 88:1101–1107

Gielen J, Ultzen T, Bontems S, Loots W, Van Schepen A, Westerbroek A, de Haan P, Van Grinsven M (1996) Coat protein-mediated protection to cucumber mosaic virus infections in cultivated tomato. Euphytica 88:139–149

Hamilton AJ, Lycett GW, Grierson D (1990) Antisense gene that inhibits synthesis of the hormone ethylene in transgenic plants. Nature 346:284–287

Hamilton AJ, Fray RG, Grierson D (1995) Sense and antisense inactivation of fruit ripening genes in tomato. Curr Top Microbiol Immunol 197:77–89

Hamilton AJ, Brown S, Han Y-H, Ishizuka M, Lowe A, Solis AGA, Grierson D, Han YH (1998) A transgene with repeated DNA causes high frequency, post-transcriptional suppression of ACC-oxidase gene expression in tomato. Plant J 15:737–746

Hamilton CM, Frary A, Xu Y-M, Tanksley SD, Zhang H-B, Xu YM, Zhang HB (1999) Construction of tomato genomic DNA libraries in a binary-BAC (BIBAC) vector. Plant J 18:223–229

Hamza S, Chupeau Y (1993) Re-evaluation of conditions for plant regeneration and *Agrobacterium*-mediated transformation from tomato (*Lycopersicon esculentum*). J Plant Sci 44:1837–1845

Hanson B, Engler D, Moy Y, Newman B, Ralston E, Gutterson N (1999) A simple method to enrich an *Agrobacterium*-transformed population for plants containing only T-DNA sequences. Plant J 19:727–734

Hareven D, Gutfinger T, Parnis A, Eshed Y, Lifschitz E (1996) The making of a compound leaf: genetic manipulation of leaf architecture in tomato. Cell 84:735–744

Hood EE, Helmer GL, Fraley RT, Chiltom M-D (1986) The hypervirulence of *Agrobacterium tumefaciens* A281 is encoded in a region of pTiBo542 outside of T-DNA. J Bacteriol 168:1291–1301

Hood EE, Gelvin SB, Melchers LS, Hoekema A (1993) New *Agrobacterium* helper plasmids for gene transfer to plants. Transgen Res 2:208–218

Horsch RB, Fry JE, Hoffman NL, Eichholtz D, Rogers SG, Fraley RT (1985) A simple and general method for transferring genes into plants. Science 227:1229–1231

Hutcheson SW (1998) Current concepts of active defense in plants. Annu Rev Phytopathol 36:59–90

Jefferson RA, Kavanagh TA, Bevan MW (1987) GUS fusions: β-glucuronidase as a sensitive and versatile gene fusion marker in higher plants. EMBO J 6:3901–3907

Jones DA, Thomas CM, Hammond-Kosack KE, Balint-Kurti PJ, Jones JDG (1994) Isolation of the tomato *Cf-9* gene for resistance to *Cladosporium fulvum* by transposon tagging. Science 266:789–793

Kim JW, Sun SSM, German TL (1994) Disease resistance in tobacco and tomato plants transformed with the tomato spotted wilt virus nucleocapsid gene. Plant Dis 78:615–621

Klann EM, Hall B, Bennett AB (1996) Antisense acid invertase (TIV1) gene alters soluble sugar composition and size in transgenic tomato fruit. Plant Physiol 112:1321–1330

Klee HJ, Hayford MB, Kretzmer KA, Barry GF, Kishore GM (1991) Control of ethylene synthesis by expression of a bacterial enzyme in transgenic tomato plants. Plant Cell 3:1187–1193

Knapp S, Larondelle Y, Rossberg M, Furtek D, Theres K (1994) Transgenic tomato lines containing Ds elements at defined genomic positions as tools for targeted transposon tagging. Mol Gen Genet 243:666–673

Koornneef M, Hanhart C, Jongsma M, Toma I, Weide R, Zabel P, Hille J (1986) Breeding of a tomato genotype readily accessible to genetic manipulation. Plant Sci 45:201–208

Koornneef M, Jongsma M, Weide R, Zabel P, Hille J (1987) Transformation of tomato. In: Nevins DJ, Jones RA (eds) Tomato biotechnology. Alan R Liss, New York, pp 169–178

Koornneef M, Bade J, Hanhart C, Horsman K, Schel J, Soppe W, Verkerk R, Zabel P (1993) Characterization and mapping of a gene controlling shoot regeneration in tomato. Plant J 3:131–141

Kuiper HA, Noteborn HPJM (1996) Food safety assessment of transgenic insect-resistant Bt tomatoes. Food safety evaluation. Proc of an OECD-sponsored workshop held on 12–15 September 1994, Oxford, UK, pp 50–57

Ling HQ, Kriseleit D, Ganal MW (1998) Effect of ticarcillin/potassium clavulanate on callus growth and shoot regeneration in *Agrobacterium*-mediated transformation of tomato (*Lycopersicon esculentum* Mill.). Plant Cell Rep 17:843–847

Lipp Joao KH, Brown TA (1993) Enhanced transformation of tomato co-cultivated with *Agrobacterium tumefaciens* C58C1Rifr::pGSFR1161 in the presence of acetocyringone. Plant Cell Rep 12:422–425

Mariani C, Gossele V, De Beuckeleer M, De Block M, Goldberg RB, De Greef W, Leemans J (1992) A chimeric ribonuclease-inhibitor gene restores fertility to male sterile plants. Nature 357:384–387

Martin GB, Brommonschenkel SH, Chunwongse J, Frary A, Ganal MW, Spivey R, Wu T, Earle ED, Tanksley SD (1993) Map-based cloning of a protein kinase gene conferring disease resistance in tomato. Science 262:1432–1436

McBride KE, Summerfelt KR (1990) Improved binary vectors for *Agrobacterium*-mediated plant transformation. Plant Mol Biol 14:269–276

McCormick S (1991) Transformation of tomato with *Agrobacterium tumefaciens*. In: Lindsey K (ed) Plant tissue culture manual. Kluwer, Dordrecht, B6:1–9c

McCormick S, Niedermeyer J, Fry J, Barnason A, Horsch R, Fraley R (1986) Leaf disc transformation of cultivated tomato (*Lycopersicon esculentum*) using *Agrobacterium tumefaciens*. Plant Cell Rep 5:81–84

Meier I, Callan KL, Fleming AJ, Gruissem W (1995) Organ-specific differential regulation of a promoter subfamily for the ribulose-1,5-bisphosphate carboxylase/oxygenase small subunit genes in to tomato. Plant Physiol 107:1105–1118

Meissner R, Jacobson Y, Melamed S, Levyatuv S, Shalev G, Ashri A, Elkind Y, Levy A (1997) A new model for tomato genetics. Plant J 12:1465–1472

Mohan R, Vijayan P, Kolattukudy PE (1993) Developmental and tissue-specific expression of a tomato anionic peroxidase (*tap1*) gene by a minimal promoter, with wound and pathogen induction by an additional 5'-flanking region. Plant Mol Biol 22:475–490

Montgomery J, Pollard V, Deikman J, Fischer RL (1993) Positive and negative regulatory regions control the spatial distribution of polygalacturonase transcription in tomato fruit pericarp. Plant Cell 5:1049–1062

Murashige T, Skoog F (1962) A revised medium for rapid growth and bioassay with tobacco tissue cultures. Physiol Plant 15:473–497

Nelson RS, McCormick SM, Delannay X, Dube P, Layton J, Anderson EJ, Kaniewaska M, Porksch RK, Horsch RB, Fraley RT, Beachy RN (1988) Virus tolerance, plant growth and field performance of transgenic tomato plants expressing coat protein from tobacco mosaic virus. Bio/Technology 6:403–409

Nevins DJ (1987) Why tomato biotechnology? A potential to accelerate the applications. In: Nevins DJ, Jones RA (eds) Tomato biotechnology. Alan R Liss, New York, pp 3–14

Ori N, Eshed Y, Paran I, Presting G, Aviv D, Tanksley S, Zamir D, Fluhr R (1997) The I2C family from the wilt disease resistance locus I2 belongs to the nucleotide binding, leucine-rich repeat superfamily of plant resistance genes. Plant Cell 9:521–532

Pfitzner AJP (1998) Transformation of tomato. In: Foster GD, Taylor SC (eds) Methods in Molecular Biology, vol 81. Plant virology protocols: from virus isolation to transgenic resistance. Humana Press, Totowa, NJ, pp 359–363

Picton S, Barton S, Bouzayen M, Hamilton AJ, Grierson D (1993) Altered fruit ripening and leaf senescence in tomatoes expressing an antisense ethylene-forming enzyme transgene. Plant J 3:469–481

Piechulla B, Merforth N, Rudolph B (1998) Identification of tomato Lhc promoter regions necessary for circadian expression. Plant Mol Biol 38:655–666

Pnueli L, Hareven D, Rounsley SD, Yanofsky MF, Lifschitz E (1994) Isolation of the tomato AGAMOUS gene TAG1 and analysis of its homeotic role in transgenic plants. Plant Cell 6:163–173

Redenbaugh K, Hiatt W, Martineau B, Emlay D (1995) Determination of the safery of genetically engineered crops. ACS Symp Ser 605. Genetically modified foods safety issues / American Chemical Society, Washington, DC, pp 72–87

Reed AJ, Kretzmer KA, Naylor MW, Finn RF, Magin KM, Hammond BG, Leimgruber RM, Rogers SG, Fuchs RL (1996) Safety assessment of 1-aminocyclopropane-1-carboxylic acid deaminase protein expressed in delayed ripening tomatoes. J Agric Food Chem 44:388–394

Ronald PC (1998) Resistance gene evolution. Curr Opin Plant Biol 1:294–298

Santino CG, Stanford GL, Conner TW (1997) Developmental and transgenic analysis of two tomato fruit enhanced genes. Plant Mol Biol 33:405–416

Schneider M, Ow DW, Howell SH (1990) The in vivo pattern of firefly luciferase expression in transgenic plants. Plant Mol Biol 14:935–947

Schuch W, Drake R, Romer S, Bramley PM (1996) Manipulating carotenoids in transgenic plants. Ann NY Acad Sci 782:1–19

Shahin EA, Sukhapinda K, Simpson RB, Spivey R (1986) Transformation of cultivated tomato by a binary vector in *Agrobacterium rhizogenes*: transgenic plants with normal phenotypes harbor binary vector T-DNA, but no Ri-plasmid T-DNA. Theor Appl Genet 72:770–777

Simons G, Groenendijk J, Wijbrandi J, Reijans M, Groenen J, Diergaarde P, van-der Lee T, Bleeker M, Onstenk J, de-Both M, Haring M, Mes J, Cornelissen B, Zabeau M, Vos P, van-der Lee T, De-Both M (1998) Dissection of the *Fusarium I2* gene cluster in tomato reveals six homologs and one active gene copy. Plant Cell 10:1055–1068

Speirs J, Lee E, Holt K, Kim Y-D, Scott NS, Loveys B, Schuch W, Kim Y-D (1998) Genetic manipulation of alcohol dehydrogenase levels in ripening tomato fruit affects the balance of some flavor aldehydes and alcohols. Plant Physiol 117:1047–1058

Stark DM, Barry GF, Kishore GM (1996) Improvement of food quality traits through enhancement of starch biosynthesis. Ann NY Acad Sci 792:26–36

Stevens AM, Rick CM (1986) Genetics and breeding. In: Atherton JG, Rudich J (eds) The tomato crop. Chapman and Hall, London, pp 35–109

Taylor IB (1986) Biosystematics of the tomato. In: Atherton JG, Rudich J (eds) The tomato crop. Chapman and Hall, London, pp 1–34

Thakur BR, Singh RK, Handa AK (1996a) Effect of an antisense pectin methylesterase gene on the chemistry of pectin in tomato (*Lycopersicon esculentum*) juice. J Agric Food Chem 44:628–630

Thakur BR, Singh RK, Tieman DM, Handa AK (1996b) Tomato product quality from transgenic fruits with reduced pectin methylesterase. J Food Sci 61:85–87,108

Theologis A (1992) One rotten apple spoils the whole bushel: the role of ethylene in fruit ripening. Cell 70:181–184

Theologis A, Oeller PW, Wong L-M, Rottmann WH, Gantz DM (1993) Use of a tomato mutant constructed with reverse genetics to study fruit ripening, a complex developmental process. Dev Genet 14:282–295

Tieman DM, Klee HJ (1999) Differential expression of two novel members of the tomato ethylene-receptor family. Plant Physiol 120:165–172

Tomato News (1998) World production: preliminary results for 1997 and forcasts for 1998. 1:5–36

Torisky RS, Kovacs L, Avdiushko S, Newman JD, Hunt AG, Collins GB (1997) Development of a binary vector system for plant transformation based on the supervirulent *Agrobacterium tumefaciens* strain Chry5. Plant Cell Rep 17:102–108

Toyoda H (1993) Transformation of tomato (*Lycopersicon esculentum* Mill.) for virus disease protection. Biotechnol Agric For 23:259–272

Ueda T, Pichersky E, Malik VS, Cashmore AR (1989) Level of expression of the tomato rbcS-3A gene is modulated by a far upstream promoter element in a developmentally regulated manner. Plant Cell 1:217–227

Van Haaren MJJ, Houck CM (1991) Strong negative and positive regulatory elements contribute to the high-level fruit-specific expression of the tomato *2A11* gene. Plant Mol Biol 17:615–630

Van Roekel JSC, Damm B, Melchers LS, Hoekema A (1993) Factors influencing transformation frequency of tomato (*Lycopersicon esculentum*). Plant Cell Rep 12:644–647

Weide R, Koornneef M, Zabel P (1989) A simple, nondestructive spraying assay for the detection of an active kanamycin resistance gene in transgenic tomato plants. Theor Appl Genet 78:169–172

Whitham S, McCormick S, Baker B (1996) The *N* gene of tobacco confers resistance to tobacco mosaic virus in transgenic tomato. Proc Natl Acad Sci USA 93:8776–8781

Zambryski P (1988) Basic processes underlying *Agrobacterium*-mediated DNA transfer to plant cells. Annu Rev Genet 22:1–30

Zambryski P, Joos H, Genetello C, Leemans J, Van Montagu M, Schell J (1983) Ti plasmid for the introduction of DNA into plant cells without alternation of their normal regeneration capacity. EMBO J 2:2143–2150

16 Transgenic Cassava (*Manihot esculenta* Crantz)

C. Schöpke[1], N.J. Taylor[2], R. Cárcamo[3], A.E. González[4], M.V. Masona[5], and C.M. Fauquet[2]

1 Introduction

Cassava (*Manihot esculenta* Crantz) plays a significant role as a carbohydrate source in many tropical countries. In 1997 the world production was an estimated 166.4 million tons of fresh root (FAO/GIEWS 1998). Due to the importance of cassava for the livelihood of millions of people, interest in this crop has increased over the last 15–20 years (Cooke and Cock 1989). It has been recognized that genetic engineering might be an efficient tool to resolve some of the problems affecting its cultivation (Roca 1989; Bertram 1990). The ability to modify a single trait without changing the principal characteristics of a given cassava cultivar is particularly advantageous in a crop where there are thousands of local cultivars (Iwanaga and Iglesias 1994; Ng et al. 1994). Although cassava breeding programs have had success in providing new cultivars to cassava growers (Jennings and Hershey 1985; Hershey and Jennings 1992), a high degree of heterozygosity, irregular flowering in many cultivars, low seed set and variable germination rates have impeded faster progress. Genetic engineering has the potential to complement traditional cassava breeding programs in several areas, such as increased resistance to pests and diseases, improved starch quality (both for bread-making and industrial purposes), modified content of cyanogenic glucosides, increased protein content and extended shelf-life of the harvested tubers (Thro et al. 1996).

In 1993 genetic transformation in cassava was limited to the recovery of chimeric embryos (Schöpke et al. 1993b). Since that time major progress has resulted in four research groups reporting the recovery of genetically transformed cassava plants (Li et al. 1996; Raemakers et al. 1996; Sarria et al. 2000; Schöpke et al. 1996). This chapter discusses cassava regeneration as it relates to transformation, and the advances that have been achieved during the last decade in cassava gene transfer technologies.

[1] Department of Botany and Plant Sciences, University of California, Riverside, California 96521, USA
[2] ILTAB/Donald Danforth Plant Science Center, 8001 Natural Bridge Road, St. Louis, Missouri 63121, USA
[3] Akkadix Corporation, 11099 North Torrey Pines Road, Ste. 200, La Jolla, California 92037, USA
[4] The Scripps Research Institute, Department of Cell Biology, 10550 North Torrey Pines Road, La Jolla, California 92037, USA
[5] University of Zimbabwe, Crop Science Department, MP 167, Harare, Zimbabwe

Biotechnology in Agriculture and Forestry, Vol. 47
Transgenic Crops II (ed. by Y.P.S. Bajaj)
© Springer-Verlag Berlin Heidelberg 2001

2 Cassava Regeneration Systems (Table 1)

2.1 Embryogenic Systems

2.1.1 Organized Embryogenic Structures

Numerous publications have described the induction of somatic embryos from various explants (summarized in Table 1). In these systems, primary explants are placed onto MS basal culture medium (Murashige and Skoog 1962) supplemented with auxins, which induce the formation of organized embryogenic structures (OES). These structures can either be regenerated directly or after the induction of secondary embryogenesis. Repetitive subculture of OES results in embryo proliferation ("secondary embryogenic cultures", Szabados et al. 1987a; "embryo clumps", Mathews et al. 1993; "cyclic embryos", Raemakers et al. 1993a). For a more detailed description of somatic embryogenesis through OES and plant regeneration in cassava see Raemakers et al. (1997a).

2.1.2 Friable Embryogenic Callus and Embryogenic Suspension Cultures

An alternative somatic embryogenic system in cassava was developed by Taylor et al. (1996). OES are produced as described above and transferred to an auxin-containing medium with GD basal salts (Gresshoff and Doy 1974) instead of MS basal salts. In some OES cultures this leads to the development of a pale yellow, highly friable embryogenic callus (FEC) that can be maintained on GD medium. This type of callus readily disperses when transferred to liquid medium and can be used to establish embryogenic suspension cultures. Mature embryos and plants have been regenerated from both the FEC and embryogenic suspension cultures (Raemakers et al. 1996, 1997b; Schöpke et al. 1996, 1997a; Taylor et al. 1996; González et al. 1998; Munyikwa et al. 1998).

Friable embryogenic callus and embryogenic suspensions were initially obtained from two cultivars that in general respond well to in vitro conditions (Taylor et al. 1996). However, Taylor et al. (1997b) expanded the range of cultivars for which embryogenic suspensions have been established to eight. Currently, embryogenic suspensions of 14 cultivars are available (Taylor, unpubl. results), and there seem to be no major barriers to increasing this number. An alternative source for FEC has been described recently (Groll et al. 1997); thin cell layers taken from internodes or petioles were reported to produce FEC, from which embryos could be regenerated.

2.1.3 Protoplasts

Cassava protoplasts have been isolated from leaves (Shahin and Shepard 1980; Mabanza and Jonard 1983; Anthony et al. 1995; McDonnell and Gray 1997),

Table 1. Plant regeneration in cassava

Explant	Regeneration pathway	Reference
Embryo axes and cotyledons from zygotic embryos	Embryogenesis via organized embryogenic structures	Stamp and Henshaw (1982)
Young leaves	"	Stamp and Henshaw (1986)
Shoot tips, young leaves	"	Szabados et al. (1987a)
Young leaves	"	Taylor et al. (1993)
Young leaves	"	Raemakers et al. (1993b)
Young leaves	"	Mathews et al. (1993)
Cotyledons from zygotic embryos	"	Konan et al. (1994a)
Meristems	"	Puonti-Kaerlas et al. (1997)
Young leaves	"	Sofiari et al. (1997)
Organized embryogenic structures	Embryogenesis via friable embryogenic callus	Taylor et al. (1996, 1997b)
Protoplasts from embryogenic suspensions	"	Sofiari et al. (1998)
Thin layer explants from stems and petioles[a]	"	Groll et al. (1997)
Callus from internodes	Adventitious shoot formation	Tilquin (1979)
Cotyledons from zygotic embryos	"	Mabanza and Jonard (1984)
Cotyledons from zygotic embryos	"	Konan et al. (1994a)
Cotyledons from somatic embryos	"	Li et al. (1998)
Shoot tips	Multiple shoot formation	Mireles and Paez de Casares (1984)
Nodal explants	"	Bhagwat et al. (1996)
Nodal explants, meristems	"	Konan et al. (1994b, 1997)
Nodal explants, meristems	"	Puonti-Kaerlas et al. (1997)

[a] Regeneration is described up to the stage of heart-shaped embryos.

embryos (Szabados et al. 1987b), and from embryogenic suspension cultures (Verdaguer et al. 1996; Sofiari et al. 1998). Shahin and Shepard reported the development of protoplast-derived callus from which shoots were regenerated as early as 1980. However, these results were not reproducible in any other laboratory. Recently, plant regeneration from cassava protoplasts was described by Sofiari et al. (1998), who utilized embryogenic suspension cultures (Taylor et al. 1996) as a source for protoplast isolation. It can be assumed that optimization of this system will make it amenable to genetic transformation, and indeed electroporation of these protoplast cultures has resulted in transgenic cassava plants (Raemakers et al. 2000).

2.2 Organogenic Systems

2.2.1 Shoot Formation from Nodal Explants

Several reports describe multiple shoot formation from nodal explants of cassava (Mabanza and Jonard 1984; Mireles and Paez de Casares 1984; Konan et al. 1994b, 1997; Bhagwat et al. 1996). The results of Bhagwat et al. (1996) and Konan et al. (1997) are interesting in the context of genetic transformation because, in addition to shoot development from existing meristems, multiple de novo shoot formation through a type of "meristematic callus" was observed by both groups. Preliminary experiments indicated that this system might be useful for *Agrobacterium*-mediated transformation (Konan et al. 1995).

2.2.2 De Novo Shoot Formation

Tilquin (1979) described de novo shoot development from internode-derived callus of cassava, but these results could not be repeated by other workers. Stamp and Henshaw (1986) observed neoformation of shoots at a low frequency on cotyledon explants derived from zygotic embryos. Cotyledons from somatic embryos can be induced to produce shoots at high frequencies, either on medium with BAP alone (Konan et al. 1994b), or with a combination of BAP and indole-3-butyric acid (IBA; Li et al. 1998).

3 Transformation

3.1 Stable Transformation of Cassava (Table 2)

Since the major focus of this review is the production of transgenic cassava plants, the stable transformation of non-regenerable cassava tissues will not be discussed here in detail. The reader is referred to the references given in Table 2.

Table 2. Stable expression of foreign genes in transformed cassava tissues and plants

Target tissue	Transf. method	Gene(s) inserted	Transgenic product	Reference
Leaf, stem, somatic embryos	A	nptII, bar, uidA	Callus	Calderón (1988)
Somatic embryos	A	nptII, uidA	Chimeric embryos	Chavarriaga et al. (1993)
Leaves	A	nptII, uidA, ACMV-CP, CsCMV-CP	Callus	Schöpke et al. (1993b)
Embryogenic suspensions	P	nptII, uidA	Plants	Schöpke et al. (1996)
Embryogenic suspensions	P	bar, luc, nptII	Plants	Raemakers et al. (1996)
Cotyledons from som. embryos	A	nptII, uidA, hpt	Plants	Li et al. (1996)
Embryogenic suspensions	A	nptII, uidA	Plants	González et al. (1998)
Embryogenic suspensions	P	luc, pat, cassava AGPase B antisense gene	Plants	Munyikwa et al. (1998)
Cotyledons from som. embryos	A	nptII, bar, uidA	Plants	Sarria et al. (2000)
Embryogenic suspensions	P	nptII, CsCMV-CP	Plants	Schöpke et al. (2000)
Embryogenic suspensions protoplasts	E	pat, luc	Plants	Raemakers et al. (2000)

A Agrobacterium-mediated transformation, *E*, electroporation, *P* particle bombardment, *ACMV-CP* African cassava mosaic virus coat protein, *AGPase B* ADP glucose pyrophosphorylase B, *bar*, *pat* phosphinotricin acetyl transferase, *CsCMV-CP* cassava common mosaic virus coat protein, *hpt* hygromycin phosphotransferase, *luc* luciferase, *nptII* neomycin phosphotransferase, *uidA*, ß-glucuronidase.

3.1.1 Transformation of Embryogenic Suspension Cultures (cv. TMS 60444)

Failure to obtain non-chimeric transgenic cassava plants using organized embryogenic structures (OES) as the target tissue for gene insertion can be attributed to the developmental process that leads to OES. Unlike in walnut (Polito et al. 1989), embryogenesis is not initiated from single cells, but by groups of cells that act together to form a new embryo (Stamp 1987; Raemakers et al. 1995). This means that, by default, any transformation procedure using OES will result initially in chimeric embryogenic tissues. Only subsequent, repetitive and prolonged induction of embryos from this chimeric tissue will eventually result in completely transformed embryos. In contrast, embryos produced through FEC and embryogenic suspensions seem to originate from single cells, and therefore the probability of obtaining fully transformed plants is much greater.

Schöpke et al. (1996) used particle bombardment to introduce a plasmid containing the *npt*II and *uid*A genes into tissues derived from embryogenic suspension cultures (Fig. 1A–F). The *npt*II gene was chosen as a selectable marker gene since assays with control tissues indicated that the antibiotic hygromycin (corresponding resistance gene: *hpt*) and the herbicide phosphinotricin (PPT; corresponding resistance gene: *bar*) were less efficient in killing non-transgenic tissue. Among the aminoglycosides against which *npt*II provides resistance, paromomycin was found to be superior to both geneticin and kanamycin.

The regeneration of paromomycin-selected tissue was achieved through differentiation and maturation of somatic embryos followed by subsequent induction of single or multiple shoots on MS medium supplemented with 4.4 µM BAP, and rooting of shoots on growth regulator-free MS medium (Fig. 1E). Southern blot hybridization analysis of four suspension lines and two plant lines confirmed stable integration of the introduced DNA. Expression of the *uid*A gene driven by the 35S promoter, as measured by histological GUS assays for the gene product, β-glucuronidase (GUS), was detected in all plant tissues, but the highest expression was found in vascular tissues and young leaves (Fig. 1F).

Raemakers et al. (1996) investigated the effect of different bombardment and culture parameters on transient and stable expression of the firefly luciferase gene (*luc*) after particle bombardment of embryogenic suspension cultures. Continuous selection and subculture of light-emitting tissue eventually resulted in cultures consisting totally of transformed tissue. Differentiation and maturation of somatic embryos occurred on an MS-based medium supplemented with a complex mixture of organic components in addition to 4.14 µM picloram and 0.43 µM adenine sulfate. LUC-positive mature embryos were multiplied through secondary embryogenesis, from which LUC-positive plants were eventually regenerated. Four shoots originating from different transformation events were analyzed by Southern blot analysis, confirming stable integration of the *luc* gene.

González et al. (1998) applied *Agrobacterium*-mediated transformation to introduce the *uid*A-intron and *npt*II genes into tissue derived from embryo-

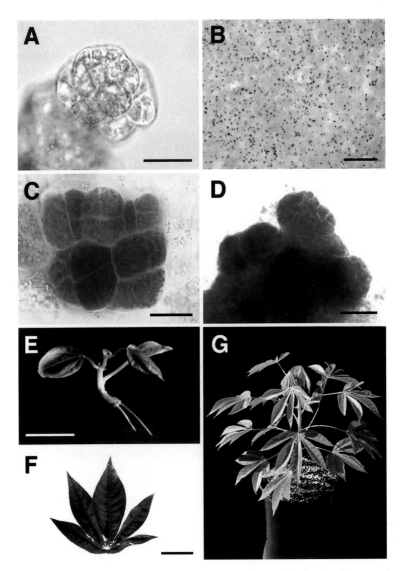

Fig. 1A–G. Cassava transformation through microbombardment of embryogenic suspension cultures. **A** Embryogenic unit containing about 15 cells from an embryogenic suspension culture of cassava. **B** Embryogenic suspension-derived tissue 3 days after bombardment with the *uid*A gene. GUS-expressing cells are stained *blue*. **C** Actively dividing, GUS-positive cells, 2 weeks after bombardment. **D** GUS-positive friable embryogenic callus 7 weeks after bombardment. **E** Regenerated plantlet 6 months after bombardment. **F** GUS expression in a young leaf of a transgenic cassava plant 12 months after bombardment. **G** Transgenic cassava plant expressing the CsCMV coat protein gene, 2 months after transfer to soil. *Bars* 50 μm in **A** and **C**, 2 mm in **B**, 100 μm in **D**, 10 mm in **E** and **F**. (**E** from Schöpke et al. 1996, **F** from Vasil 1996, reprinted by permission of Nature Publishing Co.; **G** from Schöpke et al. 2000, reprinted by permission of EMBRAPA)

genic suspension cultures. Apart from the co-culture procedures and the use of antibiotics to remove *Agrobacterium*, the selection and regeneration protocol was very similar to that used for microbombarded embryogenic tissue (Schöpke et al. 1996). Genomic DNA of five putative transgenic lines of embryogenic suspensions and two lines of regenerated plants was subjected to Southern blot analysis. In all cases, integration of the *uid*A gene was confirmed.

A cassava ADP glucose pyrophosphorylase gene (AGPase B), in antisense orientation in combination with the *luc* gene as a selectable marker, was inserted into cassava embryogenic suspensions by particle bombardment (Munyikwa et al. 1998). Selection for light-emitting tissue eventually resulted in the regeneration of 21 plants carrying the luciferase gene. Since AGPase is involved in starch formation, the presence of the AGPase antisense gene in these plants should result both in a decrease in starch accumulation and in a reduction of AGPase mRNA. Staining for starch indicated that indeed most of the transgenic plants showed a reduced starch accumulation, and Northern blot analysis with several transformants revealed a strong reduction in mRNA levels for the AGPase gene.

3.1.2 Transformation of Cotyledons from Somatic Embryos

Cotyledons from somatic embryos of cassava cv. MCol 22 have been the target for *Agrobacterium*-mediated transformation of cassava (Li et al. 1996). The plasmids used for transformation contained an *uid*A-intron gene controlled by different versions of the cauliflower mosaic virus (CaMV) 35S promoter and the *hpt* or *npt*II genes as selectable markers. Optimal transient expression of *uid*A was observed when cotyledon pieces were co-cultivated for 4 days with *Agrobacterium* strain LBA4404. After co-cultivation, explants were subcultured onto MS medium containing 2 μM cupric sulfate (Schöpke et al. 1993a), 4.4 μM BAP and 2.5 μM IBA. To select for transformed tissue the medium was supplemented with either 15 mg/l hygromycin or 20 mg/l geneticin. Both antibiotics were able to suppress shoot regeneration of non-transformed tissue and permit the formation of GUS-positive shoot primordia. In three independent experiments in which 1735 explants were transformed with the *uid*A and *hpt* genes, 30 shoots were regenerated after selection with hygromycin. Out of these, six were GUS-positive. Genomic DNA from five of the GUS-positive shoots was subjected to Southern blot analysis and was shown to contain the expected fragments. Northern blot analysis of one of these plants showed signals corresponding to the *uid*A and *hpt* transcripts.

Sarria et al. (2000) used cotyledons obtained from somatic embryos of cassava cv. M Peru 183 for transformation with the wild *Agrobacterium tumefaciens* strain CIAT 1182 in conjunction with the binary vector PGV1040. This vector contains the *npt*II, *uid*A and *bar* genes. After co-cultivation with *Agrobacterium*, the cotyledon explants were transferred to embryo induction medium with 82–164 μM PPT. Eventually, plants were regenerated from

PPT-resistant embryos. Integration of the *uid*A and *bar* genes in one of the PPT-resistant plants was verified by Southern blot analysis.

3.2 Expression of Transgenes in Cassava Tissues and Plants

The stability in time and space of the expression patterns of introduced genes in transgenic crop plants is crucial for their usefulness under field conditions. This is especially important in a vegetatively propagated crop like cassava where stable long-term transgene expression is an absolute requirement. Though transgenic cassava plants are available, information on the stability of transgene expression in this crop is still limited. Described here are results with the expression of the *uid*A gene under the control of the enhanced 35S promoter in cells, tissues and plants derived from microbombarded embryogenic suspension cultures of cassava.

GUS activity was detected using a modified histological GUS assay, originally devised by Jefferson (1987). Depending on the amount of oxidation catalysts ferri- and ferrocyanide added to the assay buffer, and on the incubation time in tissues transformed with the *uid*A gene, either only cells with a strong GUS-expression (assay A; 6.4 mM catalysts, 2 h) or all cells (assay B; 0.64 mM catalysts, 16 h) produced a dark blue stain (Schöpke et al. 1997b). Interpretations of the results of histological GUS assays with transgenic cassava tissues therefore must take into account this strong correlation between assay conditions and formation of the blue stain.

Three days after bombardment of cassava embryogenic suspension cultures with a plasmid containing the *uid*A gene (Schöpke et al. 1996) GUS expression was observed in many single cells (assay A; Fig. 1B). During development in selection medium with 25 μM paromomycin, using assay A, dark blue cell aggregates (Fig. 1C), and later FEC with dark blue cell clusters were found. When assay A was used to detect GUS expression in regenerated plants (12 months after transformation), most of the tissue stained weakly blue. Only certain cell types, i.e., guard cells (Fig. 2A), lactifers (Fig. 2C) and xylem parenchyma cells (Fig. 2D) stained dark blue (unpubl. results), giving the impression of cell-specific GUS-expression. However, when assay B was applied to the same tissues, the development of a dark blue stain was observed in all of the cells (Fig. 2B, E, F).

In a plant line transgenic for *uid*A, the pattern of GUS expression 30 months after the transformation is essentially identical to that observed 12 months after transformation. GUS is expressed in all tissues, including young leaves, stems (Fig. 2G), roots and tubers (Fig. 2H). Although gene expression was stable in transgenic cassava plants for a period of more than 2 years, long-term observations are still required to confirm the observed expression patterns.

Verdaguer et al. (1996) isolated two fragments (CVP1 and CVP2) from the promoter region of cassava vein mosaic virus (CsVMV). Although no information is available yet about long-term gene expression in cassava

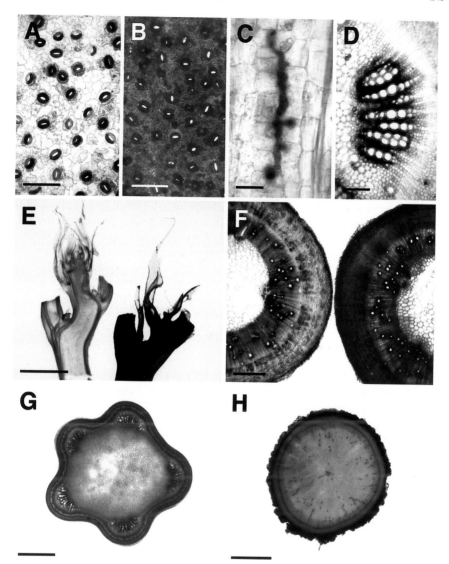

Fig. 2A–H. Cassava tissues transformed with the *uid*A gene and subjected to histological GUS assays. Two assay conditions were used: (1) assay buffer with 6.4 mM ferri/ferrocyanide, assay duration 2 h; (2) assay buffer with 0.64 mM ferri/ferrocyanide, assay duration 16 h. Under condition (1) only cells with a high GUS expression stain dark blue, while under condition (2) this is true for most of the GUS expressing cells. **A** Epidermis of a young leaf (1); only the stomata stain dark blue. **B** Epidermis of the same leaf as shown in **A** (2); most cells are dark blue. **C** Strong GUS-expression in a lactifer in a longitudinal section of a young shoot (1). **D** Vascular tissue in a close-up of a cross section of a young shoot (1). The xylem parenchyma is stained dark blue. **E** Longitudinal sections of a shoot tip. In the section on the *left* (1) only the vascular bundles are dark blue, in the section on the *right* all cells are dark blue (2). **F** Cross sections of a shoot after the onset of secondary xylem formation. Section on the *left* (1), section on the *right* (2). **G** Cross section through a young shoot, 30 months after transformation (2). **H** Cross section through a root tuber, 30 months after transformation (2). *Bars* 150 µm in **A** and **B**, 50 µm in **C**, 200 µm in **D**, 5 mm in **E**, 1 mm in **F** and **G**, 5 mm in **H**. (**A**, **B**, **C**, and **E** from Schöpke and Fauquet 1999, reprinted by permission of Oxford University Press)

plants driven by this new promoter, studies in tobacco and rice plants transformed with a CVP2-*uid*A gene construct (Verdaguer et al. 1996) have shown that the GUS expression pattern was very similar to that observed after transformation with *uid*A controlled by the frequently used 35S promoter. Strong GUS expression was observed in vascular tissues, mesophyll, and in root tips. The CsVMV promoter can therefore be used as an alternative to the 35S promoter, and the International Laboratory for Tropical Agricultural Biotechnology (ILTAB; Taylor et al. 1997a) is making it available for use in less developed countries where it is hoped to have an important impact on the improvement of tropical crops. A detailed analysis of the CsVMV promoter revealed that it is composed of different regions with distinct tissue-specific functions (Verdaguer et al. 1998). Transformation of tobacco with a series of truncated promoter: *uid*A fusion genes allowed the identification of promoter sequences that confer strong tissue-specific GUS expression in roots and vascular tissues. Ongoing investigations are aimed at using these sequences to control gene expression in cassava in a tissue-specific manner.

3.3 Transient Gene Expression – Promoter Studies (Table 3)

3.3.1 Particle Bombardment

Different promoters (35S, e35S, 4Oe35S, UBQ1) fused to the *uid*A gene were bombarded into cassava leaves of cv. Señorita with a pneumatic particle gun (Franche et al. 1991; Schöpke et al. 1993b) in order to study their efficiency in cassava tissue. Transient gene expression was measured 24 h after bombardment with a fluorometric GUS assay using methylumbelliferone glucuronide (MUG) as the substrate. The highest GUS levels were found after bombardment with e35S and 4Oe35S constructs, causing the accumulation of about three times more GUS compared with the 35S and UBQ1 promoters. Differences in activities measured fluorometrically were reflected in differences in the diameter of blue spots obtained in histological GUS assays (Jefferson 1987), i.e., higher activities measured with MUG assays corresponded to larger diameters of blue spots.

    If genetic transformation of cassava is aimed at modifying gene expression in roots and tubers, it is desirable to have a system for testing gene promoters in transient expression assays in these organs. Arias-Garzon and Sayre (1993) bombarded cassava leaves and roots of the cultivars MCol 2215 and MVen 25 with a 35S-*uid*A gene construct. Histological GUS assays on the day following bombardment revealed a relatively high number of blue spots on leaves, but virtually none in roots. Similar results were obtained with the luciferase gene (*luc*), with high expression in leaves, but low expression in roots. When the *uid*A gene coupled to a soybean root-specific glutamin synthetase promoter was employed to drive expression, no GUS was observed in either root or shoot tissue. Subsequent experiments on the effect of leaf and root extracts on gene expression and on DNA degradation led the authors to

Table 3. Transient expression of foreign genes in cassava cells and tissues

Transformation method	Target tissues/cells	Promoter(s)	Marker gene(s)	Reference
Particle bombardment	Leaves	35S, e35S, 4Oe35S, Ubq1	*uidA*	Franche et al. (1991)
"	Leaves, roots	35S, GS	*uidA, luc*	Arias-Garzon and Sayre (1993)
"	Embryogenic suspensions	e35S	*uidA*	Schöpke et al. (1997b)
"	Somatic embryos	35S, 35S with maize Adh1 intron, AHC, Act1, Spo, ßAmy, Pat		Luong et al. (1997)
"	Meristems	35S, e35S, 35SΩ, e35SΩ	*uidA, luc*	Puonti-Kaerlas et al. (1997)
"	Root cambium	35S, carrot invertase	*uidA*	Bohl et al. (1997)
Electroporation	Leaf protoplasts	35S	*uidA*	Cabral et al. (1993)
"	Somatic embryos	35S	*uidA*	Luong et al. (1995)
"	Leaves, protoplasts from embryogenic suspensions	CsVMV promoter fragments A and B	*uidA, luc*	Verdaguer et al. (1996)
Agrobacterium	Mesophyll protoplasts	35S	*uidA*	McDonnell and Gray (1997)

Abbreviations for promoters: *35S* cauliflower mosaic virus *e35S 35S* including a partial duplication of the upstream region of 35S *4Oe35S*, e35S containing a tetramer of the octopine synthase enhancer, *Ubq1* ubiquitin 1 from *Arabidopsis thaliana*, *Adh1* maize alcohol dehydrogenase 1, *AHC* maize ubiquitin, *Act1* rice actin 1, *GS* soybean glutamin synthetase, *Spo* sweet potato sporamin, *Pat* potato patatin, *ßAmy* sweet potato ß-amylase, *35SΩ 35S* containing the translational enhancer Ω of tobacco mosaic virus, *e35SΩ* e35S containing the translational enhancer Ω of tobacco mosaic virus.

conclude that cassava roots contain high levels of DNAse activity. This would reduce or eliminate transient gene expression to such a degree that cassava root tissues are unsuitable for transient gene expression studies with root-specific promoters. The fact that roots of transgenic plants transformed with *uid*A gene under the control of the 35S promoter are GUS-positive (Schöpke et al. 1996) confirms that the inhibition of transient gene expression in roots is not due to reduced expression levels in roots per se, but due to other factors.

Conditions for particle bombardment of tissue derived from embryogenic suspension cultures of cassava cv. TMS 60444 have been established and optimized by Schöpke et al. (1997b). The effect of certain parameters (bombardment pressure, particle size, number of bombardments per sample and osmotic treatment of the tissue) on transient expression of the *uid*A gene were investigated by comparing the numbers of blue spots resulting from histological GUS assays 3 days after bombardment. Blue spots were counted by employing an image processing system. This allowed the evaluation of a large number of samples in a standardized manner and in a relatively short time. The optimal bombardment parameters were found to be: 1100 psi bombardment pressure, 1.0 μm particle size, two bombardments per sample, and an osmotic treatment with 0.1 M sorbitol and 0.1 M mannitol. When used in combination, these treatments resulted in an average number of 1350 blue spots per cm^2 of bombarded sample of embryogenic suspension-derived tissue.

Luong et al. (1997) compared the effect of both constitutive and root-specific promoters (Table 3) on the expression of the *uid*A gene in microbombarded embryogenic tissues, stems, and leaves of cassava cv. MCol 1505 (syn. CMC 76). The number of blue spots that developed per microbombarded tissue sample after histological GUS assays was used as a measure for promoter activity. Gene constructs based on the 35S and the maize ubiquitin promoters caused relative high gene expression in all tissues, while the rice actin promoter permitted expression only in embryogenic tissue. The root-specific promoters Pat and βAmy resulted in high GUS expression in embryogenic tissue, and comparatively low expression in stems and leaves. The third root-specific promoter, Spo, was active at similar levels both in embryogenic tissue and leaves, but less so in stems. The insignificant level of transient gene expression found in bombarded root and tuber samples in this study is attributed to DNA degradation by these tissues, confirming the results of Arias-Garzon and Sayre (1993).

Puonti-Kaerlas et al. (1997) used a particle inflow gun (Vain et al. 1993) to investigate the efficiency of shoot meristem transformation in cassava. After bombardment with a particle preparation containing a range of sizes, particles were found in cell layers L1–L3 and deeper. Bombardment with the *uid*A gene resulted in up to 50% of bombarded meristems showing 2–8 blue spots per meristem after GUS assays. Bombardment with the *luc* gene made it possible to screen for gene expression in living tissues and thus eliminate non-expressing explants. Out of 703 explants used for bombardment, two were positive in a luciferase assay after 4 weeks and were growing normally.

Bohl et al. (1997) developed a system for transient gene expression in cassava root tissue. Root cambium explants were bombarded with a 35S-*uid*A gene construct using a particle inflow gun (Vain et al. 1993). Transient GUS assays 2–3 days after bombardment revealed many blue spots in the cambium and very few in other parts of the root. The contradicting observations by Arias-Garzon and Sayre (1993) and Luong et al. (1997) that GUS expression in microbombarded root cells controlled by the 35S promoter was very rare can be explained by the fact that in these studies whole in vitro roots were bombarded, and that the cambium therefore was not accessible to the particles. Root cambium was also bombarded with the *uid*A gene fused to the root-specific carrot invertase clone SI and SII promoters. However, in this experiment no GUS expression was observed.

3.3.2 Electroporation

Conditions for the electroporation of leaf protoplasts with the *uid*A gene were established by Cabral et al. (1993). The highest levels of GUS expression were achieved after electroporating at 1250 V/cm at 25 µF for a time of 12.8 ms. Electroporation of intact embryogenic cassava tissues with DNA as an alternative to particle bombardment and *Agrobacterium*-mediated transformation was investigated by Luong et al. (1995). Optimal transient expression of the *uid*A gene was observed when torpedo-shaped embryos, cut into pieces about 2 mm in diameter, were preincubated in electroporation cuvettes (BioRad) with 300 µl potassium aspartate buffer containing 62.5 mg/l DNA, and electroporated with one pulse at a field strength of 750 V/cm. This treatment resulted in 75% of tissue pieces producing blue spots in histological GUS assays 4 days after electroporation, at an average of 174 spots per tissue fragment. Regeneration from embryo tissues electroporated with a plasmid containing the *uid*A gene did not result in GUS-positive plantlets (Luong et al. 1997).

3.3.3 Agrobacterium-*Mediated Transformation of Protoplasts*

McDonnel and Gray (1997) transformed cassava leaf protoplasts by adding a culture of *Agrobacterium* strain LBA4404 containing the *uid*A gene to the enzyme solution for protoplast isolation. After overnight incubation, the protoplasts were separated from the bacteria by centrifugation on a density gradient. The protoplasts were then cultured in a medium supplemented with 50 mg/l carbenicillin and cefotaxim. GUS assays 1 day after the protoplast isolation resulted in GUS levels about 75 times higher than in controls. After 1 week these levels had decreased by about a factor of three. Molecular analysis is still required to demonstrate integration of the *uid*A gene into the cassava genome.

4 Cassava Transformation with Genes of Agronomic and Economic Interest

4.1 Resistance Against Viruses

Despite the absence of a system for regenerating transgenic cassava plants until very recently, research regarding genes of interest for cassava transformation has been ongoing for a number of years using tobacco as a model system. Fauquet et al. (1993) employed *Agrobacterium*-mediated transformation of *Nicotiana benthamiana*, which can be infected by both African cassava mosaic virus (ACMV) and cassava common mosaic virus (CsCMV), to study the expression of the viral coat proteins (CPs) and their ability to provide protection against the respective viruses. Plants transgenic for the ACMV-CP gene were shown to contain low levels of mRNA corresponding to the coding sequence of the ACMV-CP gene. Accumulation of the CP was detectable by western blots, but it was relatively low. Challenge of the CP positive plants with ACMV resulted in some degree of resistance at a virus concentration of 20–100 ng/ml. On the other hand, plants transgenic for the CsCMV-CP gene accumulated the CP to levels of up to 2% of total protein, and some plant lines showed a very high resistance to infection with CsCMV.

With the development of a routine genetic transformation capability at ILTAB, cassava plants expressing the CsCMV-CP gene have been produced (Fig. 1G). Recent experiments with microbombarded embryogenic suspension-derived tissues have resulted in the regeneration of more than 30 plants confirmed by PCR to contain this gene. Western blot analysis of these plant lines has revealed varying levels of expression of the CsCMV coat protein (Schöpke et al. 2000).

4.2 Modification of Starch Synthesis

The starch deposited in the tuberous roots of cassava and in the storage organs of other plant species consists mainly of two components; amylose (a linear glucose polymer) and amylopectin (a branched glucose polymer). Numerous ways to modify both starch quantity and quality have been proposed (reviewed by Visser and Jacobsen 1993; Munyikwa et al. 1997). Salehuzzaman et al. (1993) employed the antisense RNA strategy (van der Krol et al. 1989) to investigate the effect of a gene isolated from cassava on starch synthesis in transgenic plants. Potato plants were used as the test system, as cassava transformation was not possible at the time of this study. A tuber-specific cDNA library of cassava was constructed, and the cDNA for granule-bound starch synthase (GBSS), the enzyme that catalyzes the synthesis of amylose in reserve starch, was cloned. *Agrobacterium*-mediated transformation of potato plants with GBSS antisense cDNA resulted in a reduction of the amylose content of potato tubers in transgenic plants. In some cases the amylose content was below the detection level. Other genes involved in cassava starch biosynthesis have been isolated (Munyikwa et al. 1995), and recently the regeneration of cassava plants expressing the cassava AGPase B gene in antisense orientation has been reported (Munyikwa et al. 1998). In transgenic plants kept in vitro under conditions that induce starch accumulation in the shoots, 74% of the transgenic plants showed reduced starch levels compared with the controls.

4.3 Cyanogenesis

The cyanogenic glucosides linamarin and lotaustralin are found in all mature tissues of cassava, which upon wounding release hydrogen cyanide (McMahon et al. 1995). Adequate processing of cassava removes the sources for cyanide to below toxic levels, but it has been shown that in times of social or economic disturbances insufficiently processed cassava can lead to serious health problems of consumers (Banea et al. 1992). Gene technology could be used to reduce the toxicity of cassava meant for human consumption. The main source of cyanide in insufficiently processed cassava is acetone cyanohydrin (Tylleskär et al. 1992; cited in Hughes and Hughes 1995). Overexpression of the α-hydroxynitrile lyase (HNL) gene of cassava, the enzyme that catalyses the conversion of acetone cyanohydrin to the volatile compounds acetone and the hydrogen cyanide, would be an option to decrease the time needed for processing. The HNL gene has been cloned and its expression in transgenic cassava plants obtained through *Agrobacterium*-mediated transformation has been reported recently (Arias-Garzon and Sayre 2000).

5 Summary and Conclusions

Significant progress has been achieved in cassava transformation research. This has been due largely to the breakthroughs in cell culture techniques, allowing the production of morphologically competent tissues, which have also proved to be suitable as target tissues for gene transfer. The development of both embryogenic suspension cultures and organogenic cultures from somatic embryo cotyledons have facilitated the recovery of the first transgenic cassava plants and opened a new stage in cassava genetic transformation programs. No longer is it necessary to restrict investigations of gene expression to transient studies or to use model species such as tobacco and potato to evaluate the effectiveness of genes with putatively important agronomic traits. This progress is already being demonstrated, with projects underway in various laboratories to insert genes for resistance to viruses, pests and to alter starch qualities. The production of transgenic plants containing these genes will shift attention away from the technical aspects of the culture systems and regeneration protocols, towards studies on gene integration and expression, and in the near future to the performance of the transgenics under field trial conditions.

A striking aspect of any present review of cassava transformation is not only that transgenic plants can now be produced with relative ease, but that there are a number of ways in which this can be achieved. Compared with the state of the technology as late as 1994 this is significant, and compared with the large resources invested in the major temperate or tropical cash crops, such as maize, tomato, cotton and rice, the advances made in cassava over such a short time are remarkable. Transgenic cassava can be now regenerated after transformation through both particle bombardment and *Agrobacterium* infection of embryogenic suspension cultures, via organogenesis from cotyledons of somatic embryos and through regeneration from protoplasts. It remains to be seen which of these gene transfer/regeneration systems will become the favored method for producing transgenic cassava plants, or if another system, such as gene transfer and regeneration from meristems will be developed. It is very possible that all the transformation systems will have a role to play depending on the cultivar in question, the resources and expertise of the respective laboratories and the number of transgenic plant lines required.

As mentioned above, much work needs to be carried out on gene expression and to extend the transformation capability into a significant number of cultivars. There are hundreds of cassava cultivars grown throughout tropical Africa, Asia and South America. The challenge is to transform the most important cultivars from each region with ease, and to transfer this capability to the countries concerned (Taylor and Fauquet 1997). Each country or region has its own specific problems concerning cassava production. If the wish is to address these through the implementation of a transformation program, then it will become increasingly important over the next decade to establish an indigenous capability for cassava transformation within tropical countries.

Acknowledgement. We dedicate the present article to the memory of our friend and colleague Victor Masona, who unexpectedly passed away on July 22, 2000. For the last two years Dr. Masona worked actively on the production of transgenic cassava plants that are resistant to two cassava viruses present in Africa. His research was instrumental in the discovery of the biological mechanism driving the natural recombination of cassava viruses.

References

Anthony P, Davey MR, Power JB, Lowe KC (1995) An improved protocol for the culture of cassava leaf protoplasts. Plant Cell Tissue Organ Cult 42:299–302

Arias-Garzon DI, Sayre RT (1993) Tissue-specific inhibition of transient gene expression in cassava (*Manihot esculenta* Crantz). Plant Sci 93:121–130

Arias-Garzon DI, Sayre RT (2000) Genetic Engineering approaches to reducing the cyanide toxicity in cassava (*Manihot esculenta* Crantz). In: Carvalho L, Thro, AM, Vilarinhos AD (eds) Proc 4th Int Sci Meet Cassava Biotechnol Network, 3–7 November 1998, Salvador de Bahia, Brazil. Brazilian Agricultural Research Corporation – EMBRAPA, Brasilia, Brazil, pp 213–221

Banea M, Poulter NH, Rosling H (1992) Shortcuts in cassava processing and risk of dietary cyanide exposure in Zaire. Food Nutr Bull 14:137–143

Bertram RB (1990) Cassava. In: Persley GJ (ed) Agricultural biotechnology: opportunities for international development. CAB International, Wallingford, UK, pp 241–261

Bhagwat B, Vieira LGE, Erickson LR (1996) Stimulation of in vitro shoot proliferation from nodal explants of cassava by thidiazuron, benzyladenine and gibberellic acid. Plant Cell Tissue Organ Cult 46:1–7

Bohl S, Potrykus I, Puonti-Kaerlas J (1997) Searching for root-specific promoters in cassava. In: Thro AM, Akoroda MO (eds) Proc 3rd Int Sci Meet Cassava Biotechnol Network. Afr J Root Tuber Crops 2:172–175

Cabral GB, Aragão F, Monte-Neshich DC, Rech EL (1993) Transient gene expression in cassava protoplasts. In: Roca WM, Thro AM (eds) Proc 1st Int Sci Meet Cassava Biotechnol Network, Cartagena, Colombia, 25–28 August 1992. Centro Internacional de Agricultura Tropical, Cali, Colombia, pp 244–250

Calderón A (1988) Transformation of *Manihot esculenta* (cassava) using *Agrobacterium tumefaciens* and expression of the introduced foreign genes in transformed cell lines. MSc Thesis, Vrije Universiteit Brussel, Belgium, 37 pp

Chavarriaga P, Schöpke C, Sangare A, Fauquet C, Beachy RN (1993) Transformation of cassava (*Manihot esculenta* Crantz) embryogenic tissues using *Agrobacterium tumefaciens*. In: Roca WM, Thro AM (eds) Proc 1st Int Sci Meet Cassava Biotechnol Network, Cartagena, Colombia, 25–28 August 1992. Centro Internacional de Agricultura Tropical, Cali, Colombia, pp 222–228

Cooke R, Cock JH (1989) Cassava crops up again. New Sci 122:63–68

FAO/GIEWS (1998) Food outlook 2, April 1998. Internet address: http://www.fao.org/WAICENT/faoinfo/economic/giews/

Fauquet C, Schöpke C, Sangare A, Chavarriaga P, Beachy RN (1993) Genetic engineering technologies to control viruses and their application to cassava viruses. In: Roca WM, Thro AM (eds) Proc 1st Int Sci Meet Cassava Biotechnol Network, Cartagena, Colombia, 25–28 August 1992. Centro Internacional de Agricultura Tropical, Cali, Colombia, pp 190–207

Franche C, Bogusz D, Schöpke C, Fauquet C, Beachy RN (1991) Transient gene expression in cassava using high-velocity microprojectiles. Plant Mol Biol 17:493–498

González AE, Schöpke C, Taylor N, Beachy RN, Fauquet CM (1998) Regeneration of transgenic cassava plants (*Manihot esculenta* Crantz) through *Agrobacterium*-mediated transformation of embryogenic suspension cultures. Plant Cell Rep 17:827–831

Gresshoff JW, Doy CH (1974) Derivation of a haploid cell line from *Vitis vinifera* and the importance of the stage of meiotic development of anthers for haploid culture of this and other genera. Z Pflanzenphysiol 73:132–141

Groll J, Gray VM, Mycock DJ (1997) Friable embryogenic callus formation from thin cell layer explants of cassava. In: Thro AM, Akoroda MO (eds) Proc 3rd Int Sci Meet Cassava Biotechnol Network. Afr J Root Tuber Crops 2:154–158

Hershey CH, Jennings DL (1992) Progress in breeding cassava for adaptation to stress. Plant Breed Abstr 62:823–831

Hughes J, Hughes MA (1995) A molecular study of cyanogenic α-hydroxynitrile lyase from cassava (*Manihot esculenta*). In: Roca WM, Thro AM (eds) Proc 2nd Int Sci Meet Cassava Biotechnol Network, 22–26 August 1994, Bogor, Indonesia. Centro Internacional de Agricultura Tropical, Cali, Colombia, pp 414–422

Iwanaga M, Iglesias C (1994) Cassava genetic resources management at CIAT. In: Rep 1st Meet Int Network for Cassava Genetic Resources, CIAT, Cali, Colombia, 18–23 August 1992. Int Crop Network Series No 10. IPGRI, Rome, Italy, pp 77–86

Jefferson RA (1987) Assaying chimeric genes in plants: the *uid*A gene fusion system. Plant Mol Biol Rep 5:387–405

Jennings DL, Hershey CH (1985) Cassava breeding: a decade of progress from international programmes. In: Russell GE (ed) Progress in plant breeding – 1. Butterworths, London, pp 89–116

Konan NK, Sangwan RS, Sangwan-Norreel BS (1994a) Somatic embryogenesis from cultured mature cotyledons of cassava (*Manihot esculenta* Crantz) – identification of parameters influencing the frequency of embryogenesis. Plant Cell Tissue Organ Cult 37:91–102

Konan NK, Sangwan RS, Sangwan-Norreel BS (1994b) Efficient in vitro shoot regeneration systems in cassava (*Manihot esculenta* Crantz). Plant Breed 113:227–236

Konan NK, Sangwan RS, Sangwan-Norreel BS (1995) Nodal axillary meristems as target tissue for shoot production and genetic transformation in cassava (*Manihot esculenta* Crantz). In: Roca WM, Thro AM (eds) Proc 2nd Int Sci Meet Cassava Biotechnol Network, 22–26 August 1994, Bogor, Indonesia. Centro Internacional de Agricultura Tropical, Cali, Colombia, pp 276–288

Konan NK, Schöpke C, Cárcamo R, Beachy RN, Fauquet C (1997) An efficient mass propagation system for cassava (*Manihot esculenta* Crantz) based on nodal explants and axillary bud-derived meristems. Plant Cell Rep 16:444–449

Li H-Q, Sautter C, Potrykus I, Puonti-Kaerlas J (1996) Genetic transformation of cassava (*Manihot esculenta* Crantz). Nat Biotechnol 14:736–740

Li H-Q, Guo J-Y, Huang Y-W, Liang C-Y, Liu H-X, Potrykus, I, Puonti-Kaerlas J (1998) Regeneration of cassava plants via shoot organogenesis. Plant Cell Rep 17:410–414

Luong HT, Shewry PR, Lazzeri PA (1995) Transient gene expression in cassava somatic embryos by tissue electroporation. Plant Sci 107:105–115

Luong HT, Shewry PR, Lazzeri PA (1997) Transformation and gene expression in cassava via tissue electroporation and particle bombardment. In: Thro AM, Akoroda MO (eds) Proc 3rd Int Sci Meet Cassava Biotechnol Network. Afr J Root Tuber Crops 2:163–167

Mabanza J, Jonard R (1983) L'isolement et le développement in vitro des protoplastes de manioc (*Manihot esculenta* Crantz). C R Soc Biol 177:638–645

Mabanza J, Jonard R (1984) La régénération de plantes de manioc (*Manihot esculenta* Crantz) par néoformations de bourgeons à partir de cotylédons extraits de semences mûres et immatures. Bull Soc Bot Fr 131:91–95

Mathews H, Schöpke C, Cárcamo R, Chavarriaga P, Fauquet C, Beachy RN (1993) Improvement of somatic embryogenesis and plant recovery in cassava. Plant Cell Rep 12:334–338

McDonnell SL, Gray VM (1997) Transformation and culture of cassava protoplasts. In : Thro AM, Akoroda MO (eds) Proc 3rd Int Sci Meet Cassava Biotechnol Network. Afr Root Tuber Crops 2:169–172

McMahon JM, White WLB, Sayre RT (1995) Cyanogenesis in cassava. J Exp Bot 46:731–741

Mireles M, Paez de Casares J (1984) Inducción de "roseta" en yuca (*Manihot esculenta* Crantz) para la propagación multiple de la planta in vitro. Rev Fac Agron Maracay – Alcance 33:73–81

Munyikwa TRI, Chipangura B, Salehuzzaman SNIM, Jacobsen E, Visser RGF (1995) Cloning and characterization of cassava genes involved in starch biosynthesis. In: Roca WM, Thro AM (eds) Proc 2nd Int Sci Meet Cassava Biotechnol Network, 22–26 August 1994, Bogor, Indonesia. Centro Internacional de Agricultura Tropical, Cali, Colombia, pp 639–645

Munyikwa TRI, Langeveld S, Salehuzzaman SNIM, Jacobsen E, Visser RGF (1997) Cassava starch biosynthesis: new avenues for modifying starch quantity and quality. Euphytica 96:65–75

Munyikwa TRI, Raemakers CJJM, Schreuder M, Kok R, Schippers M, Jacobsen E, Visser RGF (1998) Pinpointing towards improved transformation and regeneration of cassava (*Manihot esculenta* Crantz). Plant Sci 135:87–101

Murashige T, Skoog F (1962) A revised medium for rapid growth and bioassays with tobacco tissue cultures. Physiol Plant 15:473–497

Ng, NQ, Asiedu R, Ng SYC (1994) Cassava genetic resources programme at the International Institute of Tropical Agriculture, Ibadan. In: Rep 1st Meet Int Network for Cassava Genetic Resources, CIAT, Cali, Colombia, 18–23 August 1992. Int Crop Network Series No 10. IPGRI, Rome, Italy, pp 71–76

Polito VS, McGranahan GH, Pinney K, Leslie CA (1989) Origin of somatic embryos from repetitively embryogenic cultures of walnut (*Juglans regia* L.): implications for *Agrobacterium*-mediated transformation. Plant Cell Rep 8:219–221

Puonti-Kaerlas J, Frey P, Potrykus I (1997) Development of meristem gene transfer techniques for cassava. In: Thro AM, Akoroda MO (eds) Proc 3rd Int Sci Meet Cassava Biotechnol Network. Afr J Root Tuber Crops 2:175–180

Raemakers CJJM, Amati M, Staritsky G, Jacobson E, Visser RGF (1993a) Cyclic somatic embryogenesis and plant regeneration in cassava. Ann Bot 71:289–294

Raemakers CJJM, Bessembinder J, Staritsky G, Jacobsen E, Visser RGF (1993b) Induction, germination and shoot development of somatic embryos in cassava. Plant Cell Tissue Organ Cult 33:151–156

Raemakers CJJM, Sofiari E, Jacobsen E, Visser RGF (1995) Histology of somatic embryogenesis and evaluation of somaclonal variation. In: Roca WM, Thro AM (eds) Proc 2nd Int Sci Meet Cassava Biotechnol Network, 22–26 August 1994, Bogor, Indonesia. Centro Internacional de Agricultura Tropical, Cali, Colombia, pp 336–354

Raemakers CJJM, Sofiari E, Taylor N, Henshaw G, Jacobsen E, Visser RGF (1996) Production of transgenic cassava (*Manihot esculenta* Crantz) plants by particle bombardment using luciferase activity as selection marker. Mol Breed 2:339–349

Raemakers CJJM, Jacobsen E, Visser RGF (1997a) Micropropagation of cassava (*Manihot esculenta* Crantz). In: Bajaj YPS (ed) Biotechnology in agriculture and forestry, vol 39. High-tech and micropropagation V. Springer, Berlin Heidelberg New York, pp 77–102

Raemakers CJJM, Jacobsen E, Visser RGF (1997b) Regeneration of plants from somatic embryos and friable embryogenic callus of cassava (*Manihot esculenta* Crantz). In: Thro AM, Akoroda MO (eds) Proc 3rd Int Sci Meet Cassava Biotechnol Network. Afr J Root Tuber Crops 2:238–242

Raemakers CJJM, Schreuder M, Munyikwa T, Jacobsen E, Visser R (2000) Towards a routine transformation procedure for cassava. In: Carvalho L, Thro AM, Vilarinhos AD (eds) Proc 4th Int Sci Meet Cassava Biotechnol Network, 3–7 November 1998, Salvador de Bahia, Brazil. Brazilian Agricultural Research Corporation – EMBRAPA, Brasilia, Brazil, pp 250–266

Roca WM (1989) Cassava production and utilization problems and their biotechnological solutions. In: Sasson A, Costarini V (eds) Plant biotechnology for developing countries. Proc Int Symp CTA/FAO, 26–30 June 1989, Luxembourg. The Trinity Press, UK, pp 215–219

Salehuzzaman SNIM, Jacobsen E, Visser RGF (1993) Isolation and characterization of a cDNA encoding granule-bound starch synthase in cassava (*Manihot esculenta* Crantz) and its antisense expression in potato. Plant Mol Biol 23:947–962

Sarria R, Torres E, Angel F, Chavarriaga P, Roca WM (2000) Transgenic plants of cassava (*Manihot esculenta*) with resistance to Basta obtained by *Agrobacterium*-mediated transformation. Plant Cell Rep 19:339–344

Schöpke C, Fauquet CM, Paterson HF (1999) Introduction of materials into living cells. In: Lacey A (ed) Light microscopy in biology – a practical approach, 2nd edn. Oxford University Press, Oxford, pp 373–397

Schöpke C, Chavarriaga P, Fauquet C, Beachy RN (1993a) Cassava tissue culture and transformation: improvement of culture media and the effect of different antibiotics on cassava. In: Roca WM, Thro AM (eds) Proc 1st Int Sci Meet Cassava Biotechnol Network, Cartagena, Colombia, 25–28 August 1992. Centro Internacional de Agricultura Tropical, Cali, Colombia, pp 140–145

Schöpke C, Franche C, Bogusz D, Chavarriaga P, Fauquet C, Beachy RN (1993b) Transformation in cassava (*Manihot esculenta* Crantz). In: Bajaj YPS (ed) Biotechnology in agriculture and forestry, vol 23. Plant protoplasts and genetic engineering IV. Springer, Berlin Heidelberg New York, pp 273–289

Schöpke C, Taylor N, Cárcamo R, Konan NK, Marmey P, Henshaw GG, Beachy RN, Fauquet C (1996) Regeneration of transgenic cassava plants (*Manihot esculenta* Crantz) from microbombarded embryogenic suspension cultures. Nat Biotechnol 14:731–735

Schöpke C, Cárcamo R, Beachy RN, Fauquet C (1997a) Plant regeneration from transgenic and non-transgenic embryogenic suspension cultures of cassava (*Manihot esculenta* Crantz). In: Thro AM, Akoroda MO (eds) Proc 3rd Int Sci Meet Cassava Biotechnol Network. Afr J Root Tuber Crops 2:194–195

Schöpke C, Taylor NJ, Cárcamo R, Beachy RN, Fauquet C (1997b) Optimization of parameters for particle bombardment of embryogenic suspension cultures of cassava (*Manihot esculenta* Crantz) using computer image analysis. Plant Cell Rep 16:526–530

Schöpke C, Masona MV, Taylor NJ, Cárcamo R, Ho T, Beachy RN, Fauquet CM (2000) Production and characterization of transgenic cassava plants expressing the coat protein gene of cassava common mosaic virus. In: Carvalho L, Thro AM, Vilarinhos AD (eds) Proc 4th Int Sci Meet Cassava Biotechnol Network, 3–7 November 1998, Salvador de Bahia, Brazil. Brazilian Agricultural Research Corporation – EMBRAPA, Brasilia, Brazil, pp 236–243

Shahin EA, Shepard JF (1980) Cassava mesophyll protoplasts: isolation, proliferation, and shoot formation. Plant Sci Lett 17:459–465

Sofiari E, Raemakers CJJM, Kanju E, Danso K, Van Lammeren AM, Jacobsen E, Visser RGF (1997) Comparison of NAA and 2,4-D induced somatic embryogenesis in cassava. Plant Cell Tissue Organ Cult 50:45–56

Sofiari E, Raemakers CJJM, Bergervoet JEM, Jacobsen E, Visser RGF (1998) Plant regeneration from protoplasts isolated from friable embryogenic callus of cassava. Plant Cell Rep 18:159–165

Stamp JA (1987) Somatic embryogenesis in cassava: the anatomy and morphology of the regeneration process. Ann Bot 59:451–459

Stamp JA, Henshaw GG (1982) Somatic embryogenesis in cassava. Z Pflanzenphysiol 105:183–187

Stamp JA, Henshaw GG (1986) Adventitious regeneration in cassava. In: Withers LA, Alderson PG (eds) Plant tissue culture and its agricultural applications. Butterworths, London, pp 149–157

Szabados L, Hoyos R, Roca WM (1987a) In vitro somatic embryogenesis and plant regeneration of cassava. Plant Cell Rep 6:248–251

Szabados L, Narváez J, Roca WM (1987b) Techniques for isolation and culture of cassava (*Manihot esculenta* Crantz) protoplasts. Working Document no 23. Biotechnol Res Unit, CIAT, Cali, Colombia, 42 pp

Taylor NJ, Fauquet CM (1997) Transfer of rice and cassava gene biotechnologies to developing countries. Biotechnol Int 1:239–246

Taylor NJ, Clarke C, Henshaw GG (1993) The induction of somatic embryogenesis in fifteen African and one South American cassava cultivars. In: Roca WM, Thro AM (eds) Proc 1st Int Sci Meet Cassava Biotechnol Network, Cartagena, Colombia, 25–28 August 1992. Centro Internacional de Agricultura Tropical, Cali, Colombia, pp 134–139

Taylor NJ, Edwards M, Kiernan RJ, Davey C, Blakesley D, Henshaw GG (1996) Development of friable embryogenic callus and embryogenic suspension culture systems in cassava (*Manihot esculenta* Crantz). Nat Biotechnol 14:726–730

Taylor NJ, Beachy RN, Fauquet CM (1997a) The international laboratory for tropical agricultural biotechnology. Developing plant gene techniques for food security in developing countries. In: Jordan C, Persaud B, Zimmerman H (eds) Global aid. Hanson and Cooke, London, pp 97–100

Taylor NJ, Kiernan RJ, Davey C, Henshaw GG, Blakesley D (1997b) Improved procedures for the production of embryogenic tissues across a range of African cassava cultivars and the implications for genetic transformation. In: Thro AM, Akoroda MO (eds) Proc 3rd Int Sci Meet Cassava Biotechnol Network. Afr J Root Tuber Crops 2:200–204

Thro AM, Beachy RN, Bonierbale M, Fauquet C, Henry G, Henshaw GG, Hughes MA, Kawano K, Raemakers CJJM, Roca W, Schöpke C, Taylor N, Visser RGF (1996) International research on biotechnology of cassava (tapioca, *Manihot esculenta* Crantz) and its relevance to Southeast Asian economies: a cassava biotechnology network review. Asian J Trop Biol 2:1–30

Tilquin JP (1979) Plant regeneration from stem callus of cassava. Can J Bot 57:1761–1763

Tylleskär T, Banea M, Bikangi J, Cooke RD, Poulter NH, Rosling H (1992) Cassava cyanogens and konzo, an upper motoneuron disease found in Africa. Lancet 339:208–211

Vain P, Keen N, Murillo J, Rathus C, Nemes C, Finer JJ (1993) Development of the particle inflow gun. Plant Cell Tissue Organ Cult 33:237–246

Van Der Krol AR, Mol JNM, Stuitje AR (1989) Modulation of eukaryotic gene expression by complementary RNA or DNA sequences. Biotechniques 6:958–976

Vasil IK (1996) Milestones in crop biotechnology-transgenic cassava and *Agrobacterium*-mediated transformation of maize. Nat Biotechnol 14:702–703

Verdaguer B, De Kochko A, Beachy RN, Fauquet C (1996) Isolation and expression in transgenic tobacco and rice plants, of the cassava vein mosaic virus (CVMV) promoter. Plant Mol Biol 31:1129–1139

Verdaguer B, De Kochko A, Fux C, Beachy RN, Fauquet C (1998) Functional organization of the cassava vein mosaic virus (CsVMV) promoter. Plant Mol Biol 37:1055–1067

Visser RGF, Jacobsen E (1993) Towards modifying plants for altered starch content and composition. Trends Biotechnol 11:63–68

17 Transgenic Banana (*Musa* Species)

L. Sági[1], S. Remy[1], J.B.P. Hernández[1], B.P.A. Cammue[2], and R. Swennen[1]

1 Introduction

1.1 Importance

The term banana is meant here to cover dessert, cooking and beer bananas as well as plantains. Banana is probably the most important fruit crop in the world with an annual production of more than 80 million tons (Anonymous 1995). In many of the 120, mostly less-developed banana producing countries, banana is locally consumed fresh, prepared by cooking or processed for chips and other food products. In fact, it is (after rice, milk and wheat) the fourth major food source for the Third World and also ranks fourth (after rice, wheat and maize) in terms of gross value of production.

Some 10% of banana production is exported and serves as dessert for many more millions of people. Export banana trade has a gross value of more than four billion US$ (Anonymous 1996), and banana cultivation is thus of great socio-economic importance for many less-developed countries.

1.2 Major Constraints in Banana Production

Most cultivated landraces are parthenocarpic and sterile triploids which prevents elaboration of successful hybridization programmes. In addition, the relatively long life cycle of this crop makes even simple field testing experiments last for at least 4 to 5 years. These obstacles make it very difficult for the breeders to produce banana cultivars that are resistant to their numerous virulent diseases and pests.

The most devastating disease is undoubtly the Sigatoka complex caused by the fungal pathogens *Mycosphaerella fijiensis* (Morelet) Deighton (black sigatoka) and *M. musicola* Leach ex Mulder (yellow sigatoka). Sigatoka disease presents the most serious problem in large plantations where yield losses may reach up to 30–50% (Stover 1983; Mobambo et al. 1993). Annual

[1] Laboratory of Tropical Crop Improvement, Catholic University of Leuven, Kard. Mercierlaan 92, 3001 Heverlee, Belgium
[2] FA Janssens Laboratory of Genetics, Catholic University of Leuven, Kard. Mercierlaan 92, 3001 Heverlee, Belgium

Biotechnology in Agriculture and Forestry, Vol. 47
Transgenic Crops II (ed. by Y.P.S. Bajaj)
© Springer-Verlag Berlin Heidelberg 2001

costs of fungicide spraying control in plantations range between US$ 600 and 1800 per ha.

The second major fungal disease is Panama disease or banana wilt caused by *Fusarium oxysporum* Schlecht. f. sp. *cubense* (E.F. Smith) Snyd. and Hansen. A new race presently threatens the banana export industry in the subtropics (Ploetz 1990).

Presently, the most serious viral disease affecting banana is bunchy top disease, partially because no resistant source is known in banana germplasm. The causal agent, banana bunchy top virus (BBTV) is a multicomponent single-stranded DNA virus which probably represents a new group of plant viruses (Harding et al. 1993). Recently, banana streak badnavirus (Lockhart 1986), banana bract mosaic potyvirus (Rodoni et al. 1997) and cucumber mosaic cucumovirus have caused increasing damage to banana.

Among the migratory endoparasitic nematodes, which are the most damaging and widespread ones [e.g. the root-lesion nematodes, *Pratylenchus* spp., and the spiral nematode, *Helicotylenchus multicinctus* (Cobb) Golden], the burrowing nematode, *Radopholus similis* (Cobb) Thorne is the most dangerous. It causes steadily increasing problems in commercial plantations in Latin America.

2 Genetic Transformation

2.1 A Brief Review of Banana Transformation

The development of embryogenic cell suspensions (ECSs) and of regenerable protoplast cultures in banana (reviewed by Panis et al. 1994) have made it possible to transfer foreign genes into this crop (reviewed by Sági et al. 1995a). Recently, the first transgenic banana plants have been produced by particle bombardment of ECSs (Sági et al. 1995b,c). Also, *Agrobacterium*-mediated transformation of meristematic tissues has been reported to result in regeneration of transgenic banana plants (May et al. 1995).

At present, three systems appear to be feasible for banana transformation (Table 1): (1) introduction of DNA into regenerable, cell suspension-derived

Table 1. Summary of various studies on banana transformation by different gene transfer techniques

Reference	Explant/culture used	Vector/method used	Observations/Remarks
Sági et al. (1994)	Protoplast	pBI221/electroporation	2–5% Transient expr.
Sági et al. (1995b,c)	Embryogenic cell suspension (ECS)	pAHC27/particle bombardment	High frequency of transgenic plants
May et al. (1995)	In vitro meristem	pIBT141/*Agrobacterium*	Chimeric plants?
Remy et al. (1998)	ECS	pFAJ3084/particle bomb.	Co-transformation
Pérez Hernández et al. (1998)	ECS	pFAJ3000/*A. tumefaciens*	More efficient than particle bombardment

protoplasts by electroporation (Sági et al. 1994), (2) particle bombardment using an in-house developed particle gun device which has been optimized for transformation of ECSs (Sági et al. 1995b,c; Remy et al. 1998), and (3) *Agrobacterium*-mediated transformation of in vitro meristems (May et al. 1995) or ECSs (Pérez Hernández et al. 1998).

2.2 Materials and Methods

Plant Material. Embryogenic cell suspension lines of the cooking banana cultivar Bluggoe (*Musa* spp., ABB group), the plantain cultivar Three Hand Planty (AAB group), and the Cavendish dessert banana cultivar Williams (AAA group) were maintained and subcultured weekly in MS medium (Murashige and Skoog 1962) supplemented with 5µM 2,4-D and 1µM zeatin (ZZ medium) as described by Dhed'a et al. (1991). *In vitro* plants of the cultivars Bluggoe and Pome were micropropagated on MS medium containing 2.5µM benzyladenine.

Protoplast Isolation. Isolation and purification of protoplasts from ECSs was done according to Panis et al. (1993). The purified protoplasts were resuspended in electroporation buffer (Sági et al. 1994) at a concentration of 1.25×10^6 protoplasts per ml. Protoplasts were counted using a modified Neubauer hemocytometer. Complete removal of the cell wall was confirmed by Calcofluor white staining while the viability of freshly isolated or electroporated protoplasts was assessed by staining either with fluorescein diacetate or Evans' blue.

Transformation Procedures. Protoplast transformation by electroporation was carried out as described by Sági et al. (1994). Briefly, a 800-µl aliquot containing 10^6 protoplasts in electroporation buffer was placed into cuvettes of 0.4cm gap. After addition of plasmid DNA to a concentration of 60µg/ml, cuvettes were stored on ice for 10min, then electroporated with a 960µF capacitor at a field strength of 800V/cm. After electroporation, the cuvettes were placed on ice for 10min and then for 10min at room temperature. Protoplasts were diluted in ZZ medium supplemented with 0.55M mannitol to a concentration of 10^5 protoplasts per ml and incubated in the dark at 24°C. The following controls were used: (1) samples electroporated with pUC19 DNA, (2) samples electroporated without plasmid DNA, (3) non-electroporated samples incubated with plasmid DNA.

For particle bombardment of suspension cells, a flowing helium gun was constructed and the procedure is described in detail in Sági et al. (1995b). Briefly, for one bombardment approximately 1µg plasmid DNA was precipitated according to Perl et al. (1992) onto 500µg gold (Bio-Rad) or M-17 tungsten (Sylvania) particles and applied into a Sartorius syringe filter unit. For co-transformation of multiple gene constructs, equimolar amounts of plasmid DNAs were precipitated. Suspension cells were collected 4 to 6 days after subculture and 30 to 100µl settled cell volume (approximately 25–60mg fresh weight of cells) was used for bombardment. Particles were accelerated in vacuum at a pressure of 5–8 bars. Cells were then cultured in ZZ medium for 1 or 2 days and assayed for transient GUS expression.

Transient β-glucuronidase (GUS) Expression Assay. For histochemical in situ assays, protoplasts were collected 48 h after electroporation, resuspended in 50 mM sodium phosphate buffer, pH 7.0, and incubated for periods ranging from overnight up to 10 days at 37 °C in the presence of 1 mM X-Gluc (5-bromo-4-chloro-3-indolyl-β-D-glucuronide) as described by Jefferson (1987). Protoplast transformation frequency was quantitated by counting blue-stained protoplasts according to Sági et al. (1994). For bombarded cells, transient GUS expression frequencies were expressed as numbers of blue foci per shot averaged over three to six replicates per treatment. Cultures of *Escherichia coli* were used as positive controls for all GUS assays.

Regeneration and Characterization of Transgenic Plants. Embryogenic suspension cells were cultured on solidified ZZ medium for 7–10 days without selection, then transferred to solid selective ZZ medium usually containing 50 mg/l hygromycin or geneticin and subcultured every 2 weeks for 2 months. Actively growing surviving colonies were proliferated and regenerated according to Dhed'a et al. (1991).

Screening of putative transgenic plants by polymerase chain reaction (PCR) was performed with primers specific to a portion of the selectable marker gene and/or to the promoter region. Both in vitro and greenhouse plants were screened using a capillary PCR cycler. For Southern hybridization analysis, total DNA isolated from young leaves was transferred after restriction enzyme digestion and gel electrophoresis to nylon membranes and hybridized with digoxigenin-probes labelled by PCR. Chemiluminescent detection of the Southern blots was performed according to Neuhaus-Url and Neuhaus (1993).

For the introduction of multiple genes, the frequency of co-transformations was evaluated in three independent experiments (Table 2). In experiment 1, the selectable marker gene (gene A) and an antifungal peptide gene (gene B) were transformed into embryogenic cells of the plantain cultivar Three Hand Planty in a linked position, i.e., the two genes were present on the

Table 2. Co-transformation frequencies of linked and unlinked foreign genes in transgenic plants of the plantain cultivar Three Hand Planty obtained by PCR screening. (Remy et al. 1998)

Gene combination	Co-transformation frequency (%) in experiment no.		
	1	2	3
Linked genes (same plasmid)	95.0	89.2	100.0
(No. of lines tested)	(40)	(37)	(14)
Unlinked gene (diff. plasmids)	NA[a]	83.8	71.4
(No. of lines tested)	(NA)	(37)	(14)

[a] NA, not applicable.

same plasmid. In experiments 2 and 3, the plasmid with genes A and B was co-transformed with another plasmid that carried a different antifungal protein gene (gene C). Gene C was thus not linked to genes A or B. Transgenic shoots were then regenerated from all three experiments and screened by PCR for the presence/absence of each of the three foreign genes. Selected plants from each experiment were analyzed by DNA blot hybridization in order to confirm integrative transformation of these genes.

The number of plants carrying both genes A and gene B allows for the calculation of co-transformation frequencies of linked genes in all the three independent experiments according to the following equation: $[(A + B)^+/A^+] \times 100(\%)$. Similarly, in experiments 2 and 3, the proportion of plants containing both gene A and gene C indicates the co-transformation frequencies of unlinked genes.

2.3 Results and Discussion

2.3.1 Transient GUS Expression (TGE) in Banana Protoplasts by Electroporation

Electroporation conditions were established for transient expression of chimeric *gusA* gene constructs in banana protoplasts isolated from ECSs of the cooking banana cultivar Bluggoe. When using a 960 μF capacitor, the optimized parameters were as follows: (1) electric field strength, 800 V/cm; (2) ASP-electroporation buffer, containing 70 mM potassium-aspartate, 5 mM calcium-gluconate, 5 mM MES, and 0.55 M mannitol (pH 5.8) as described by Tada et al. (1990); (3) polyethylene glycol (PEG) concentration, 5%; (4) heat shock, 45 °C for 5 min before addition of PEG; (5) protoplast parameters, highly viable protoplasts isolated from 1-week-old ECSs, and (6) chimeric gene constructs for optimized expression. The maximum frequency of DNA introduction as detected by an in situ histochemical TGE assay amounted to 2–5% of total protoplasts.

Besides providing a fast method for determination of promoter strength in banana (Sági et al. 1994, 1995c), this protocol can also prove to be suitable for the production of transgenic banana using the plant regeneration protocol previously described (Panis et al. 1993).

2.3.2 Particle Bombardment of ECS

Our in-house developed particle gun was found to deliver DNA-coated particles into embryogenic cells of a wide range of banana cultivars (Table 3) at a high frequency as measured by a histochemical assay of *gusA* reporter gene expression (Sági et al. 1995a,b). This observation provided the basis for selecting stably transformed cells and regenerating transgenic banana plants. Two selective agents have been used successfully so far, i.e. hygromycin and geneticin. Putative transgenic shoots were regenerated on media containing

Table 3. Transient and stable transformation of various banana cultivars by particle bombardment

Cultivar (genome group)	Transient transformation[a]	Stable
Grande Naine (AAA)	Yes (1)	ND
Williams (AAA)	Yes (2)	Yes (2)
Three Hand Planty (AAB)	Yes (2)	Yes (2)
Bluggoe (ABB)	Yes (2)	Yes (2)
Cardaba (ABB)	Yes (1)	ND
Monthan (ABB)	Yes (1)	ND
M. balbisiana Tani (BB)	Yes (1)	ND

[a] (1) L. Sági et al. (unpubl.); (2) Sági et al. (1995b,c).

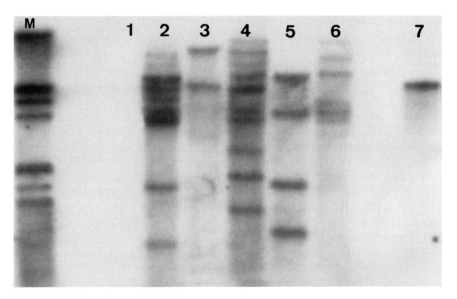

Fig. 1. Southern blot hybridization analysis of total DNA isolated from independent transgenic plants of the plantain cultivar Three Hand Planty with a digoxigenin-labelled probe specific to the *neo* gene (Remy et al., in prep.). *M* Size marker, λ/*Hin*dIII; *lane 1* untransformed control plant; *lanes 2–6* transformed plants; *lane 7* positive control, linearized plasmid

high concentrations (50–100 mg/l) of these antibiotics which were toxic and lethal to untransformed cells and shoots. In addition, histochemical GUS assays demonstrated uniform expression of the introduced *gusA* gene in various parts of the transgenic plants including leaves, pseudostems and roots. Finally, PCR and Southern blot hybridization analyses (Fig. 1) confirmed the presence and integration of all introduced genes in the banana genome. Stable integration of foreign genes was also demonstrated in subsequent micropropagated progeny by histochemical staining, RT-PCR and Southern hybridization.

These three lines of evidence, i.e. (1) the antibiotic-resistant phenotype, (2) stable expression of the *gusA* reporter gene in various tissues, and (3) the presence and stable integration of the introduced genes in transformed plants, prove that the regenerated plants are indeed transgenic.

Molecular Characterization of Transgenic Plants. With this technology, close to 900 independent transgenic lines have been recently produced from the plantain Three Hand Planty and the Cavendish dessert banana cultivar Williams. A large-scale molecular and biochemical characterization of these lines confirmed that the vast majority of them contained the foreign genes. For instance, PCR screening of 776 putative transgenic plants revealed that more than 90% of them contained the selectable marker gene (Table 4). No significant differences were found in this respect between the two cultivars (92.5% for Three Hand Planty and 98.1% for Williams) and the two genes used for selection (89.6% for *hph* and 94.1% for *neo*). Similar results were obtained by Southern blot hybridization (Table 5, Fig. 1) which confirmed that the majority of the plants analyzed had the selectable marker gene incorporated in their genome. Finally, in preliminary ELISA experiments, expression of

Table 4. Overview of PCR-screening experiments in putative transgenic plants of two banana cultivars. (Remy et al., in prep.)

Cultivar	Screening for		Total
	hph	*neo*	
Three Hand Planty	110/124 (88.7)	511/547 (93.4)	621/671 (92.5)
Williams	11/11 (100.0)	92/94 (97.9)	103/105 (98.1)
Total	121/135 (89.6)	603/641 (94.1)	724/776 (93.3)

Total DNA isolated from putative transgenic plants was amplified by PCR using primers specific for either the hygromycin phosphotransferase gene (*hph*) or the neomycin phosphotransferase gene (*neo*). For each tested cultivar-gene combination, the number of positive plants per the number of plants analyzed is shown. In parentheses, relative frequency of PCR-positive plants is expressed in percentages.

Table 5. Overview of Southern hybridization analysis experiments in transgenic plants of two banana cultivars. (Remy et al., in prep.)

Cultivar	Selectable marker gene		Total
	hph	*neo*	
Three Hand Planty	13/15	29/32	42/47
Williams	1/1	7/8	8/9
Total	14/16	36/40	50/56

Total DNA isolated from independent transgenic plants was hybridized to a probe specific for either the hygromycin phosphotransferase gene (*hph*) or the neomycin phosphotransferase gene (*neo*). For each tested cultivar-gene combination, the number of positive plants per the number of plants analyzed is shown.

the *neo* gene was reproducibly detected in all the PCR-positive samples assayed so far.

More than 200 independent lines have been transferred to a greenhouse for further analyses and for observation of off-types in the vegetative phase. A significant part of these lines has now been multiplied and prepared for testing in the field.

Co-transformation of Linked and Unlinked Foreign Genes. Long-term and multiple disease resistance by molecular improvement is likely to be achieved by integrating numerous genes with different modes of action into the plant genome. Technically, this can be done in several consecutive steps or simultaneously. Particle bombardment of embryogenic cell cultures, the present method of choice for genetic transformation of banana, relies on the time-consuming development of highly embryogenic cultures from a number of target cultivars which makes consecutive transformations not practical. On the other hand, simultaneous gene transfer can be performed by co-precipitation of a mixture of chimeric gene constructs onto microparticles before bombardment. Therefore, several bombardment experiments were performed to assess co-transformation frequencies of linked and unlinked genes (Remy et al. 1998).

PCR screening of transgenic plants revealed that, as one might expect, linked genes co-existed at a high frequency that ranged between 90 and 100% in the different experiments (Table 2). Similarly, as expected, the unlinked gene showed a lower co-transformation frequency than the linked genes. However, this frequency was still remarkably high, in the range of 70 to 80%, probably due to efficient co-precipitation of the two plasmids onto the microparticles. As in the results obtained by Hadi et al. (1996), this observation indicates that simultaneous bombardment of different plasmid molecules may be a convenient way for the introduction and co-expression of multiple genes into crop plants.

2.3.3 Agrobacterium-*mediated Transformation*

The *Agrobacterium*-based banana transformation system utilizes apical meristematic tissues (May et al. 1995). Meristems prepared from 100 in vitro plantlets each of the dessert banana cultivars Grands Naine and Williams were wounded by particle bombardment using uncoated particles. After a 4-day recovery period, these meristems were co-cultivated with *Agrobacterium tumefaciens* harboring the plant transformation vector pTOK233 (Hiei et al. 1994) in the presence of acetosyringone, an inducer of the *Agrobacterium* virulence genes. This transformation vector contains a chimeric *gusA* gene with an intron in the coding sequence (Ohta et al. 1990) which prevents expression of the *gusA* reporter gene in the bacteria. After 3 days of co-cultivation, no TGE was observed, while after 6 days of co-cultivation with *A. tumefaciens*, 2 and 3% of the Grande Naine and Williams explants showed TGE, respectively. This observation suggests that T-DNA transfer to banana may indeed occur, though at a low frequency.

In order to evaluate the reasons for the possible recalcitrance of this system to *Agrobacterium* we investigated two early steps in the plant-bacterium interaction, i.e. (1) chemotaxis to exudates of different excised or extensively wounded tissues, and (2) attachment of agrobacteria to single cells or different tissues. Chemotaxis was studied by a swarm agar plate assay that is described in detail elsewhere (Pérez Hernández et al. 1999). Our results demonstrated that excised leaf, corm, root and in vitro proliferating tissues from different banana cultivars were able to elicit a positive chemotactic reaction of *A. tumefaciens* that could be enhanced by extensive wounding (Fig. 2). In addition, physical binding of agrobacteria to banana cells and tissues was confirmed by light and fluorescent microscopy as well as by scanning electron microscopy (Fig. 3). On the basis of these observations it can be concluded that the interaction of *A. tumefaciens* with banana is not inhibited in these early steps which may be a basis for the development of efficient *Agrobacterium*-mediated transformation protocols in this crop. Indeed, recently we have been able to obtain a high frequency of transient GUS expression after

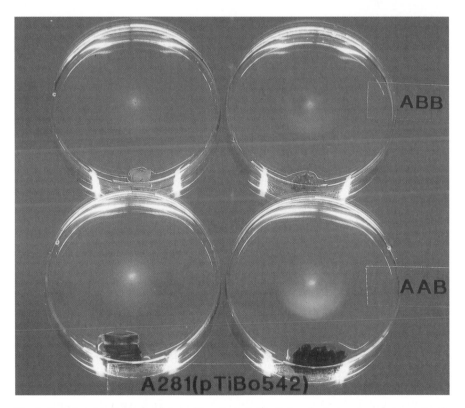

Fig. 2. Swarm agar plates showing positive chemotaxis of *Agrobacterium tumefaciens* towards excised (on the *left* side) and wounded (on the *right* side) explants from corms of the cooking banana Bluggoe (*upper row*) and from roots of the cultivar Pome (*bottom row*). (Pérez Hernández et al. 1999)

Fig. 3. Scanning electron micrograph of *Agrobacterium tumefaciens* attached to root hairs of the banana cultivar Pome reveals massive aggregation in a cap form over the root tip; *Bar* 20μm. (Pérez Hernández et al. 1999)

Agrobacterium-mediated transformation of embryogenic cell suspension cultures and have regenerated the first transgenic shoots (Pérez Hernández et al. 1998).

3 Applications and Prospects

Since the most significant damage to banana production is caused by fungal pathogens and host resistance genes remain to be identified, heterologous genes encoding proteins with antifungal activity (Broekaert et al. 1997) are the primary targets for expression in banana. Perhaps the most promising candidates are the recently described new types of antifungal proteins (AFPs) which are stable, cysteine-rich small peptides isolated from seeds of diverse plant species (Broekaert et al. 1992; Cammue et al. 1992; Terras et al. 1992; Cammue et al. 1995; Osborn et al. 1995). These AFPs have a broad antifungal spectrum, and show high in vitro antifungal activity to *Mycosphaerella fijiensis* and *Fusarium oxysporum*, the main fungal pathogens in banana, while they exert no toxicity to human or banana cells (Cammue et al. 1993). One of the

AFPs isolated from radish has recently been expressed in transgenic tobacco which resulted in an increased resistance to *Alternaria longipes* (Terras et al. 1995). It is expected that the expression of these AFPs in transgenic banana plants combined with cytological and ultrastructural analysis of the fungus-banana interaction (Beveraggi et al. 1993) in planta will lead to a more detailed understanding of the infection process and symptom development. Furthermore, high and tissue-specific expression of AFPs may result in the production of commercially acceptable fungal disease-resistant bananas.

Engineering resistance to banana bunchy top virus (BBTV) is another obvious objective of banana molecular breeding. Recently, Harding et al. (1991) and Thomas and Dietzgen (1991) reported the association of small ssDNA with isolated virus-like particles as well as with the bunchy top disease and its transmission as demonstrated using a cloned DNA probe and virion-specific monoclonal antibodies. The BBTV genome has six different ssDNA components which have now been cloned and sequenced: each of these components but one contains one large open reading frame (ORF; Burns et al. 1995). One of these ORFs has been identified by Harding et al. (1993) as the putative replicase gene which can be used to prevent BBTV replication in banana. Components of the BBTV genome have also been recently isolated by another research group (Yeh et al. 1994) who confirmed the above findings. It appears, however, that there are two strains of BBTV: an Asian and a South Pacific strain (Karan et al. 1994; Xie and Hu 1995) and, therefore, different constructs may be required to produce resistance to these two strains.

The recent cloning and sequencing of the C-terminal region of the coat protein and the 3′ terminal untranslated region of banana bract mosaic virus (Bateson and Dale 1995) opens the opportunity for using pathogen-derived resistance to this virus, too. Transformation experiments are now in progress in this direction in the authors' laboratory.

4 Present Status

More than 100 transgenic lines have been micropropagated for testing the stability of the transgenes in different field environments. Also, close to 200 individual plants have been transferred to the greenhouse for the evaluation of morphological uniformity and of the stability of transgene integration in subsequent vegetative progenies. Preliminary results obtained with the Cavendish type cultivar Williams indicate that transgene integration is stable as the Southern hybridization pattern has not changed up to the fourth vegetative generation (Swennen et al. 1998).

To date no transgenic bananas have entered the phase of field testing. This is basically due to the lack of biosafety guidelines and/or functional regulatory bodies in many tropical countries where banana is cultivated. This situation appears to be a major bottleneck to progress in transgenic field testing of tropical crops, especially in Sub-Saharan Africa.

Table 6. Strategies for molecular improvement of banana by in planta expression of foreign genes to control pathogens and pests

Pathogen or pest	Target genes encoding
Viruses (bunchy top, bract mosaic, etc.)	Replicase, coat protein
Fungi (*Mycosphaerella, Fusarium*)	Antifungal peptides, hydrolytic enzymes
Nematodes (*Radopholus, Pratylenchus*, etc.)	Protease inhibitors, lectins

5 Summary and Conclusion

As for many other crops, particle bombardment and co-cultivation with *Agrobacterium tumefaciens* are the most efficient gene delivery methods for banana transformation. Using particle bombardment of embryogenic cell suspension cultures of cultivars representing the genetic diversity of banana (Table 3), hundreds of transgenic plants have been regenerated with confirmed integration and expression of the introduced genes. More recently, *Agrobacterium*-mediated transformation has also been found to be compatible with banana cells and tissues.

The described transformation techniques have great potential for molecular improvement in banana, especially in controlling the major diseases and pests of this crop. Introduction of available genes with agronomic importance (Table 6) may allow the assessment of various strategies in banana for protection against viruses, fungal pathogens and nematode pests. One of the most promising among these strategies is the expression in banana of antifungal peptides or proteins isolated from various plant species which exert fungistatic activity on *Mycosphaerella fijiensis* and *Fusarium oxysporum*, the causal agents of black sigatoka and Panama disease, respectively. Another practical example is the introduction of the coat protein gene of banana bract mosaic virus into banana. It is expected that in the near future this list of applications will be further extended.

Acknowledgements. Financial support from the International Network for Improvement of Banana and Plantain (INIBAP) through a grant of the Belgian Administration for Development Cooperation (BADC/ABOS) and from the CFC/FAO/World Bank Banana Improvement Project is acknowledged. Technical contributions and assistance from Isabelle François, Inge Holsbeeks, An Buyens and Nico Smets is appreciated.

References

Anonymous (1995) FAO production yearbook, 1994, vol 48. Food and Agriculture Organization of the United Nations, Rome, 243 pp

Anonymous (1996) FAO trade yearbook, 1995, vol 49, Food and Agriculture Organization of the United Nations, Rome, 383 pp

Bateson MF, Dale JL (1995) Banana bract mosaic virus: characterisation using potyvirus specific degenerate PCR primers. Arch Virol 140:515–527

Beveraggi A, Mourichon X, Sallé G (1993) Study of host-parasite interactions in susceptible and resistant bananas inoculated with *Cercospora fijiensis*, pathogen of black leaf streak disease.

In: Ganry J (ed) Genetic improvement of banana and plantain for resistance to diseases and pests. CIRAD/INIBAP/CTA, Montpellier, France, pp 171–192

Broekaert WF, Marien W, Terras FRG, De Bolle MFC, Proost P, Van Damme J, Dillen L, Claeys M, Rees SB, Vanderleyden J, Cammue BPA (1992) Antimicrobial peptides from *Amaranthus caudatus* seeds with sequence homology to the cysteine/glycine-rich domain of chitin-binding proteins. Biochemistry 31:4308–4314

Broekaert WF, Cammue BPA, De Bolle M, Thevissen K, De Samblanx G, Osborn RW (1997) Antimicrobial peptides from plants. Crit Rev Plant Sci 16:297–323

Burns TM, Harding RM, Dale JL (1995) The genome organization of banana bunchy top virus: analysis of six ssDNA components. J Gen Virol 76:1471–1482

Cammue BPA, De Bolle MFC, Terras FRG, Proost P, Van Damme J, Rees SB, Vanderleyden J, Broekaert WF (1992) Isolation and characterization of a novel class of plant antimicrobial peptides from *Mirabilis jalapa* L. seeds. J Biol Chem 267:2228–2233

Cammue BPA, De Bolle MFC, Terras FRG, Broekaert WF (1993) Fungal disease control in *Musa*: application of new antifungal proteins. In: Ganry J (ed) Genetic improvement of banana and plantain for resistance of diseases and pests. CIRAD/INIBAP/CTA, Montpellier, France, pp 221–225

Cammue BPA, Thevissen K, Hendriks M, Eggermont K, Goderis IJ, Proost P, Van Damme J, Osborn RW, Guerbette F, Kader J-C, Broekaert WF (1995) A potent antimicrobial protein from onion (*Allium cepa* L.) seeds showing sequence homology to plant lipid transfer proteins. Plant Physiol 109:445–455

Dhed'a D, Dumortier F, Panis B, Vuylsteke D, De Langhe E (1991) Plant regeneration in cell suspension cultures of the cooking banana cv. 'Bluggoe' (*Musa* spp. ABB group). Fruits 46:125–135

Hadi MZ, McMullen MD, Finer JJ (1996) Transformation of 12 different plasmids into soybean via particle bombardment. Plant Cell Rep 15:500–505

Harding RM, Burns TM, Dale JL (1991) Virus-like particles associated with banana bunchy top disease contain small single-stranded DNA. J Gen Virol 72:225–230

Harding RM, Burns TM, Hafner G, Dietzgen RG, Dale JL (1993) Nucleotide sequence of one component of the banana bunchy top virus genome contains a putative replicase gene. J Gen Virol 74:323–328

Hiei Y, Ohta S, Komari T, Kumashiro T (1994) Efficient transformation of rice (*Oryza sativa* L.) mediated by *Agrobacterium* and sequence analysis of the boundaries of the T-DNA. Plant J 6:271–282

Jefferson RA (1987) Assaying chimeric genes in plants: the GUS gene fusion system. Plant Mol Biol Rep 5:387–405

Karan M, Harding RM, Dale JL (1994) Evidence for two groups of banana bunchy top virus isolates. J Gen Virol 75:3541–3546

Lockhart BEL (1986) Purification and serology of a bacilliform virus associated with banana streak disease. Phytopathology 76:995–999

May GD, Rownak A, Mason H, Wiecko A, Novak FJ, Arntzen CJ (1995) Generation of transgenic banana (*Musa acuminata*) plants via *Agrobacterium*-mediated transformation. Bio/Technology 13:486–492

Mobambo KN, Gauhl F, Vuylsteke D, Ortiz R, Pasberg-Gauhl C, Swennen R (1993) Yield loss in plantain from black sigatoka leaf spot and field performance of resistant hybrids. Field Crops Res 35:35–42

Murashige T, Skoog F (1962) A revised medium for rapid growth and bioassays with tobacco tissue cultures. Physiol Plant 15:473–497

Neuhaus-Url G, Neuhaus G (1993) The use of the nonradioactive digoxigenin chemiluminescent technology for plant genomic Southern blot hybridization: a comparison with radioactivity. Transgen Res 2:115–120

Ohta S, Mita S, Hattori T, Nakamura K (1990) Construction and expression in tobacco of a β-glucuronidase (GUS) reporter gene containing an intron within the coding sequence. Plant Cell Physiol 31:805–813

Osborn RW, De Samblanx GW, Thevissen K, Goderis I, Torrekens S, Van Leuven F, Attenborough S, Rees S, Broekaert WF (1995) Isolation and characterisation of plant defensins from seeds of Asteraceae, Fabaceae, Hippocastaneaceae and Saxifragaceae. FEBS Lett 368:257–262

Panis B, Van Wauwe A, Swennen R (1993) Plant regeneration through somatic embryogenesis from protoplasts of banana (*Musa* spp.). Plant Cell Rep 12:403–407

Panis B, Sági L, Swennen R (1994) Regeneration of plants from protoplasts of *Musa* species (banana). In: Bajaj YPS (ed) Biotechnology in agriculture and forestry, vol 29. Plant protoplasts and genetic engineering V. Springer, Berlin Heidelberg New York, pp 100–112

Pérez Hernádez JB, Remy S, Galán Saúco V, Swennen R, Sági L (1998) Chemotaxis, attachment and transgene expression in the *Agrobacterium*-mediated banana transformation system. Med Fac Landbouww Univ Gent 63/4b:1603–1606

Pérez Hernández JB, Remy S, Galán Saúco V, Swennen R, Sági L (1999) Chemotactic movement and attachment of *Agrobacterium tumefaciens* to banana cells and tissues. J Plant Physiol 155:245–250

Perl A, Kless H, Blumenthal A, Galili G, Galun E (1992) Improvement of plant regeneration and GUS expression in scutellar wheat calli by optimization of culture conditions and DNA-microprojectile delivery procedures. Mol Gen Genet 235:279–284

Ploetz RC (1990) *Fusarium* wilt of banana. APS Press, St Paul, Minnesota

Remy S, François I, Cammue BPA, Swennen R, Sági L (1998) Co-transformation as a potential tool to create multiple and durable resistance in banana (*Musa* spp.). Acta Hortic 461:361–365

Rodoni BC, Ahlawat YS, Varma A, Dale JL, Harding RM (1997) Identification and characterization of banana bract mosaic virus in India. Plant Dis 81:669–672

Sági L, Remy S, Panis B, Swennen R, Volckaert G (1994) Transient gene expression in electroporated banana (*Musa* spp., cv. 'Bluggoe', ABB group) protoplasts isolated from regenerable embryogenic cell suspensions. Plant Cell Rep 13:262–266

Sági L, Remy S, Verelst B, Swennen R, Panis B (1995a) Stable and transient genetic transformation of banana (*Musa* spp.) protoplasts and cells. In: Bajaj YPS (ed) Biotechnology in agriculture and forestry, vol 34. Plant protoplasts and genetic engineering VI. Springer, Berlin Heidelberg New York, pp 214–227

Sági L, Panis B, Remy S, Schoofs H, De Smet K, Remy S, Swennen R, Cammue BPA (1995b) Genetic transformation of banana and plantain (*Musa* spp.) via particle bombardment. Bio/Technology 13:481–485

Sági L, Remy S, Verelst B, Panis B, Cammue BPA, Volckaert G, Swennen R (1995c) Transient gene expression in transformed banana (*Musa* spp., cv. 'Bluggoe') protoplasts and embryogenic cell suspensions. Euphytica 85:89–95

Stover RH (1983) Effet du *Cercospora* noir sur les plantains en Amérique Centrale. Fruits 38:326–329

Swennen R, Van den Houwe I, Remy S, Sági L, Schoofs H (1998) Biotechnological approaches for the improvement of Cavendish bananas. Acta Hortic 490:415–423

Tada Y, Sakamoto M, Fujimura T (1990) Efficient gene introduction into rice by electroporation and analysis of transgenic plants: use of electroporation buffer lacking chloride ions. Theor Appl Genet 80:475–480

Terras FRG, Schoofs HME, De Bolle MFC, Van Leuven F, Rees SB, Vanderleyden J, Cammue BPA, Broekaert WF (1992) Analysis of two novel classes of plant antifungal proteins from radish (*Raphanus sativus* L.) seeds. J Biol Chem 267:15301–15309

Terras FRG, Eggermont K, Kovaleva V, Raikhel N, Osborn R, Kester A, Rees SB, Vanderleyden J, Cammue BPA, Broekaert WF (1995) Small cysteine-rich antifungal proteins from radish (*Raphanus sativus* L.): their role in host defense and their constitutive expression in transgenic tobacco leading to enhanced resistance to a fungal disease. Plant Cell 7:573–588

Thomas JE, Dietzgen RG (1991) Purification, characterization and serological detection of virus-like particles associated with banana bunchy top disease in Australia. J Gen Virol 72:217–224

Xie WS, Hu JS (1995) Molecular cloning, sequence analysis, and detection of banana bunchy top virus in Hawaii. Phytopathology 85:339–347

Yeh HH, Su HJ, Chao YC (1994) Genome characterization and identification of viral-associated dsDNA component of banana bunchy top virus. Virology 198:645–652

18 Transgenic Common Bean (*Phaseolus vulgaris*)

F.J.L. ARAGÃO and E.L. RECH

1 Introduction

The common bean has been grown on more than 12 million ha and constitutes the most important food legume for direct consumption of more than 500 million people in Latin America and Africa (FAO 1995). Beans have been a very important source of protein and calories for these two continents. The world annual production is about 8.5 million metric tons. Production occurs in a wide range of cropping systems and diverse environments in Latin America, Africa, the Middle East, China, Europe, the United States and Canada (Schwartz and Pastor-Corrales 1989). In Latin America and Africa, beans have been primarily cultivated on small farms with few purchased inputs, in association with other crops.

Despite its nutritional importance, productivity has been declining in some regions. The average national yield is about 600 kg/ha, but beans possess the potential to yield over 4000 kg/ha. Countries from Latin America and Africa yield 500 to 600 kg/ha while West Asia and North America yield 1100 to 1500 kg/ha. The main limiting factors are: poor agronomic practices, diseases, insects, nutritional deficiencies, soils, climate constraints, lack of improved cultivars and weed competition. Consequently, there is considerable interest in the introduction of agronomically useful traits into beans by breeding and genetic engineering.

Desirable characteristics have been searched for within germplasm banks, with considerable success. For self-pollinated crops such as the common bean, the breeding methods most commonly used are bulk, pedigree, back cross, and their modifications. In general, inter-specific hybridization and exploration of genetic variability is very limited within the genus *Phaseolus*. In addition, some barriers to hybridization and genetic recombination have been observed within *P. vulgaris*, with great consequences for genetic and breeding studies (Shii et al. 1981; Grafton et al. 1983; Rabakoarihanta and Baggett 1983; Gepts and Bliss 1985). Another fact which has to be considered, is that both inter- and intraspecific hybridization produce hybrid plants with undesirable characteristics. Thus, an extra period in the breeding program is necessary to obtain a commercial cultivar. Despite this drawback, the utilization of sexual

Embrapa Recursos Genéticos e Biotecnologia Parque Estação Biológica, Final Av. W3 Norte 70.770-900 Brasíia, DF, Brazil hybridization has resulted in several important examples of gene transfer, on which many of our current cultivars have been based.

With the advent of recombinant DNA technology, genes can be cloned from any organism, independent of the kingdom. These genes can be introduced and expressed in crop plants. Thus, the source of variability available for crop improvement has been expanded to all living organisms. Indeed, this technology, associated with sexual breeding methods, may accelerate the production of plants with useful traits.

2 Genetic Transformation

2.1 Gene Transfer in *Phaseolus*

The application of plant genetic engineering in bean was limited by the absence of an efficient methodology of introduction, integration and expression of cloned genes into the plant genome. Considerable advances have been achieved in methodologies for plant transformation, mainly using *Agrobacterium*, electroporation and biolistics.

The susceptibility of bean to *Agrobacterium* has been demonstrated, and some transgenic tissues, such as callus, meristems, cotyledon and hypocotyl, have been achieved (McClean et al. 1991; Franklin et al. 1993; Lewis and Bliss 1994; Brasileiro et al. 1996; Table 1). However, it has not been possible to regenerate transgenic plants. Mariotti et al. (1989) reported the production of transgenic bean plants through the utilization of the *Agrobacterium* system. However, there was no molecular evidence for genetic transformation or progeny analysis. Recently, Dillen et al. (1997) have described a methodology to transform *P. accutifolius* utilizing *Agrobacterium tumefaciens*.

Using the electroporation of protoplasts and tissues, it has been possible to introduce and express the *gus* (*uidA*) gene in bean cells. However, due to the impossibility of regenerating plants from the transformed cells, no transgenic plants have been obtained (Crepy et al. 1986; Dillen et al. 1995).

During the last decade, efforts to achieve an efficient methodology for common bean transformation have been obstructed due to the difficulties in regenerating plants from transformed cells. Numerous attempts have been made to regenerate bean plants from several types of isolated cells and tissues. Although no satisfactory results have been achieved, some methodologies have described shoot organogenesis (through multiple shoot induction) of the apical and axillary meristems from bean embryonic axis (McClean and Grafton 1989; Malik and Saxena 1992; Mohamed et al. 1992, 1993; Aragão et al. 1996; Aragão and Rech 1997). This background information, in association with the biolistic process, allowed us to recover transformed bean plants.

The possibility of combining the *Agrobacterium* and biolistics systems has been recently proposed (Brasileiro et al. 1996). It would associate the high efficiency of the *Agrobacterium* system to integrate T-DNA in a susceptible host

Table 1. Summary of various studies conducted to introduce and express foreign genes in *Phaseolus vulgaris*

Reference	Explant	Method[a]	Progeny	Observations
Mariotti et al. (1989)	Stem	Agr	No	Expression of the *gus* gene
McClean et al. (1991)	Cotyledonary node, hypocotyl	Agr	No	Expression of the *nptII* gene
Genga et al. (1991)	Embryonic axis, cotyledon	B	No	Expression of the *gus* gene
Becker et al. (1991)	Cotyledonary and leaf nodes.	Agr	No	Expression of the *gus* and *nptII* genes
Aragão et al. (1992)	Embryonic axis	B	No	Transient expression of the 2S protein from Brazil nut (*be 2s2*) and *gus* genes
Franklin et al. (1993)	Leaf disc, hypocotyl	Agr	No	Transgenic callus expressing the *gus* and *nptII* genes
Aragão et al. (1993)	Embryonic axis, leaf, cotyledon	B	No	Factors influencing gene introduction by the biolistic process; expression of the *gus* gene
Russel et al. (1993)	Embryonic axis	B	Yes, F_1	Transgenic navy beans with a frequency of 0.03%; expression of the *gus*, *pat* and *BGMV* coat protein genes
Lewis and Bliss (1994)	Seedling	Agr	No	Expression of the *gus* genes
Grossi de Sá et al. (1994)	Embryonic axis	B	No	Transient expression of the *gus* gene
Dillen et al. (1995)	Seedling	Elec	No	Transient expression of the *gus* gene
Brasileiro et al. (1996)	Embryonic axis, seedling	Agr; B	No	Expression of the *gus* gene. Experiments in combining the *Agrobacterium* and biolistic process systems
Aragão et al. (1996)	Embryonic axis	B	Yes, F_3	Transgenic common beans with a frequency of 0.9%. Expression of the 2S protein, *gus*, *nptII*, and *AL1-AL2-AL3* and *BC1* from BGMV genes
Kim and Minamikawa (1996)	Embryonic axis	B	Yes, F_1	Transgenic plants expressing the *gus* genes
Nagl et al. (1997)	Cotyledonary node	Agr	No	Expression of the *gus*, *nptII* and *pat* genes
Aragão and Rech (1997)	Embryonic axis	B	Yes, F_1	Transgenic plants. Morphology influencing transformation
Kim and Minamikawa (1997)	Embryonic axis	B	Yes, F_1	Expression of the *gus* gene
Aragão et al. (1998)	Embryonic axis	B	Yes, F_3, F_4	Expression of the *AC1*, *TrAP*, *REn* and *BC1* genes from BGMV
Aragão et al. (1999)	Embryonic axis	B	Yes, F_5	Expressions of the gene *be2s2* from Brazil nut 2S protein

[a] Agr = *Agrobacterium*; B = biolistic; Elec = electroporation.

with the capacity of introducing genetic information into any living cell by the biolistic process. Our studies have demonstrated that it is possible to combine *Agrobacterium*/biolistic systems to transform bean apical meristems (Brasileiro et al. 1996). The micro-wounds caused by the microparticle bombardment (biolistic) can greatly enhance the frequency of *Agrobacterium*-mediated transformation. However, we have been unable to recover transgenic bean plants. Further investigation will be necessary in order to confirm the practical utility of the combined techniques for bean transformation.

Early efforts to achieve transgenic bean through the biolistic process have shown the possibility of introducing and expressing genes in any type of tissue (Aragão et al. 1992, 1993). Russel et al. (1993) have obtained transgenic navy bean plants using an electrical-discharge particle acceleration device. Nevertheless, the efficiency of this transformation protocol was extremely low, i.e., 0.03% germ line transformed plants per bombarded embryonic axis. Recently, we have developed an efficient and cultivar-independent bean transformation system based on microparticle bombardment of the apical meristematic region followed by multiple shoot induction (Aragão et al. 1996; Aragão and Rech 1997).

2.2 Transformation of *P. vulgaris* by Bombardment of Meristematic Cells

To achieve a transgenic plant regenerated from a single cell or a group of cells, there are two basic and essential requirements: the cells should be competent for (1) transformation and (2) regeneration.

The biolistic process has allowed the bombardment of intact meristem cells, which, followed by cytokinin-induced shoot organogenesis, has resulted in plants whose germ line has been transformed. The system developed allows the regeneration of transgenic bean plants without selection (Aragão and Rech 1997).

One of the advantages of the system, based on the exploitation of the biolistic process, is the short period of in vitro culture, which may reduce the possibilities of somaclonal variation (Irvine et al. 1991) and other abnormalities such as aneuploids (De Block 1993; Son et al. 1993). In addition, the process has allowed significant advances in the studies of gene expression in any type of cellular tissue and the in situ evaluation of promoters which are spatially and/or temporally regulated in specific tissue (Grossi de Sá et al. 1994).

The apical meristem has been the target of several basic studies. Most of them have investigated the function of different meristematic cell structures. However, molecular studies are still limited, possibly due to the difficulties in obtaining specific types of cells in large enough quantities (Hara 1995).

Studies on the development of the apical meristematic region of beans are rare. Consequently, there is some confusion regarding the terms and definitions employed to describe some tissues in the apical region. According to most of the concepts, the apical region is constituted by the meristem itself, the primordia of the lateral organs and the mature region (Cutter 1965). The bean apical meristematic region is composed of the apical meristem, the leaf primordia and the primary leaves (Fig. 1).

The different parts of the meristem have been defined by two distinct concepts: layers and zones (Medford 1992). The first defines three distinct cell layers (L1, L2 and L3; Satina et al. 1940). Layer L1 is the most external and

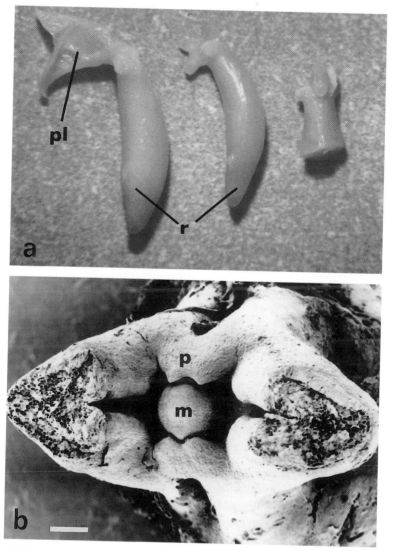

Fig. 1. a Embryonic axis is excised from bean seeds, the apical meristem is exposed by removing the primary leaves (*pl*), and the rootlet (*r*) is cut. **b** The embryonic axes are placed with the apical meristematic region (*m*) directed upwards in the bombardment medium, immediately prior to the bombardment. *Bar* 100 μm; *p* primordia leaf. (F.J.L. Aragão, unpubl.)

forms the epidermis of the differentiated regions. Layer L2 divides preferentially in the anticlinal plane (perpendicular to the surface) and in the periclinal plane (parallel to the surface), when the organs are formed. Layer L3 divides in the anticlinal and periclinal plane. The second concept is based on meristem zones (Esaú 1977; Steeves and Sussex 1989; Lyndon 1990), with one central zone, a peripheral zone and a rib zone. The central zone includes all three layers. The cells in this zone usually divide with a lower frequency. Despite being the origin of other cells of the apical meristem, this type of cells is not permanent (Ruth et al. 1985; Lyndon 1990). All the peripheral cells have a function in the formation of the meristem lateral region.

The utilization of the apical meristematic region to clonally multiply plants was first demonstrated by Morel (1960), studying *Cymbidium* micropropagation. Since then, significant advances in the development of several micropropagation techniques have been achieved. The induction of multiple shooting in the apical and lateral meristematic region in bean has been demonstrated (Kartha et al. 1981; Martins and Sondahl 1984). Neoformation of shoots in the meristems has been induced by cytokinin (N^6 benzylaminopurine, BAP). The induction of morphogenesis in bean apical meristems has been achieved with success, through the culture of mature embryos in high doses of different cytokinins, such as kinetin, zeatin and BAP (McClean and Grafton 1989; Franklin et al. 1991; Malik and Saxena 1992; Mohamed et al. 1992, 1993; Aragão et al. 1996; Aragão and Rech 1997). Compounds that present effects similar to cytokinin, such as TDZ (thidiazuron) and CPPU (forclhorofenuron; Mohamed et al. 1992), have also been evaluated. These compounds are not cytokinins, however, they can induce the accumulation of the endogenous cytokinins of the plant, through the inhibition of cytokinin oxidase (Hare and Van Staden 1994). However, the utilization of BAP has been the most effective.

Several studies have demonstrated that the differentiated de novo shoots originate from the sub-epidermal layers (L2 and L3) of the apical meristem, though the L1 layer could participate in their formation (McClean and Grafton 1989; Franklin et al. 1991; Malik and Saxena 1992; Mohamed et al. 1992). These shoots have been formed in the peripheral regions of the apical meristem (Aragão and Rech 1997).

The biolistic process can de defined as a universal process to introduce macromolecules into any living cell. In plants, the possibility of directly introducing desirable genes into meristems is not limited by genotype or cultivar (Sanford 1990; Christou 1993). However, the morphology of the explant utilized during the bombardment process may greatly influence the successful recuperation achievement of transgenic bean plants (Aragão and Rech 1997). In some cultivars, the embryonic axes revealed that the apical meristematic region is partially exposed and only the central region could be visualized. Thus, the number of meristematic cells which could be reached by the microparticle coated-DNA, will be drastically reduced. Consequently, the efficiency of transformation could also be reduced. The cells located in the central zone of the meristematic region have been indicated as the center of the morphological and histological activities of the shoot apex (Medford 1992; Hara 1995).

However, several studies have shown that de novo shoot differentiation in embryos of bean cultivated on cytokinins appeared in the peripheral layers of the meristematic ring (McClean and Grafton 1989; Franklin et al. 1991; Malik and Saxena 1992; Aragão and Rech 1997). Thus, based on these concepts, cultivars with a non-exposed apical meristematic region are not suitable for biolistic transformation, since removal of the leaf primordia is not practical.

Though a great number of shoots have been formed after 15 days culture on MS basal medium containing BAP, only a small number of shoots have developed as much as 1 cm. When these shoots were excised, those remaining have elongated, maintaining a constant number of shoots per axis during a 3-month period of cultivation. This apical dominance-like effect may suggest a regulatory response mechanism (Aragão and Rech 1997).

3 Methodology

3.1 Embryonic Axes Preparation

Explant preparation was carried out essentially as described by Aragão et al. (1993). Basically, mature seeds of common bean (cv. Olathe or Carioca) were surface sterilized in 1% sodium hypochlorite for 20 min, and rinsed three times in sterile distilled water. The seeds were soaked in distilled water for 16–18 h. Then, the embryonic axes were excised from the seeds and the apical meristems were exposed by removing their primary leaves (Fig. 1). Subsequently, the rootlets were excised, and the embryonic axes were surface sterilized again in 0.1% sodium hypochlorite for 10 min. The prepared embryonic axes were rinsed three times with sterile distilled water. Ten to fifteen embryonic axes were placed with the apical region directed upwards in a 5-cm dish containing the bombardment medium: MS medium (Murashige and Skoog 1962), supplemented with 44.3 µM benzylaminopurine (BAP), 3% sucrose and 0.7% Phytagel (Sigma), immediately prior to the bombardment.

3.2 Preparation of Microparticles

3.2.1 Particle Sterilization

Weigh 60 mg of M-10 tungsten (Sylvania Inc., GTE Chemical-metals, Towanda, PA, USA) or gold (Aldrich 32,658-5) microparticles, and place in a microcentrifuge tube. Add 1 ml of 70% ethanol and vortex vigorously. Soak the particles for 15 min, centrifuge at 12000 rpm for 5 min and wash three times with distilled water. After the last wash, resuspend the microparticles in 1 ml glycerol 50% (v/v). The tungsten particle suspension can be stored at room temperature for 1–4 weeks (prolonged storage can led to oxidation of particle surfaces).

3.2.2 DNA Precipitation onto Microparticles

DNA was bound to tungsten M-10 or gold microparticles by mixing sequentially in a microcentrifuge tube: 50 µl microparticles (60 mg.ml^{-1} in 50% glycerol), 5 µl (1 µg.µl^{-1}) of plasmid DNA (containing the genes for useful traits, *nptII* (*neo*) gene for kanamycin selection and *gus* gene for transformation screening), 50 µl CaCl$_2$ (2.5 M) and 20 µl spermidine free-base (100 mM). After 10 min of incubation, the DNA-coated microparticles were centrifuged at 15 000 g for 10 s and the supernatant was removed. The pellet was washed with 150 µl of 70% ethanol and then with absolute ethanol. The final pellet was resuspended in 24 µl of absolute ethanol and sonicated for 2 s, just before use. Aliquots of 3 µl were spread onto the carrier membranes (Kapton, 2 mil, DuPont) and allowed to evaporate in a desiccator, under 12% relative humidity.

3.3 Microparticle Bombardment

A high pressure helium-driven particle acceleration device built in our laboratory (Aragão et al. 1996) and/or the PDS-1000 (BIORAD), as described by Sanford et al. (1991) was used for bombardment. The relative humidity in the biolistic laboratory has to be maintained at 50%, the gap distance from shock wave generator to the carrier membrane was 8 mm; the carrier membrane flying distance to the stopping screen was 13 mm; the DNA-coated microparticles flying distance to the target was 80 mm; the vacuum in the chamber was 27 in. of Hg and the helium pressure utilized in all experiments was 1200 psi.

3.4 Plant Growth: Selection, Shoot Culture and Acclimatization

After bombardment, the embryonic axes were cultivated for 1 week at 28 °C with a 16 h photoperiod (50 µmols m^{-2} s^{-1}) on the bombardment medium. Then, the germinated embryonic axes were transferred to MS medium containing 44.3 µM BAP, 100 mg l^{-1} kanamycin and 0.8% agar (Fisher). After 2 weeks, the embryonic axes developing shoots were transferred to MS medium without kanamycin and BAP to allow shoot elongation. As soon as the shoots reached 2- to 4-cm in length, they were excised at the stem basal level. A 1-mm section of the basal stem and the leaf tip was excised and assayed for the expression of the *gus* gene, as described by McCabe et al. (1988). Shoots expressing the *gus* gene were individually cultured on MS medium, with no growth regulators, 0.8% agar (Fisher), and the sucrose level was reduced to 1%. Once the plantlets were rooted, they were transferred to a plastic pot containing autoclaved soil:vermiculite (1:1) and covered with a plastic bag for a week. Plantlets were transferred to soil and allowed to set seeds. The temperature in the greenhouse was maintained at 25 °C and the relative humidity above 80%.

Table 2. Sequence of the oligonucleotides utilized as primers for PCR analysis (Aragão et al. 1996)

Gene	Position	Sequence (5′ → 3′)	Expected size of fragment (bp)
nptII	60	GAGGCTATTCGGCTATGACTG	410
	470c	TCGACAAGACCGGCTTCCATC	
gus	251	TTGGGCAGGCCAGCGTATCGT	420
	671c	ATCACGCAGTTCAACGCTGAC	

3.5 Progeny Analysis

The transgenes were detected in transformed mother-plants (R_0) and their progenies (R_1 to R_3) by polymerase chain reaction (PCR) and Southern blot.

For PCR analysis, DNA was isolated from leaf disks according to Edwards et al. (1991) and the reactions were carried out according to Aragão et al. (1996). The primers utilized to amplify a fragment within the *gus* and *nptII* coding sequences are listed in Table 2. For Southern blot analysis, genomic DNA was isolated according to Dellaporta et al. (1983) and hybridization was carried out as previously described by Sambrook et al. (1989).

4 Results and Discussion

We have developed an efficient and reproducible system to routinely achieve transgenic bean plants. The bean transformation system was based upon the development of a tissue culture protocol of multiple shoot induction, shoot elongation and rooting. In 21 independent experiments, 3079 embryonic axes were bombarded and 377 (9%) induced axes were obtained. After histochemical analysis of the basal stem and leaf tip for the expression of the *gus* gene, 27 (7%) of the generated shoots were *gus* positive. The average frequency of transformation (the total number of putative transgenic plants divided by the total number of bombarded embryonic axes) was 0.9%. Recently, Russel et al. (1993) have been able to achieve transgenic bean plants using an electrical particle acceleration device. However, the frequency of transgenic plants obtained was much lower (0.03%). In addition, their described tissue culture protocol was time consuming, involving several temperature treatments and medium transfers of the bombarded embryos prior to recovery of transgenic shoots.

All GUS-positive shoots were transferred to MS with 1% sucrose. As soon as the plantlets developed vigorous roots, they were acclimated and transferred to soil. All plants presented a normal phenotype, were fertile and have set pods and seeds. Histochemical GUS assays performed on leaves of all transgenic plants revealed intense enzymatic activity. The analysis of the R_0,

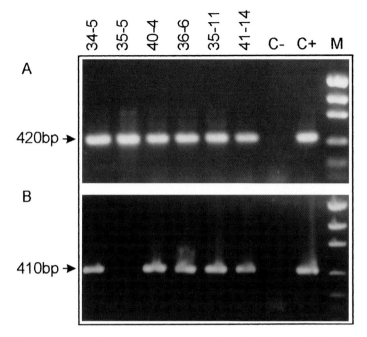

Fig. 2. PCR analysis of transformed bean plants containing the (**A**) *gus* and (**B**) *nptII* genes. The plants are indicated above the lanes, *C–* non-transformed plants, C+ positive control (pBI426), *M* molecular size standard (DNA mass-BRL). The expected fragments are indicated by an *arrow*. (Aragão et al. 1996)

R_1, R_2 and R_3 generations was conducted by detecting the introduced foreign genes by PCR (Fig. 2).

In order to evaluate the integration of the introduced foreign genes, Southern blot analysis of the genomic DNA from the R_3 generation of the transgenic bean plants was conducted. The results revealed the presence of a small number of integrated copies of the introduced genes in most transgenic lines (Fig. 3).

The progeny (R_1 generation) of the nine self-fertilized transgenic plants were screened by *gus* gene expression PCR analysis. Chi-square (χ^2) analysis indicated that most of them segregated in a Mendelian fashion (3:1; Table 3). The other two plants (26-16 and 26-2) revealed a non-Mendelian ratio (Table 3). Chimerism is frequent in transgenic plants obtained by the biolistic process (Christou 1990). However, this non-Mendelian ratio obtained in some plants was not due to chimerism, since the positions of transgenic seeds were randomly distributed on the mother plant (R_0). These results might suggest that the inserted foreign genes may cause some de-stabilization of the chromosome structure and a poor transgene transmission to the progeny. Moreover, an insertion mutation of an essential gene required for ovule fecundation and/or development might account for this aberrant inheritance.

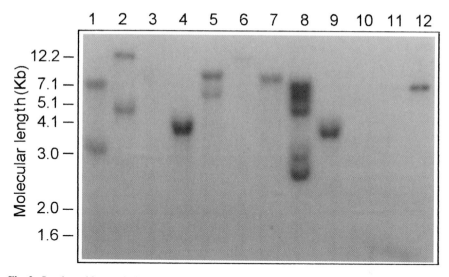

Fig. 3. Southern blot analysis of representative putative transformed lines in the R₃ generation. DNAs were digested with *Hin*dIII, transferred to a nylon membrane and probed with an internal fragment of the *nptII* gene. *Lanes 1-9* Different transformed lines: 10-33 (*lane 1*), 26-16 (*lane 2*), 26-1 (*lane 3*), 31-29 (*lane 4*), 34-5 (*lane 5*), 35-11 (*lane 6*), 36-6 (*lane 7*), 40-4 (*lane 8*), 41-14 (*lane 9*). *Lane 10* Non-transformed plant, *lane 11* pBI426 (10 pg), *lane 12* pBI426 (50 pg). Molecular size markers are indicated on the *left*. (Aragão et al. 1996)

Table 3. Segregation analysis of self-fertilized transgenic plants in the R₁ generation (Aragão et al. 1996)

R₀ plants	R₁ generation[a]		Segregation ratio	χ^2	P[b]
	Positive	Negative			
34-5	29	9	3:1	0.03	>0.8
35-11	20	6	3:1	0.05	>0.8
36-6	34	11	3:1	0.03	>0.8
40-4	15	6	3:1	0.14	>0.7
41-14	32	10	3:1	0.03	>0.8
26-16	6	51	3:1	126.3	<0.001
26-2	6	66	3:1	170.6	<0.001
10-33	28	8	3:1	0.14	>0.7
31-29	23	7	3:1	0.04	>0.8

[a] Data are based on histochemical assay for *gus* gene expression and PCR analysis.
[b] *P* is the probability that the observed ratios reflect the expected segregation ratio.

5 Present Status of Transgenic Bean Plants

We have transformed dry bean to introduce some useful traits. Transgenic bean (third and fourth generations) containing the genes *Rep-TrAP-REn* and *BC1* from bean golden mosaic virus, in antisense orientation were challenged against the virus. Two transgenic lines had both delayed and attenuated viral symptoms (Aragão et al. 1998). Plants containing the *be2s2* gene of the methionine-rich 2S protein from Brazil nut were analyzed in the fifth generation. In some transgenic lines, the methionine content was increased 14–23% over the control (Aragão et al. 1999).

The transgenic bean lines are being used in our bean-breeding program in order to evaluate the gene expression in different genetic backgrounds in both greenhouse and field conditions. Transgenic plants in the fifteenth generation are expressing the introduced genes stably.

6 Concluding Remarks

To achieve transgenic bean plants by particle bombardment of the apical meristematic region, there are three biological factors which must be evaluated. First, the morphology of the shoot apex is critical. Some cultivars have their apical meristematic region partially covered by leaf primordia (Aragão and Rech 1997). This problem could be solved by utilizing an appropriate cultivar within the desirable bean group or breeding each desirable line, aiming to achieve plants with small leaf primordia. Second, multiple shoot induction in the apical meristematic region must be feasible. In general, cytokinin or cytokinin-like compounds have been successfully utilized to culture bean embryos and promote shoot organogenesis (McClean and Grafton 1989; Franklin et al. 1991; Malik and Saxena 1992; Mohamed et al. 1992). Other cytokinins [6-furfuryllaminopurine (kinetin), isopentenylaminopurine (IPA) and zeatin] could be tested. In addition, changes in medium composition and incubation of the explants in the dark may enhance multiple shoot initiation (Mohamed et al. 1992). Third, the rooting of the shoots is essential for transfer of plants to soil. It is strongly influenced by the shoot length. Usually, shoots 1–2 cm long cultivated on medium with a reduced level of sucrose ($\leq 1\%$) formed adventitious roots within 10 days (Mohamed et al. 1992; Aragão et al. 1996; Aragão and Rech 1997).

These results are of particular importance to the understanding of the basic process of integration of foreign genes in bean, which, in turn, may form the foundation for the effective and practical use of genetic engineering to introduce traits such as protein quality and disease resistance in this important legume.

Acknowledgments. We would like to thank Prof. Linda S. Caldas for critical reading of the manuscript. This work was supported by Empresa Brasileira de Pesquisa Agropecuária (EMBRAPA) and Conselho Nacional de Desenvolvimento Científico e Tecnológico (CNPq), Brazil.

References

Aragão FJL, Rech EL (1997) Morphological factors influencing recovery of transgenic bean plants (*Phaseolus vulgaris* L.) of a carioca cultivar. Int J Plant Sci 158:157–163

Aragão FJL, Sá MFG, Almeida ER, Gander ES, Rech EL (1992) Particle bombardment-mediated transient expression of a Brazil nut methionine-rich albumin in bean (*Phaseolus vulgaris* L.). Plant Mol Biol 20:357–359

Aragão FJL, Sá MFG, Davey MR, Brasileiro ACM, Faria JC, Rech EL (1993) Factors influencing transient gene expression in bean (*Phaseolus vulgaris*) using an electrical particle acceleration device. Plant Cell Rep 12:483–490

Aragão FJL, Barros LMG, Brasileiro ACM, Ribeiro SG, Smith FD, Sanford JC, Faria JC, Rech EL (1996) Inheritance of foreign genes in transgenic bean (*Phaseolus vulgaris* L.) co-transformed via particle bombardment. Theor Appl Genet 93:142–150

Aragão FJL, Ribeiro SG, Barros LMG, Brasileiro ACM, Maxwell DP, Rech EL, Faria JC (1998) Transgenic beans (*Phaseolus vulgaris* L.) engineered to express viral antisense RNAs showed delayed and attenuated symptoms to bean golden mosaic geminivirus. Mol Breed 4: 491–499

Aragão FJL, Barros LMG, Sousa MV, Grossi de Sá MF, Almeida ERP, Gander ES, Rech EL (1999) Expression of a methionine-rich storage albumin from the Brazil nut (*Bertholletia excelsa* H.B.K. Lecythidaceae) in transgenic bean plants (*Phaseolus vulgaris* L. Fabaceae). Genetics Mol Biol 22:445–449

Becker J, Vogel T, Iqbal J, Nagl W (1994) *Agrobacterium* mediated transformation of *Phaseolus vulgaris*. Adaptation of some conditions. Annu Rep Bean Improv Coop USA 37:127–128

Brasileiro ACM, Aragão FJL, Rossi S, Dusi DMA, Barros LMG, Rech EL (1996) Susceptibility of common and tepari bean to *Agrobacterium* spp. strains and improvement of *Agrobacterium*-mediated trasformation using microprojectile bombardment. J Am Soc Hortic Sci 121:810–815

Christou P (1990) Morphological description of transgenic soybean chimeras created by the delivery, integration and expression of foreign DNA using electric discharge particle acceleration. Ann Bot 66:379–386

Christou P (1993) Particle gun mediated transformation. Curr Opin Biotechnol 4:135–141

Crepy L, Barros LMG, Valente VRN (1986) Callus production from leaf protoplasts of various cultivars of bean (*Phaseolus vulgaris* L.). Plant Cell Rep 5:124–126

Cutter EG (1965) Recent experimental studies of the shoot apex and shoot morphogenesis. Bot Rev 31:7–113

De Block M (1993) The cell biology of transformation: current state, problems, prospects and the implications for plant breeding. Euphytica 71:1–14

Dellaporta SL, Wood J, Hicks JB (1983) A plant DNA minipreparation: version II. Plant Mol Biol Rep 1:19–21

Dillen W, Engler G, Van Montagu M, Angenon G (1995) Electroporation-mediated DNA delivery to seedling tissues of *Phaseolus vulgaris* L. (common bean). Plant Cell Rep 15:119–124

Dillen W, De Clercq J, Groossens A, Van Montagu M, Angenon G (1997) *Agrobacterium*-mediated transformation of *Phaseolus acutifolius* A. Gray. Theor Appl Genet 94:151–158

Edwards K, Johnstone C, Thompson C (1991) A simple and rapid method for the preparation of plant genomic DNA for PCR analysis. Nucleic Acids Res 19:1349

Esau K (1997) Anatomy of seed plants. John Willey, New York

FAO – Food and Agriculture Organization of the United Nations (1995) FAO Production Yearbook 1994. FAO statistics series 125, vol 48, Rome

Franklin CI, Trieu TN, Gonzales RA, Dixon RA (1991) Plant regeneration from seedling explants of green bean (*Phaseolus vulgaris* L.) via organogenesis. Plant Cell Tissue Organ Cult 24:199–206

Franklin CI, Trieu TN, Cassidy BG, Dixon RA, Nelson RS (1993) Genetic transformation of green bean callus via *Agrobacterium* mediated DNA transfer. Plant Cell Rep 12:74–79

Genga A, Cerjotti A, Bollini R, Bernacchia G, Allavena A (1991) Transient gene expression in bean tissues by high velocity microprojectile bombardment. J Genet Breed 45:129–134

Gepts P, Bliss FA (1985) F1 hybrid weakness in the common bean: differential geographic origin suggests two gene pools in cultivated bean germphasm. J Hered 76:447–450

Grafton KF, Wyatt JE, Weiser GC (1983) Genetics of a vicescent foliage mutant in beans. J Hered 74:385

Grossi de Sá M, Weinberg DF, Rech EL, Barros LMG, Aragão FJL, Holmstroem KO, Gander ES (1994) Functional studies on a seed-specific promoter from a Brazil nut 2S gene. Plant Sci 103:189–198

Hara N (1995) Developmental anatomy of the three-dimensional structure of the vegetative shoot apex. J Plant Res 108:115–125

Hara PD, Van Staden J (1994) Cytokinin oxidase: biochemical features and physiological significance. Physiol Plant 91:128–136

Irvine JE, Benda GTA, Legendre BL, Machado GRJ (1991) The frequency of marker changes in sugarcane plants generated from callus culture. 2. Evidence for vegetative and genetic transmission, epigenetic effects and chemical disruption. Plant Cell Tissue Organ Cult 26: 115–125

Kartha KK, Pahl K, Leung NL, Mroginski LA (1981) Plant regeneration from meristems of grain legumes: soybean, cowpeas, peanut, chickpea and bean. Can J Bot 59:1671–1679

Kim JW, Minamikawa T (1996) Transformation and regeneration of Frebch bean plants by the particle bombardment process. Plant Sci 117:131–138

Kim JW, Minamikawa T (1997) Stable delivery of a canavalin promoter-β-glucuronidase gene fusion into French bean by particle bombardment. Plant Cell Physiol 38:70–75

Lewis ME, Bliss FA (1994) Tumor formation and β-glucuronidase expression in *Phaseolus vulgaris* inoculated with *Agrobacterium tumefaciens*. J Am Soc Hortic Sci 119:361–366

Lyndon RF (1990) Plant development: the cellular basis. Unwin Hyman, Winchester, Massachusetts

Malik KA, Saxena PK (1992) Regeneration in *Phaseolus vulgaris* L.: high-frequency induction of direct shoot formation in intact seedlings by N-benzylaminopurine and thidiazuron. Planta 186:384–389

Mariotti D, Fontana GS, Santini L (1989) Genetic transformation of grain legumes: *Phaseolus vulgaris* L. and *P. coccineus* L. J Genet Breed 43:77–82

Martins IS, Sondahl MR (1984) Axillary bud development from nodal cultures of bean seedlings (*Phaseolus vulgaris* L.). Turrialba 34:157–161

McCabe DE, Swain WF, Martinell BJ, Christou P (1988) Stable transformation of soybean (*Glycine max*) by particle acceleration. Bio/Technology 6:923–926

McClean P, Grafton KF (1989) Regeneration of dry bean (*Phaseolus vulgaris*) via organogenesis. Plant Sci 60:117–122

McClean P, Chee P, Held B, Simental J, Drong RF, Slightom J (1991) Susceptibility of dry bean (*Phaseolus vulgaris* L.) to *Agrobacterium* infection: transformation of cotyledonary and hypocotyl tissues. Plant Cell Tissue Organ Cult 24:131–138

Medford JI (1992) Vegetative apical meristems. Plant Cell 4:1029–1039

Mohamed MF, Read PE, Coyne DP (1992) Plant regeneration from in vitro culture of embryonic axis explants in common and terapy beans. J Am Soc Hortic Sci 117:332–336

Mohamed MF, Coyne DP, Read PE (1993) Shoot organogenesis in callus induced from pedicel explants of common bean (*Phaseolus vulgaris* L.). J Am Soc Hortic Sci 118:158–162

Morel G (1960) Producing virus-free cymbidiums. Am Orchid Soc Bull 29:495–497

Morikawa H, Iida A, Yamada Y (1989) Transient expression of foreign genes in plant cells and tissues obtained by simple biolistic device (particle-gun). Appl Microbiol biotechnol 31:320–322

Murashige T, Skoog F (1962) A revised medium for rapid growth and bioassays with tobacco tissue cultures. Physiol Plant 15:473–497

Nagl W, Ignacimuthu S, Becker J (1997) Genetic engineering and regeneration of *Phaseolus* and *Vigna*. State of the art and new attempts. J Plant Physiol 150:625–644

Rabakoarihanta A, Baggett JR (1983) Inheritance of leaf distortion tendency in bush lines of beans, *Phaseolus vulgaris* L., of blue lake background. J Am Soc Hortic Sci 108:351–354

Russel DR, Wallace KM, Bathe JH, Martinell BJ, MacCabe DE (1993) Stable transformation of *Phaseolus vulgaris* via electric-discharge mediated particle acceleration. Plant Cell Rep 12:165–169

Ruth J, Klekowski EJ, Stein OL (1985) Impermanent initials of the shoot apex and diplontic selection in a juniper chimera. Am J Bot 72:1127–1135

Sambrook J, Fritsch EF, Maniatis T (1989) Molecular cloning: a laboratory manual. Cold Spring Harbor Laboratory Press, Cold Spring Harbor, New York

Sanford JC (1990) Biolistic plant transformation. Physiol Plant 79:206–209

Sanford JC, Devit MJ, Russel JA, Smith FD, Harpending PR, Roy MK, Johnston SA (1991) An improved, helium-driven biolistic device. Technique 1:3–16

Santina S, Blakeslee AF, Avery AG (1940) Demonstration of the three germ layers in the shoot apex of *Datura* by means of induced polyploidy in periclinal chimeras. Am J Bot 27:895–905

Schwartz HF, Pastor-Corrales MA (eds) (1989) Bean production problems in the tropics. CIAT, Cali, 726 pp

Shii CT, Mok MC, Mok DW (1981) Developmental controls of morphological mutants of *Phaseolus vulgaris* L.: differential expression of mutant loci in plant organs. Dev Genet 2:279–290

Son SH, Moon HK, Hall RB (1993) Somaclonal variation in plants regenerated from callus culture of hybrid aspen (*Populus alba* L. x *P. gradidentata* Michx.). Plant Sci 90:79–84

Steeves TA, Sussex IM (1989) Patterns in plant development. Cambridge University Press, New York

Vasil IK (1994) Molecular improvement of cereals. Plant Mol Biol 25:925–937

19 Genetic Transformation of Pea (*Pisum sativum*)

N.V. Malysheva[1], Z.B. Pavlova[1], N.S. Chernysh[1], L.V. Kravchenko[2],
Y.N. Kislin[3], V. Chmelev[4], and L.A. Lutova[1]

1 Introduction

Several approaches have been used to study plant tissue differentiation processes. Wounding, such as organ explantation in vitro (e.g., Lutova and Zabelina 1988) and exogenous phytohormone application (Gamborg et al. 1974; Kartha et al. 1974; Malmberg 1979; Hussey and Gunn 1984; Rubluo et al. 1984; Ezhova et al. 1985) showed that the main morphogenetic processes were directed by auxins and cytokinins (auxin/cytokinin ratio) and depended on internal factors (plant genotype, age and tissue), which can determine the endogenous phytohormonal balance of plant tissue and its susceptibility to phytohormones. Transformation with *Agrobacterium tumefaciens* and *A. rhizogenes* strains, which introduce bacterial phytohormonal genes into the plant genome, allows the study of plant morphogenetic responses to directed changing of the endogenous hormonal balance (Medford et al. 1989; Sitbon et al. 1991). Transformation can also be used as an inducing condition to reveal differences between genetic forms in traits which characterize plant hormonal status. This approach reveals forms with atypical responses to transformation, i.e., potential hormonal mutants.

Garden pea (*Pisum sativum* L.) is a suitable and interesting object for studying different morphogenetic processes. As a leguminous plant, it can be a model for studying nodule formation and symbiotic nitrogen fixation. Pea is an important crop (used as food and forage) rich in protein, so elaboration of pea transformation and regeneration systems could contribute finally to crop improvement.

[1] St. Petersburg State University, Dept. of Genetics, Universitetskaya emb., 7/9, 199034 St. Petersburg, Russia
[2] All-Russian Institute of Agricultural Microbiology, Lab. of Biotechnology, sh. Podbelskiy 3, 189620 Pushkin 8, St. Petersburg, Russia
[3] All-Russian Institute of Plant Protection, Lab. of Plant Growth Regulators, sh. Podbelskiy 3, 189620 Pushkin 8, St. Petersburg, Russia
[4] All-Russian Plant Research Institute, Lab. of Biochemistry, Isaakievskaya sq. 2, 199000 St. Petersburg, Russia

Biotechnology in Agriculture and Forestry, Vol. 47
Transgenic Crops II (ed. by Y.P.S. Bajaj)
© Springer-Verlag Berlin Heidelberg 2001

2 Pea Regeneration and Genetic Transformation

2.1 Tissue Culture and Regeneration

Numerous studies have been conducted on various aspects of tissue culture, regeneration, and genetic transformation of pea (see Griga and Novak 1990; de Kathen and Jacobsen 1993; Atkins and Smith 1997). The regeneration processes have been shown to be strongly genotype-dependent. Pea genetic forms differ in callus formation capacity, shoot formation (Rubluo et al. 1984; Ezhova et al. 1985; Lutova and Zabelina 1988; Lutova et al. 1994, 1996), root formation in callus cultures, and rooting of the shoots (Lutova and Zabelina 1988). The dependence of regeneration capacity on plant genotype must reflect the differences in endogenous phytohormonal status and hormone susceptibility between pea genetic forms (Lutova and Zabelina 1988).

Pea regeneration capacity was also shown to be tissue-specific (Kunakh et al. 1984; Ezhova et al. 1985; Lutova and Zabelina 1988; Mallick and Rashid 1989) and to depend on tissue age (e.g., Hussey and Gunn 1984; Ezhova et al. 1985; Kysely and Jacobsen 1990).

Pea regenerants were obtained on media with phytohormones via organogenesis from pre-existing meristems (e.g., Kartha et al. 1974), organogenesis de novo from pea explant tissues and primary calli (Rubluo et al. 1984; Lutova and Zabelina 1988) and somatic embryogenesis in primary callus (Jacobsen and Kysely 1985; Kysely and Jacobsen 1990). Shoot formation was induced in long-term calli obtained from pea stem apices, epicotyls, explants of stems, leaves and roots (Malmberg 1979; Hussey and Gunn 1984; Kunakh et al. 1984; Ezhova et al. 1985; Gostimsky et al. 1985).

2.2 Pea Transformation

2.2.1 Determinants of Agrobacterium Host Range

Plant responses to transformation with *Agrobacterium* strains and regeneration capacity in vitro in general depend on common factors (plant genotype, tissue, developmental stage, exogenous phytohormones). Specific interaction of products of bacterial and plant genes is important at all stages of the transformation process: bacterium attachment to the plant cell, virulence induction, T-DNA processing and transfer, its integration and expression.

Bacterial chromosomal genes are necessary for bacterium attachment to the plant cell (Thomashow et al. 1987). The peculiarities of the plant cell wall are possibly influenced by phytohormones (Lowe and Krul 1991).

Agrobacterial plasmid determinants of the host range are located in T-DNA (sequence delimited by 25 bp direct repeats which is transferred into plant genome) and the vir region (regulating the transfer of T-DNA into plant cells; see Binns 1990). The peculiarities of the vir region effect pea transformation frequency much more than the chromosomal background of the

Agrobacterium (De Kathen and Jacobsen 1990). The induction of bacterial *vir* genes demands excretion of the specific phenol compounds acetosyringone (AS) and hydroxyacetosyringone by plant-wounded cells (Stachel et al. 1985). The ability to synthesize these compounds and their concentration in the wound region depend on the plant genotype. Several strains of *A. tumefaciens* possess chemotaxis to certain sugars and carry the gene for periplasmic galactose-binding protein (Cangelosi et al. 1990). Successful transformation induction must be promoted by a high concentration of indole acetic acid (IAA) at the plant wound, determined by expression of the bacterial gene *iaaP* located in the vir region, the *Agrobacterium* chromosomal gene controlling IAA synthesis from tryptophan (Liu et al. 1982; see Chernin and Avdienco 1985) and also by endogenous IAA contents in the plant tissue (Lowe and Krul 1991).

T-DNA genes controlling phytohormone (auxin and cytokinin synthesis or, in the case of *A. rhizogenes*, susceptibility to phytohormones (e.g., Nilsson et al. 1993), were shown to determine *Agrobacterium* host range (Hoekema et al. 1984; Yanofsky et al. 1985). Host specificity and transformation frequency can be influenced by pretreatment with phytohormones as shown in grape (Lowe and Krul 1991). The authors suggest that the endogenous hormonal balance of some plants hosts could functionally complement the oncogenic function of pathogens with narrow host ranges.

Several facts suggest dependence of plant morphogenetic responses to transformation on plant hormonal status and/or plant susceptibility to hormones. Tobacco mutants, non-susceptible to auxin and unable to form roots, had modified reactions to inoculation with wild-type *Agrobacterium* and mutant strains (Tourneur et al. 1985). *Agrobacterium* strains, inducing teratomas (tumors with shoot-like structures) on tobacco, caused unorganized tumors on pea (Puonti-Kaerlas et al. 1989).

Plant hormonal balance and susceptibility of plant tissues to phytohormones are complicated traits determined by hormone metabolism, the system of regulation of the free hormone pool (Olsson et al. 1990), hormone influence on metabolism of the other hormone(s) and on susceptibility to other hormone(s) (Klee and Estelle 1991), the system of hormone signal reception, and finally transduction (Prasad and Jones 1990; Ryan and Farmer 1991). Host genes participating in the plant response to transformation are unknown, but the data obtained on grape and pea suggest polygenic control of plant transformation traits (Lowe and Krul 1991; Robbs et al. 1991).

2.2.2 *Internal Factors Affecting Pea Responses to Transformation*

Pea intraspecific variability of morphogenetic responses to transformation with *Agrobacterium* strains has been demonstrated (Hobbs et al. 1989; De Kathen and Jacobsen 1990; Robbs et al. 1991). Pea genetic forms with atypical responses to transformation were revealed: for example, those forming tumors instead of "hairy" roots after inoculation with an *A. rhizogenes* strain and those forming roots in response to *A. tumefaciens* strains (Robbs et al.

1991; Lutova and Sharova 1993; Lutova et al. 1994, 1996; Pavlova et al. 1998).

The pea line resistant to *Agrobacterium* did not differ from the suscepti-ble ones in the first steps of the transformation process (binding, chemotaxis, bacterial ability to survive in plant wounds), suggesting that pea resistance to *Agrobacterium* depends on the endogenous hormonal balance and the change resulting from transformation (Robbs et al. 1991). Pea resistance to crown gall (*A. tumefaciens*) was inherited as a polygenic trait in this study.

Pea morphogenetic response to transformation and transformation effi-ciency depended on the *Agrobacterium* strain (Puonti-Kaerlas et al. 1990; De Kathen and Jacobsen 1990). *A. tumefaciens* strain A281 was shown to be super-virulent and induced large tumors on legumes, including pea (Jin et al. 1987; Puonti-Kaerlas et al. 1989). The interaction of certain plant genotype with certain *Agrobacterium* strain also determined transformation characteristics (Hobbs et al. 1989; Puonti-Kaerlas et al. 1990).

Pea transformation with *A. tumefaciens* and *A. rhizogenes* strains was shown to be tissue-specific (Hobbs et al. 1989; De Kathen and Jacobsen 1990) and to depend on the age of inoculated plant tissue: young seedlings were more susceptible to transformation than mature plants (Puonti-Kaerlas et al. 1989).

2.2.3 Obtaining Transgenic Pea Regenerants

Lines with a great ability to regenerate and root provide success in genetic engineering (Atkins and Smith 1997). In several studies, transgenic pea calli resistant to antibiotic hygromycin or kanamycin could not be induced to regenerate (Puonti-Kaerlas et al. 1989; Lulsdorf et al. 1991; Table 1). In these studies pea stem explants and microcolonies derived from protoplasts were transformed with *A. tumefaciens* non-oncogenic strains carrying marker genes (*hpt*, hygromycin phosphotransferase, or *nptII*, neomycin phosphotransferase).

Transformation of axenic cultures of pea shoots and epicotyls with dis-armed *A. tumefaciens* strains allowed culture of hygromycin-resistant calli, which were induced to shoot formation and then to rooting by means of hormone application (kanamycin-resistant calli failed to form shoots; Puonti-Kaerlas et al. 1990). The introduced gene was inherited as a dominant Mendelian trait (Puonti-Kaerlas 1991). In another study, transformation of pea epicotyl segments and nodal explants and further organogenesis induction resulted in chimeric plantlets composed of both transformed and non-transformed cells (De Kathen and Jacobsen 1990). In this experiment imma-ture and highly meristematic explants showed quite low susceptibility to trans-formation. Nodal explants (thin segments of node with excised leaves and axillary buds) were successfully used for pea transformation with binary vectors (carrying *nptII* and *hpt* genes) and further organogenesis induction (Nauerby et al. 1991). The transformation, regeneration, and rooting capaci-ties depended greatly on the plant genotype, since in general the yield of pea

Table 1. Summary of pea genetic transformation studies

Reference used	Explant	Vector	Observation/Remarcs
1. Hobbs et al. (1989)	(Whole plant), leaf explants	Wild type *A.t.* strains	Intraspecific variability in tumor formation
2. Puonti-Kaerlas et al. (1989)	(Whole plant), shoot epicotyl explants	Wild type *A.t.* strains; *A.t.* constructs: disarmed control vector, vectors, containing gene *NPTII*, "shooty" strains, carrying T-DNA gene 4, "rooty" strain with T-DNA genes 1 and 2; vector with Ri-plasmid	Kanamycin-resistant calli, strain-dependent tumor formation, root formation in response to Ri strain, no shoot formation
3. De Kathen and Jacobsen (1990)	Epicotyl, nodal explants from etiolated seedlings	Wild type *A.t.* strains; *A.t.* constructs: binary vector with genes *NPTII* and *GUS*, cointegrative vector with gene *HPT*	Genotype- and strain-dependent tumor formation, transgenic kanamycin-resistant regenerants
4. Puonti-Kaerlas et al. (1990)	Epicotyl and shoot explants	Disarmed *A.t.* transconjugant strains: with gene *HPT*; with genes *NPTII*, *CAT*; with genes *NPTII* and metothrexate resistance	Transgenic flowering plants, carrying gene of hygromycin resistance
5. Zubko et al. (1990)	Stem and leaf explants	*A.t.* vector "shooty" with T-DNA gene 4 and *NPTII*	Organogenic calli on hormone-free medium, transgenic shoots and seeds
6. Lulsdorf et al. (1991)	Stem explants	*A.t.* strains with genes *NPTII*, *HPT*	Transformed calli, resistant to hygromycin and/or kanamycin, no regenerants
7. Nauerby et al. (1991)	Nodal explants	Disarmed *A.t.* binary vectors with genes *NPTII* and *GUS*	Transgenic regenerants, expressing gene *GUS*
8. Robbs et al. (1991)	Axenic seedlings	Wild type *A.t.* strains, *A.rh.* strain; the Ti-pasmid-cured strain as a control	Intraspecific variability in tumor and root formation; tumorigenesis in cultivar, resistance

Table 1. *Continued*

Reference used	Explant	Vector	Observation/Remarcs
			to *A.t.*, was not blocked in early stages of infection; resistance to crown gall was inherited as a quantitative trait
9. Schaerer and Pilet (1991)	Intact roots, root explants, protoplasts	Wild-type *A.t.* strains; disarmed *A.t.* strain; oncogenic *A.t.* binary vector with gene; wild-type *A.rh.* strains	Transformed tissues (calli, tumors, roots), no regenerants
10. Lutova and Sharova (1993)	(Intact axenic seedlings), internodal and leaf explants	Wild-type *A.t.* and *A.rh.* strains; *A.rh.* strain with gene *NPTII*	Intraspecific variability in response to transformation, genotypes with a typical response, transformed tumor and root tissues
11. Schroeder et al. (1993)	Sections of the embryonic axis of immature seeds	*A.t.* binary vector with *BAR* and *NPTII* genes	Transgenic plants, seeds; Mendelian inheritance of *BAR* and *NPTII* genes
12. Shade et al. (1994)	Explants of the embryonic axis of immature seeds	Disarmed *A.t.* vector, carrying gene of α-amylase inhibitor (αAI-Pv) from common bean under strong seed-specific promoter, *GUS, BAR, NPTII* genes	Transgenic seeds (from transgenic plants) resistant to bruchid beetles (*Callosobruchus maculatus* and *C. chinensis*), expressing gene αAI-Pv (α-amylase inhibitor of the common bean that is toxic to the larvae)
13. Chowrira et al. (1995)	Intact nodal meristems	DNA electroporation, *GUS* gene	Transgenic plants, stable integration and expression of *GUS* gene in electroporated tissues and offspring
14. Grant et al. (1995)	Explants of immature cotyledons	*A.t.* binary vector with *NPTII* and *BAR* genes	Transgenic regenerants, stable inheritance and expression of transgenes

Table 1. *Continued*

Reference used	Explant	Vector	Observation/Remarcs
15. Lutova et al. (1996), Pavlova et al. (1998)	Internodal explants	Wild-type *A.t.* and *A.rh.* strains	Differences in phytohormonal contents between genotypes, differing in response to transformation, influence of exogenous phytohormones on phenotype of transformed tissues, genetic analysis of transformation phenotypic traits
16. Bean et al. (1997)	Lateral cotyledonary meristems of germinating seeds	*A.t.* binary vectors with *BAR* gene	Transgenic regenerants, stable inheritance of transgene in Mendelian manner
17. Grant et al. (1998)	Immature cotyledons	*A.t.* binary vectors with *NPTII* gene	Transgenic plants with *NPTII* gene, stable inheritance of the transgene
18. Lurquin et al. (1998)	Slices of embryonic axes of mature growing seeds	*A.t.* binary vectors with *GUS* gene	GUS-positive transgenic plants

Abbreviations: *A.t.*, *Agrobacterium tumefaciens*; *A.rh. Agrobacterium rhizogenes;* NPTII, neomycin phosphotransferase, resistance to kanamycin; HPT, hygromycin phosphotransferase, resistance to hygromycin; GUS, β-glucuronidase; CAT, chloramphenicol acyltransferase (chloramphenicol resistance); BAR, phosphinothricin acetyltransferase, resistance to the herbicide Basta (glufosinate = bialophos = phosphinothricin).

transgenic regenerants was low. Transformation of embryonic axis explants of immature pea embryos or half-embryos produced transgenic pea plants containing a gene conferring resistance to the herbicide phosphinothricin (*bar*), selective marker for kanamycin resistance and reporter β-glucuronidase gene (Schroeder et al. 1993; Lurquin et al. 1998). The introduced *bar* gene has stably inherited as a monogenic dominant trait (3 bar +:1 bar −) (Schroeder et al. 1993). In recent studies, transformation has been performed in planta (so that the need for tissue culture was avoided); pea meristematic cells from intact nodal meristems transformed by means of electroporation produced transgenic shoots (expressing reporter gene *gus* – β-glucuronidase) and seed (Chowrira et al. 1995). Developing cotyledon segments were also used for transformation (Grant et al. 1995, 1998).

Pea was transformed with other constructs in addition to marker genes. For example, transgenic pea seeds expressing the α-amylase inhibitor gene (αAI-Pv) of the common bean were obtained (Shade et al. 1994). The product of this bean gene is toxic to the larvae of bruchid beetles feeding on seeds. So pea transformation with the gene αAI-Pv driven by a strong seed-specific promoter resulted in pea seeds (from transgenic regenerants) resistant to pests. Pea transformation was performed as described by Schroeder et al. (1993).

In order to study the biological action of phytohormones, pea epycotyl and shoot explants were transformed with constructed *Agrobacterium* vectors carrying single phytohormonal genes. "Rooty" strain (pGV3304) containing the intact T-DNA genes 1 and 2 (auxin genes) produced tumors in pea, as did the "shooty" strain (pGV2298), containing the intact T-DNA gene 4 (cytokinin gene; Puonti-Kaerlas 1991) in contrast to results obtained with tobacco (Joos et al. 1983), suggesting that the levels of phytohormones produced by the inserted genes were not sufficient to induce regeneration processes in pea, or that these hormones were inactivated, or that pea tissue susceptibility to phytohormones differed from other species.

However, in the other study, transgenic pea plants with a changed phytohormonal balance and high shoot formation capacity were obtained after transformation of stem and leaf explants with the "shooty" mutant of *A. tumefaciens* (pGV2206), carrying substitution in genes of auxin synthesis (Zubko et al. 1990). In preliminary experiments, shoot formation from original untransformed cell lines was induced on medium containing cytokinin BAP (benzylaminopurine) with low frequency. Transformation resulted in production of actively regenerating cell lines on hormone-free medium, giving rise to transgenic pea shoots which were grafted onto plants of the original genotypes and formed seeds. Thus, certain progress has been achieved in elaboration of foreign gene introduction into pea genomes with further regeneration of transgenic plants. However, such transgenic plants have been obtained with low frequency (<1–5%; Atkins and Smith 1997).

This study is devoted to the investigation of inheritance of pea differentiation characteristics (regeneration and transformation capacities) and their dependance on plant hormonal balance.

3 Material and Methods

3.1 Plant Material and Bacterial Strains

We used pea lines 32, 15, 9, 3, 4, Sprint-2 and cultivars Sparkle, Rondo, Finale, and also mutant lines with changed nodulation capacity: mutant lines unable to form nodules Nod⁻, *sym5*, and *sym8* obtained from Sparkle cultivar (Guinel and LaRue 1991; kindly supplied by Prof. Th.A. LaRue, Boyce Thompson Institute, USA), line Sprint-2 Nod⁻-2 (*sym8*) produced from line Sprint-2

(Borisov 1992; kindly supplied by Dr. A.Yu. Borisov, All-Russian Institute of Agricultural Microbiology, St. Petersburg); supernodulating forms nod3 (Jacobsen and Feenstra 1984; kindly provided by Dr. E. Jacobsen, Agricultural University, The Netherlands) and RisFixC (Novak et al. 1993; kindly provided by Dr. K. Engvild, Riso National Laboratory, Denmark).

Pea lines and cultivars used in our work were received from the gene collection of the All-Russian Research Institute of Agricultural Microbiology. The lines differ in morphological and physiological traits, including symbiotic nitrogen fixation and nodulation capacities (Table 2).

Table 2. Plant material

Genetic form	Origin	Height	Vegetation period	Red color (presence +/absence−)	Nodule formation and nitrogenase activity (n.a.)[c]
Line 3	Local cv.[a] of Arhangelskaya region (K3302[b])	High	Medium	+	Numerous nodules, low n.a. (Nod[+] Fix[±])
Line 4	Local cv.* of Arhangelskaya region (K3302)	High	Medium and long	+	Numerous medium and large nodules, high n.a. (Nod[+] Fix[+])
Line 9	Crupp Pelushke cv. (K5970),	High	Long	+	Numerous medium and large nodules, high n.a. (Nod[+] Fix[+])
Line 15	*Germany* cv., Great Britain (K6264)	Low	Medium	−	Numerous medium and large nodules, medium n.a. (Nod[+] Fix[+])
Line 32	Local cv. of Novgorodskaya region	High	Medium and long	+	Numerous, very large nodules, high n.a. (Nod[+] Fix[++])
cv. Spar kle	T.A. LaRue collection, USA	Low	Short	−	Numerous, small nodules predominantly on the lateral roots (Nod[+] Fix[+])
Line E2	Sparkle cv., (chemical mutagenesis), mutation in sym5 gene (1st chromosome)	Short intern odes	Short	−	No or few nodules, ethylene- and t-dependent resistance to nodulation (Nod[−]) (Guinel and LaRue 1991)
Line R25	Sparkle cv., (γ-rays), mutation in sym8 gene (6th chromosome	Short intern odes	Short	−	No nodules (Nod[−])
Line Sprint -2	From Institute of Cytology and Genetics, Russia	Low	Very short	−	Medium nodules predominantly on the medium roots

Table 2. *Continued*

	(Berdnikov et al. 1989)				(Nod$^+$ Fix$^+$)
Line S-2 Nod$^-$-2	L. S-2 (EMS-treatment mutation in sym8 gene (6th chromosome)	Low	Very short	–	No nodules (Nod$^-$)
cv. Rondo	E. Jacobsen collection, The Netherlands	Low	Medium	–	Medium nodulation ability and medium n.a. (Nod$^+$ Fix$^+$)
nod$_3$	Rondo cv., (chemical mutagenesis)	Low	Medium	–	Nodulation ability is much higher than in parental cv. (Nod^{++} Fix$^+$) (Jacobsen and Feenstra 1984)
cv. Finale	K. Engvild collection, Denmark	Low	Medium	–	Medium nodulation ability and medium n.a. (Nod$^+$ Fix$^+$)
RisFix C	Finale cv., (chemical mutagenesis)	Low	Medium	–	Nodulation ability is much higher than in parental cv. (Nod^{++} Fix$^+$) (Novak et al. 1993)

[a] Cultivar (cv.) from which the line was bred.
[b] Number according to the All-Russian Plant Research Institute catalogue.
[c] Nitrogenase activity was detected according to modified acetylene reduction assay (ARA;

To study the inheritance of differentiation traits we obtained hybrids F_1 and F_2 between lines. *A. tumefaciens* strain A281 and *A. rhizogenes* strain 8196 were obtained from Prof. M. Ondrej (Institute of Molecular Biology, Academy of Sciences of the Czech Republic, Ceske Budejovice).

3.2 Plant Germination

Pea seeds were sterilized for 4 min in concentrated sulfuric acid followed by five washes in sterile distilled water. They were germinated on the hormone-free MS medium (Murashige and Skoog 1962) containing Difco agar 8 g/l. The seedlings were grown in closed jars for 14 days at 18–20 °C under fluorescent light with a 12-h photoperiod.

3.3 Regeneration Capacity Assessment

Internodal explants were used to study regeneration capacity. The basal MS medium was supplemented with the cytokinin BAP (6-benzylaminopurine, Sigma, St. Louis) and/or the auxin NAA (α-naphtaleneacetic acid, Sigma) at the following concentrations: (1) 1 mg/l BAP, (2) 1 mg/l NAA, (3) 1 mg/l BAP

+ 2 mg/l NAA, (4) 5 mg/l BAP + 2 mg/l NAA. Emerging shoots were transferred to the rooting medium half-strength MS with 1% sucrose and 0.1 mg/l
NAA). The regeneration capacity was assessed as described earlier (Lutova et
al. 1996).

To obtain the organogenic calli from immature pea embryos, a modified
procedure by Schroeder et al. (1993) was used. The embryonic tissue was
placed on MS medium containing 5 mg/l BAP and 2 mg/l NAA. Emerging
shoots were excised and transferred to the rooting medium.

3.4 *Agrobacterium* Transformation

Internodal explants of 14-day-old axenic seedlings were inoculated with
Agrobacterium suspension as described earlier (e.g., Lutova et al. 1996). The
apical ends of stem explants were immersed in the solid MS (or MS + 1 mg/l
BAP + 2 mg/l NAA) medium. Assessment (40 days after co-cultivation) followed conventional procedures (see Lutova and Sharova 1993).

3.5 High-Performance Liquid Chromatography (HPLC) Analysis of Auxin and Its Derivatives (and Possible Precursor)

Indole acetic acid (IAA) and its derivatives were extracted from the roots of
14- to 20-day-old axenic plantlets with 70% ethanol. Further purification procedures are described by Kravchenko et al. (1994). IAA and its derivatives
were analyzed using HPLC apparatus (Jusco 900, Japan) connected to a UV
detector ($\lambda = 220$ nm). We used a mixture of water:acetonitrile:acetic acid
[83:17:0.2 (v/v)] as an eluant. LiChrosorb C18 with 5-μm particle size was
used as a sorbent; the sorbent was loaded into the stainless steel column (250
× 4.6 mm). The elution rate was 0.9 ml/min at a column temperature of 33 °C.
The control probe contained a mixture of indole aldehyde (IAld), indole lactic
acid (ILA), indole carbonic acid (ICA) and indole acetic acid (IAA) dissolved
in methanol and all at a concentration of 25 mg/ml.

3.6 HPLC Analysis of Cytokinins

A Beckman HPLC unit was used to determine cytokinin content in 20-day-
old axenic seedlings of cv. Sparkle and lines Sprint-2, 32, 9, and 4. The characteristics of the unit have been described elsewhere (Lutova et al. 1996).
Cytokinin detection in green parts of the plant used procedures described
earlier (Kislin 1991).

The following phytohormones (from Serva Company, St. Louis) were
used as standard: zeatin (Z), zeatin riboside (ZR), dihydro-zeatin (DZ),
dihydrozeatin-riboside (DZR), 6-(γ,γ-dimethylallyl-amino)-purine (DAP),
6-γ,γ-dimethylallyladenosine (DAPR). The hormones moved in the following
order: ZR, DZR, Z, DZ, DAP, DAPR.

3.7 Enzyme-Linked Immunosorbent Assay (ELISA) Analysis of Cytokinins

Cytokinins were extracted from the green parts of axenic 14- to 20-day-old plantlets with 70% ethanol. Then extracts were evaporated to the water phase at 45 °C under vacuum. The water phase was alkalined with 2 N NaOH to pH 8–9 and extracted with hexane. The hexane phase was extracted twice with butanol. The resulting extracts were evaporated to dryness, dissolved in 400 ml ethanol, cleaned for 30 min with polyvinyl pyrrolidone and passed through membrane filters (pore size 0.45 μm). Kits for analysis of zeatin (Z), zeatin riboside (ZR), dihydrozeatin (DZ), dihydrozeatin riboside (DZR) were performed by "Uralinvest" Co. (Ufa, Russia).

4 Results and Discussion

4.1 Variability in Regeneration Capacity

The pea lines showed diversity in callus and root formation capacities on the hormone-free medium (Lutova and Zabelina 1988; Lutova et al. 1994) and on medium with auxin NAA (1 mg/l). Cytokinin supplement (1 mg/l BAP) did not stimulate regeneration in the pea genetic lines tested (Lutova et al. 1996) Both cytokinin BAP (1 or 5 mg/l) and auxin NAA (2 mg/l) application induced callus formation with high frequency and genotype-dependent shoot formation with low frequency. Shoots rooted with low frequency on the rooting medium. Shoot formation and rooting capacities decreased greatly after subculture. Line 32 showed the highest regeneration capacity in all the experiments on media with different hormone constitutions and on hormone-free medium.

Use of the immature pea embryos as the source of explants (Schroeder et al. 1993) allowed a significant increase in pea regeneration capacity. Though the frequency of shoot formation from immature embryos was high, intraspecific variability in this trait was observed: line 4 formed shoots with the lowest frequency; line Sprint-2, cv. Sparkle and line E2 obtained from cv. Sparkle showed the highest shoot formation capacity. We obtained rooted regenerants, although rooting capacity of regenerants was extremely low. So this method of shoot formation induction may be used to obtain further transgenic regenerants.

4.2 Responses to Transformation

Pea morphogenetic responses to transformation with *Agrobacterium* strains were highly genotype-specific (Fig. 1; Lutova and Sharova 1993; Lutova et al. 1996; Pavlova et al. 1998). The results of transformation were consistent with the phenotypic responses to wounding and exogenous phytohormones. For

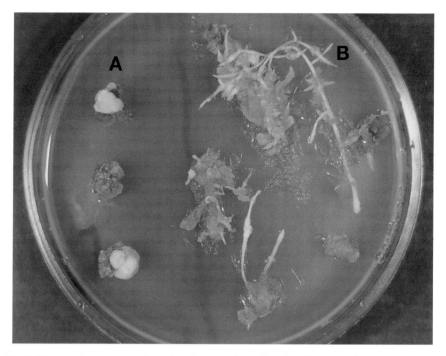

Fig. 1. Phenotypic responses of pea line 9 to transformation. **A** with *A. tumefaciens* strain A281 (tumor formation); **B** with *A. rhizogenes* strain 8196 (root formation)

example, line 32 showed the highest root formation capacity in vitro and formed roots even after transformation with tumor-inducing *A. tumefaciens* strain A281 (Tables 3, 4). This must reflect the influence of genotype-dependent endogenous hormone status (hormone content, tissue sensitivity to phytohormones, etc. on morphogenetic processes.

Pea nodulation mutants did not differ from their original parents in response to transformation (Table 4).

The differences revealed between pea genotypes in root and tumor for-mation, induced by transformation, allowed genetic analysis of these traits (Pavlova et al. 1998). Tumor formation in response to *A. rhizogenes* strain 8196 was inherited as a polygenic trait in crossings between tumor-forming line 15 and line 4. The atypical root formation in response to *A. tumefaciens* strain A281 was inherited as a dominant train (in crossings between root-forming line 32 and line 9). Reciprocal differences in root formation of F_2 hybrids were observed. If the maternal line was 32, 3:1 segregation ($\chi^2 = 0.3$, $P > 0.5$) was observed in F_2 progeny (taking into account the penetrance), suggesting monogenic differences between these lines in root formation. We named the gene controlling this trait *Rtf* (root forming).

We analyzed an effect of exogenous phytohormones (at concentrations inducing callus formation and regeneration in untransformed explants) applied after explantation, on transformation traits of pea genotypes

Table 3. Tumor and root formation of pea lines transformed by the different *Agrobacterium* strains (MSO medium)

Line	Strain	Number of explants	Tumor formation		Root formation	
			(%)	Mark[a]	(%)	Mark
3	A281	37	100 − 3.0	3	0.0 + 2.6	0
	8196	37	0.0 + 2.6	0	80.0 + 6.6	2
4	A281	19	52.6 + 11.8	2	0.0 + 5.0	0
	8196	111	0.0 + 0.9	0	54.1 + 4.7	2
9	A281	64	73.4 + 5.5	2	0.0 + 1.5	0
	8196	74	0.0 + 1.3	0	70.0 + 5.3	3
15	A281	34	100 − 2.9	3	0.0 + 2.9	0
	8196	30	66.7 + 8.6	3	70.0 + 8.4	2
32	A281	99	55.6 + 5.1	2	15.2 + 3.6	2
	8196	52	51.9 + 6.9	2	65.4 + 6.7	2

[a] Mark 0 absence of tumors or roots; mark 1 small tumor (up to 2 mm) or single roots; mark 2 medium tumor (3–5 mm) or numerous roots; mark 3 large tumor (>5 mm) or vigorous root formation.

Table 4. Phenotypical effects of transformation of pea lines and cultivars and their nodulation mutants with *A. tumefaciens* and *A. rhizogenes* strains. Untransformed control explants (treated with medium without bacteria) of tested pea genotypes formed small calli on explant surfaces and sometimes single roots without hairs; the control explants soon died on hormone-free medium

Plant genotype	*A. tumefaciens* (strain A281)	*A. rhizogenes* (strain 8196)
Sparkle	Large tumors	Large tumors, single roots
Nod⁻ E2	Large tumors	Large tumors, single roots
Nod⁻ R25	Large tumors	Large tumors, single roots
Rondo	Large tumors	Small tumors, roots
Nod⁺⁺ nod3	Large tumors	Large tumors, roots
Finale	Large tumors, rare roots	Small tumors, single roots
Nod⁺⁺ RisFixC	Large tumors, rare roots	Large tumors, single roots
Sprint-2	Tumors	Roots
Sprint-2 Nod⁻-2	Tumors	Roots
3	Large tumors	Roots
4	Large tumors	Roots
9	Large tumors	Roots
15	Large tumors	Tumors, roots
32	Large tumors, single roots	Tumors, roots

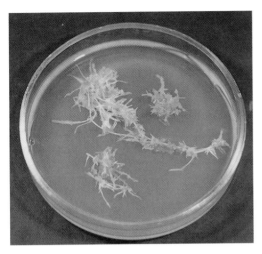

Fig. 2. Hormone-independent growth of rooty culture of line 32 transformed with strain 8196

inoculated with *A. rhizogenes* (Pavlova et al. 1998). Exogenous hormones increased the differences between pea lines in tumor and root formation. Auxin and cytokinin supplement increased the percentage of tumor formation in line 9 and cv. Sparkle and decreased the tumor formation frequency in line 32. Exogenous hormones completely suppressed root formation in cv. Sparkle, decreased the percentage of root formation in lines 9 and 4, and insignificantly increased root formation in line 32. Therefore, the reaction of line 32 to exogenous auxin and cytokinin applied with *A. rhizogenes* transformation is also non-standard.

Long-term transformed tumors and roots (Fig. 2), maintained on the hormone-free medium, were replaced with green callus after transfer to the medium with cytokinin BAP. Auxin NAA induced root formation in the tumorous cultures (Sparkle-8196), but the effect of NAA was less than the influence of BAP on the rooty transformed cultures (Lutova et al. 1996). Exogenous phytohormone supplements made it possible to obtain phenocopies of typical morphogenetic responses of plant tissues to transformation with certain *Agrobacterium* strains.

Thus, exogenous hormone application confirmed the role of phytohormones in the creation of transformation phenotype.

4.3 Analysis of Contents of Auxin, Its Derivatives and Possible Precursor in Pea Forms

HPLC analysis showed that pea genotypes differed both quantitatively and qualitatively in their content of auxin derivatives (Table 5). IAld, the product of oxidative decarboxilation of IAA, was found in the roots of lines 32 and 15 only. The IAld content of line 32 was four times higher than the IAld content of line 15. ICA, one of the terminal products of IAA oxidative decarboxyla-

Table 5. Concentrations of IAA derivatives in the roots of various pea genotypes, shown by HPLC analysis (μg/g fresh weight)

Auxin derivative[a]	9	32	4	15	cv. Sparkle	Nod⁻ E2
ILA	0.0114	0.010	–	0.027	0.002	0.002
ICA	–	–	–	0.228	0.101	0.142
IAld	–	0.211	–	0.056	–	–
IAA	–	–	–	–	–	–

[a] *ILA*, indole lactic acid; *IAld*, indolealdehyde; *ICA*, indole carbonic acid; *IAA*, indole-3-acetic acid.

Table 6. Variability of pea genotypes in cytokinin spectra and contents (μg/g fresh weight) shown by HPLC and ELISA analyses

Cytokinin[a]	9		32		4		15 E	cv. Sparkle		Nod⁻ E2 E	Sprint-2 HPLC
	E[b]	HPLC	E	HPLC	E	HPLC		E	HPLC		
DZ	+[c]	0.2184	–	–	+	0.3819	–	–	–	+	–
DZR	+	0.0601	+	0.2198	+	0.1904	–	+	0.0266	+	–
Z	–	–	+	0.4202	–	–	–	+	0.0707	+	0.2591
ZR	–	–	–	–	–	–	–	–	–	+	+
DAP	nt	0.0064	nt	–	nt	–	nt	nt	0.0040	nt	–
DAPR	nt	0.0112	nt	0.0049	nt	0.0324	nt	nt	0.0237	nt	0.0726

[a] *Z*, zeatin; *ZR*, zeatin riboside; *DZ*, dihydrozeatin; *DZR*, dihydrozeatin riboside; *DAP*, 6-(γ, γ-dimethylallylamino)-purine; *DAPR*, 6-γ,γ – dimethylallyladenosine.
[b] *E*, ELISA analysis.
[c] Minimum detection level is 0.5 ng aliquot; *nt*, not tested.

tion, was found only in line 15 roots. One can suppose that IAA degradation in line 32 is stopped at the IAld stage, probably because of lack of the enzyme controlling conversion from IAld to ICA.

　　ILA, one of the possible precursors of IAA synthesis, was found in the roots of all lines except line 4. ILA and ICA contents of Nod⁻ mutant E2 and parental cv. Sparkle were almost identical. IAA was not found in the roots of all lines studied.

4.4　Study of Cytokinin Spectra of Pea Forms

Data from the immunological assay (ELISA) of four cytokinins (Z, ZR, DZ, DZR) correlate well with previous HPLC results (Lutova et al. 1996; Pavlova et al. 1998; Table 6). None of the studied cytokinins were found in the plantlets of line 15. Probably, the main cytokinins of this line are isopentenyladenines. The most typical cytokinin of the studied pea genotypes was DZR, whereas ZR was much rarer. Only Nod⁻ mutant E2 contained all of the studied purines. The Nod⁻ mutant E2 differs from parental cultivar Sparkle in its cytokinin spectra. Thus, mutations affecting nodule formation can change the plant hormonal spectrum. Pea genetic forms demonstrating different (contrasting) mor-phogenetic responses to transformation differed in the set of main endoge-nous cytokinins: Z, DZ, ZR, DZR, DAP, DAPR (Table 6).

HPLC and ELISA analysis of auxin derivative content and the cytokinin spectrum showed that forms with the typical responses to transformation (lines 9 and 4) possess almost identical phytohormonal sets (Tables 5, 6). Genotypes with non-standard reactions to the transformation (lines 32 and 15, cv. Sparkle and its mutant E2) differed strongly in hormonal balance from lines 9, 4, and from each other.

So, the type of phytohormonal spectrum reflected the phenotypic effect of transformation.

5 Summary and Conclusion

Pea has been genetically transformed with *A. tumefaciens* and *A. rhizogenes* strains. Electroporation with DNA was also used. The transgenic tissues, expressing introduced marker genes (antibiotic resistance, tolerance to herbicides) and/or a reporter gene (β-glucuronidase) were obtained. Several approaches, such as suing immature embryos and nodes as a source of explants, made it possible to obtain transgenic pea regenerants; transgenes showed a Mendelian pattern of inheritance. Genes of agricultural importance such as the α-amylase inhibitor from common bean introduced into the pea genome gave pea seeds with resistance to bruchid beetles. In order to study developmental processes, separate agrobacterial phytohormonal genes were introduced into the pea genome.

Different pea genotypes, including nodulation mutants, were used to study the phytohormone-dependent differentiation processes. Intraspecific pea vari-ability in morphogenetic processes (callus, tumor, shoot, and root formation), induced by wounding, exogenous phytohormones, and transformation with *Agrobacterium* strains, reflected the differences between pea lines in endoge-nous phytohormonal balance and/or hormone metabolism systems (and maybe also sensitivity to hormones).

The shoot formation capacity of the pea lines studied was low, but use of a modified method by Schroeder et al. (1993) increased the yield of regener-ants, which is important for obtaining transgenic plants. In general, regenera-tion processes in vitro were consistent with morphogenetic responses to transformation. Changing of the transformed tissue phenotype after exoge-nous phytohormone treatment confirmed our suggestions concerning the hormone balance of lines with atypical responses to transformation. The quan-tities of several cytokinins, auxin, its derivatives and the possible auxin pre-cursor in different pea forms were analyzed using HPLC and ELISA. The quantitative and qualitative differences in the phytohormonal spectra in dif-ferent pea genetic forms have been shown. Lines with atypical responses to transformation significantly differed in hormonal spectra from those with typical responses.

Revealing forms with atypical responses to transformation allowed genetic analysis of transformation morphogenetic traits which reflect the peculiarities of plant hormone status. Tumor formation in response to transformation with *A. rhizogenes* was inherited as a polygenic trait. Root formation in response to transformation with *A. tumefaciens* was inherited as a dominant trait with incomplete penetrance (reciprocal differences in F_2 progeny were observed). The results of F_2 segregation in one direction of the crosses suggest a monogenous dominant nature for this trait (or monogenic differences between tested lines in this trait. We named the gene controlling this trait *Rtf* (root forming).

Pea transformation characteristics as well as regeneration capacity were shown to be highly genotype-specific. Resistance to crown gall disease was shown to be inherited as a polygenic trait. In our work, atypical morphogenetic responses to transformation with the wild-type *Agrobacterium* strains indicated genotypes with unusual hormonal contents or sensitivity to phytohormones, which was confirmed by HPLC and ELISA analyses and by the influence of exogenous auxins and cytokinins on the transformation characteristics of pea genotypes. The differences revealed between pea lines in root and tumor formation enabled genetic analysis of these traits.

Acknowledgments. These investigations were supported by grants from the Russian Fund for Fundamental Investigations (RFFI), Frontiers of Russian Genetics, and International Soros Science Education Program (ISSEP).

References

Atkins CA, Smith MC (1997) Genetic transformation and regeneration of legumes. In: Legocki A, Bothe H, Puhler A (eds) Biological fixation of nitrogen for ecology and sustainable agriculture. NATO ASI Ser G, vol 39. Springer, Berlin Heidelberg New York, pp 283–304

Bean SJ, Gooding PS, Mullineaux PM, Davies DR (1997) A simple sytem for pea transformation Plant Cell Rep 16:513–519

Binns AN (1990) *Agrobacterium*-mediated gene delivery and the biology of host range limitations. Physiol Plant 79:135–139

Borisov AYu (1992) Obtaining and characterization of symbiotic mutants of pea (*Pisum sativum* L.). PhD Thesis, St Petersburg, 128 pp (in Russian)

Brewin NJ, Ambrose MJ, Downie JA (1993) Root nodules, *Rhizobium* and nitrogen fixation. In: Casey R, Davies DR (eds) Peas: genetics, molecular biology and biotechnology. CAB International, Wallingford

Cangelosi GA, Ankenbauer RG, Nester EV (1990) Sugars induce the *Agrobacterium* virulence genes through a periplasmic binding protein and a transmembrane signal protein. Proc Natl Acad Sci USA 87:6708–6712

Chernin LS, Avdienco ID (1985) Plasmid phytohormone genes and their role in plant oncogenesis. Russian J Mol Biol 19(4):869–888

Chowrira GM, Akella V, Lurquin PF (1995) Electroporation-mediated gene transfer into intact nodal meristems in planta. Generating transgenic plants without in vitro tissue culture. Mol Biotechnol 3(1):17–23

De Kathen A, Jacobsen H-J (1990) *Agrobacterium*-mediated transformation of *Pisum sativum* L. using binary and cointegrate vectors. Plant Cell Rep 9:276–279

De Kathen A, Jacobsen H-J (1993) Transformation of pea (*Pisum sativum* L.). In: Bajaj YPS (ed) Biotechnology in agriculture and forestry, vol 23. Plant protoplasts and genetic engineering IV. Springer, Berlin Heidelberg New York, pp 331–347

Ezhova TA, Bagrova AM, Gostimskii SA (1985) Shoot formation in calluses from stem tips, epicotyls, internodes and leaves of different pea genotypes. Russian J Plant Physiol 32:409–414

Gamborg OL, Constabel F, Shyluk JP (1974) Organogenesis in callus from shoot apices of *Pisum sativum*. Physiol Plant 30:125–128

Gostimsky SA, Bagrova AM, Ezhova TA (1985) Revealing and cytological analysis of variability, appearing when plants are regenerating from pea tissue culture. Theses of the Academy of Sciences of the USSR 283(4):1007–1011 (in Russian)

Grant JE, Cooper PA, Mcara AE, Rrew TJ (1995) Transformation of peas (*Pisum sativum* L.) using immature cotyledons. Plant Cell Rep 15:254–258

Grant JE, Cooper PA, Gilpin BJ, Hoglund SJ, Reader JK, Pither-Joyce MD, Timmerman-Vaughan GM (1998) Kanamycin is effective for selecting transformed peas. Plant Sci 139:159–164

Griga M, Novak EJ (1990) Pea (*Pisum sativum* L.). In: Bajaj YPS (ed) Biotechnology in agriculture and forestry, vol 10. Legumes and oilseed crops I. Springer, Berlin Heidelberg New York, pp 65–69

Guinel FC, and LaRue TA (1991) Light microscopy study of nodule initiation in *Pisum sativum* L. cv. Sparkle and its low-nodulating mutant E2 (*sym5*). Plant Physiol 97:1206–1211

Hawes MC, Robbs SL, Pueppke SG (1989) Use of a root trumorigenesis assay to detect genotypic variation in susceptibility of 34 cultivars of *Pisum sativum* to crown gall. Plant Physiol 90:180–184

Hobbs SLA, Jacobsen JA, Mahon JD (1989) Specifity of strain and genotype in the susceptibility of pea to *Agrobacterium tumefaciens*. Plant Cell Rep 8:274–277

Hoekema A, de Pater BS, Fellinger AJ, Hooykaas PJJ, Schilperoort RA (1984) The limited host range of an *Agrobacterium tumefaciens* strain extended by a cytokinin gene from a wide host range T-region. EMBO J 3:2485–2490

Hussey G, Gunn HV (1984) Plant production in pea (*Pisum sativum* L. cv. puget and Upton) from long-term callus with superficial meristems Plant Sci Lett 37:143–148

Jacobsen E, Feenstra WJ (1984) A new pea mutant with efficient nodulation in the presence of nitrate. Plant Sci Lett 33:337–344

Jacobsen H-J, Kysely W (1985) Induction of in vitro regeneration via somatic embryogenesis in pea (*Pisum sativum*) and bean (*Phaseolus vulgaris*). Gen Manipul Plant Breed Proc Int Symp Berl (West), Sept, 1985, pp 445–448

Jin S, Komaki T, Gordon MP, Nester EV (1987) Genes responsible for the supervirulence phenotype of *Agrobacterium tumefaciens* A281. J Bacteriol 169:4417–4425

Joos H, Inze D, Caplan A, Sormann M, Van Montagu M, Schell J (1983) Genetic analysis of T-DNA transcripts in nopaline crown galls. Cell 32:1057–1067

Karssen CM, Groot SPC, Koorneef M (1987) Hormone mutants and seed dormancy in *Arabidopsis* and tomato. In: Thomas H, Grierson D (eds) Developmental mutants in higher plants. SEB Seminar Series 32. Cambridge Univ Press, Cambridge, pp 119–133

Kartha KK, Gamborg OL, Constabel F (1974) Regeneration of pea (*Pisum sativum* L.) plants from shoot apical meristems. Z Pflanzenphysiol 72:172–176

Kislin YeN (1991) Changes of the cytokinin content in the corn leaves damaged by grass aphid Russian. J Plant Physiol Biochem 23:602–605

Klee H, Estelle M (1991) Molecular genetic approaches to plant hormone biology. Annu Rev Plant Physiol Plant Mol Biol 42:529–551

Knauf V, Panagopoulos CG, Nester EW (1982) Genetic factors controlling the host range of *Agrobacterium tumefaciens*. Phytopathology 72:1545–1549

Korber H, Strizhov N, Staiger D, Feldwisch J, Olsson O, Sandberg G, Palme K, Schell J, Koncz C (1991) T-DNA gene 5 of *Agrobacterium* modulates auxin response by autoregulated synthesis of a growth hormone antagonist in plants. EMBO J 10(13):3983–3991

Kravchenko LV, Leonova EI, Tikhonovich IA (1994) Effect of root exudates of non-legume plants on the response of auxin production by associated diazotrophs. Microbe Releases 2:267–271

Kunakh VA, Voityuk LI, Alkhimova EG, Alpatova LK (1984) Production of callus tissues and induction of organogenesis in *Pisum sativum*. Russian J Plant Physiol 31(30):542–548

Kysely W, Jacobsen H-J (1990) Somatic embryogenesis from pea embryos and shoot apices. Plant Cell Tissue Organ Cult 20:7–14

Lowe BA, and Krul WR (1991) Physical, chemical, developmental and genetic factors that modulate the *Agrobacterium-Vitis* interaction. Plant Physiol 96:121–129

Lulsdorf MM, Rempel H, Jackson JA, Baliski DS, Hobbs SLA (1991) Optimizing the production of transformed pea (*Pisum sativum* L.) callus using disarmed *Agrobacterium tumefaciens* strains. Plant Cell Rep 9:479–483

Lurquin PF, Cai Z, Stiff CM, Fuerst PE (1998) Half-embryo cocultivation technique for estimating the susceptibility of pea and lentil cultivars to *Agrobacterium tumefaciens*. Mol Biotechnol 9:175–179

Lutova LA, Sharova NV (1993) Study of pea response to transformation with *Agrobacterium tumefaciens* and *Agrobacterium rhizogenes* strains. Russian J Genet 29:1157–1172

Lutova LA, Zabelina YaK (1988) Callus and shoot in vitro formation in different forms of peas *Pisum sativum* L. Russian J Genet 24:1632–1640

Lutova LA, Bondarenko LV, Buzovkina IS, Levashina EA, Tikhodeev ON, Khodzhaiova LT, Sharova NV, Shishkova SO (1994) The influence of plant genotype on regeneration process. Russian J Genet 30:928–936

Lutova LA, Sharova NV, Kislin YeN (1996) The role of phytohormones in response to transformation of pea lines, differing in symbiotic characteristics, and their hybrids. Symbiosis 21:61–80

Mallick MA, Rashid A (1989) Induction of multiple shoots from cotyledonary node of grain legumes, pea and lentil. Biol Plant 31:230–232

Malmberg RL (1979) Regeneration of whole plants from callus culture of diverse genetic lines of *Pisum sativum* L. Planta 146:143–144

Medford J, Horgan R, El-Sawi Z, Klee HJ (1989) Alterations in endogenous cytokinins in transgenic plants using a chimeric isopentenyl transferase gene. Plant Cell 1:403–404

Murashige T, Skoog F (1962) A revised medium for rapid growth and bioassays for tobacco tissue cultures. Physiol Plant 115:473–479

Nauerby B, Madsen M, Christiansen J, Wyndaele R (1991) A rapid and efficient regeneration system for pea (*Pisum sativum*) suitable for transformation. Plant Cell Rep 9(12):696–679

Nilsson O, Crozier A, Schmulling T, Sandberg G, Olsson O (1993) Indole-3-acetic acid homeostasis in transgenic tobacco plants expressing *A. rhizogenes rolB* gene. Plant J 3(5):681–689

Novak K, Skrdleta V, Nemcova M, Lisa L (1993) Behavior of pea nodulation mutants as affected by increasing nitrate level. Symbiosis 15:195–206

Olsson O, Sitbon F, Sundberg B, Sandberg G (1990) IAA and IAA-conjugate content of transgenic tobacco plants carrying the *iaaM* and *iaaH* genes from *Agrobacterium tumefaciens*. Physiol Plant 79(2):27

Pavlova ZB, Malysheva NV, Kravchenko LV, Chmelev V, Lutova LA (1998) Response of pea genotypes to *Agrobacterium* as a means of probing their endogenous hormone level. Plant Sci 133:167–176

Prasad P, Jones AM (1991) Identification and characterization of a novel auxin receptor. Plant Physiol Suppl 91(1):17

Puonti-Kaerlas J (1991) Tissue culture and genetic transformation of pea (*Pisum sativum* L.). Acta Universitatis Upsaliensis, Comprehensive Summaries of Uppsala Dissertations from the Faculty of Science 340, Uppsala, 47 pp

Puonti-Kaerlas J, Stabel P, Eriksson T (1989) Transformation of pea (*Pisum sativum* L.) by *Agrobacterium tumefaciens*. Plant Cell Rep 8:321–324

Puonti-Kaerlas J, Eriksson T, Engstrom P (1990) Production of transgenic pea (*Pisum sativum* L.) plants by *Agrobacterium tumefaciens*-mediated gene transfer. Theor Appl Genet 80:246–252

Robbs SL, Hawes MC, Lin H-J, Pueppke SG, Smith LY (1991) Inheritance of resistance to crown gall in *Pisum sativum*. Plant Physiol 95:52–57

Rubluo A, Kartha KK, Mroginski LA, Dyck J (1984) Plant regeneration from pea leaflets cultured in vitro and genetic stability of regenerants. J Plant Physiol 117:119–130

Ryan CA, Farmer EE (1991) Oligosaccharide signals in plants: a current assessment. Annu Rev Plant Physiol Mol Biol 42:651–674

Schaerer S, Pilet P-E (1991) Roots, explants and protoplasts from pea transformed with strains of *Agrobacterium tumefaciens* and *rhizogenes*. Plant Sci 78:247–258

Schroeder D, Schotz AH, Wardley-Richardson J, Spencer D, Higgins ThJV (1993) Transformation and regeneration of two cultivars of pea (*Pisum sativum* L.). Plant Physiol 101:751–757

Shade RE, Schroeder HE, Pueyo JJ, Tabe LM, Murdock LL, Higgins TJV, Chrispeels MJ (1994) Transgenic pea seeds expressing the α-amylase inhibitor of the common bean are resistant to bruchid beetles. Biotechnology 12(8):793–796

Sitbon F, Sundberg B, Olsson O, Sandberg G (1991) Free and conjugated indoleacetic acid (IAA) content in transgenic tobacco plants expressing the *iaaM* and *iaaH* IAA biosynthesis genes from *Agrobacterium tumefaciens*. Plant Physiol 95:480–485

Stachel SE, Messens E, Van Montagu M, Zambryski P (1985) Identification of the signal molecules produced by wounded plant cells that activate T-DNA transfer in *Agrobacterium tumefaciens*. Nature 318:624–628

Thomashow MF, Karlinsey JE, Marks JR, Hurlbert RE (1987) Identification of a new virulence locus in *Agrobacterium tumefaciens* that affects polysaccharide composition and plant cell attachment. J Bacteriol 169:3209–3216

Tikhonovich IA, Alisova SM, Chetkova SA, Berestetku OA (1987) Enhancement of nitrogen fixation efficiency in pea by selection of lines for the nitrogenase activity. Agric Biol 2:29–34 (in Russian)

Tourneur J, Jouanin L, Muller J-F, Caboche M (1985) A genetic approach to the study of the mechanism of action of auxin in tobacco. Susceptibility of an auxin resistant mutant to *Agrobacterium* transformation. Proc 3rd Annu Symp Plant Biol Keystone, April 1985, New York, pp 791–797

Yanofsky M, Lowe B, Montoya A, Rubin R, Krul W, Gordon M, Nester E (1985) Molecular and genetic analysis of factors controlling host range in *Agrobacterium tumefaciens*. Mol Gen Genet 201:237–246

Zubko EI, Kuchuk NV, Tumanova LG, Vironskaya NA, Gleba YuYu (1990) Genetic transformation of pea plants mediated by *Agrobacterium tumefaciens*. Biopol Cell 6:77–80 (in Russian)

20 Transgenic Potato (*Solanum tuberosum* L.)

D.R. ROCKHOLD, M.M. MACCREE, and W.R. BELKNAP

1 Introduction

Potato (*Solanum tuberosum* L.) is a very important source of human nutrition for people in temperate climates around the world. According to the FAO, it is the fourth most important world food crop, exceeded in production only by wheat, maize, and rice. The FAO reports that 18.4 million hectares of potatoes were harvested worldwide in 1996. However, both pre- and post-harvest disorders (Hooker 1981) significantly reduce net yield, and therefore both farm profitability and food supplies suffer. In recent decades, potato breeders and growers have developed cultural and agronomic practices aimed at reducing losses to microbial pests or weaknesses of existing cultivars. While the production of certified, virus-free seed tubers using tissue culture-based technology represents an important tool in limiting virus-dependent losses, addressing other sources of commodity loss will require the introduction of improved potato genotypes. To that end, a number of programs have emerged that attempt to bring about cultivar improvement by traditional breeding. These programs around the world are working to breed desirable traits from different potato cultivars and related *Solanum* species into new commercial selections. In North America, these programs examine hundreds of thousands of seedlings from sexual crosses each year; yet commercialization of a new varietal selection is rare. The effectiveness of standard breeding practices in generating improved cultivars is limited by inherent genetic properties – low fertility, tetraploidy, and high heterozygosity – common to commercially viable strains. Horticultural practices, e.g., commercial vegetative propagation, coupled with susceptibility to *Agrobacterium*-mediated transformation, make the potato an ideal candidate for improvement via molecular genetic techniques. Using a biotechnological approach, it is possible to introduce desirable traits into existing cultivars, thereby creating new selections capable of producing more or better food for an all-too-hungry world.

As is true for most commercially produced farm commodities, potatoes are the subject of constant research to produce varieties with better agronomic properties such as insect and disease resistance, lower or higher levels of solids, improved yield, improved cold storage, cold tolerance in the field or altered

[1] United States Department of Agriculture Agricultural Research Service, Western Regional Research Center, 800 Buchanan Street, Albany, California 94710, USA

Biotechnology in Agriculture and Forestry, Vol. 47
Transgenic Crops II (ed. by Y.P.S. Bajaj)
© Springer-Verlag Berlin Heidelberg 2001

biochemical characteristics. In our program, we are attempting to make a variety of changes to commercial cultivars. Among them are increased bruise resistance, lower levels of steroidal glycoalkaloids, resistance to late blight, (*Phytophthora infestans*) and bacterial soft rot, (*Erwinia carotovora*). Other researchers around the world seek to alter the level of solids or produce potato varieties that produce insecticidal compounds designed to make the plants resistant to the Colorado potato beetle (*Leptinotarsa decemlineata*).

In this chapter, we present an overview of methods and procedures now in use to bring about genetic transformation of potato, field performance of genetically engineered plants, and discuss the potential of genetic engineering for potato improvement.

2 *Agrobacterium*-Mediated Transformation of Potato

2.1 A Brief Overview

Agrobacterium-mediated transformation is being used worldwide to introduce a wide variety of transgenes into potato. By the end of 1996, there had been 231 potato related applications for field release and interstate movement of transgenic potato plants in the US. Of these, 175 were submitted by industrial concerns (USDA, Animal and Plant Health Inspection Service, worldwide web). The remaining 56 applications were submitted by US government and academic laboratories. In these permits, research into application of molecular genetics to the improvement of potato cultivars focuses primarily on resistance to insect pests, e.g., Colorado potato beetle, or on the manipulation of metabolic characteristics such as altered carbohydrate metabolism, resistance to blackspot bruise, or pesticide tolerance. Another significant proportion of the permit applications is directed toward the development of plants that are resistant to microbial diseases.

2.2 Plant Materials

Many investigators (Table 1) have used *Agrobacterium*-mediated transformation in a variety of protocols to regenerate transgenic plants following inoculation of potato tissue. Various explant sources have been used for genetic transformation of potato. Many researchers report that in vitro grown plantlets are used to provide either leaf disks (De Block 1988) or stem segments (Ooms et al. 1987) as target tissue in transformation experiments while others use axenic microtubers (Snyder and Belknap 1993; Kumar 1995). Discs made from field-grown tubers or in vitro-grown microtubers have been useful in transforming the commercially important cultivars Bintje, Desiree, Escort, Golden Wonder, Lemhi Russet, Pentland Dell, Ranger Russet, and Russet Burbank. Regeneration of plants from tuber discs was reported to lead to a lower level of somaclonal variation than that observed in plants derived from other

Table 1. Summary of various studies conducted on transformation of potato

Year	Author(s)	Explant source/ culture used	Vector/method used	Remarks
1983	Ooms et al.	*In vitro* shoots	Oncogenic *A. tumefaciens*	cv. Maris Bard
1986	An et al.	Leaf/stem	pGA472	Used feeder layer
1987	Hanish ten Cate et al.	Tuber disk	*A. rhizogenes*	cv. Bintje
1987	Chung and Sims	Tuber disk	Oncogenic *A. tumefaciens*	
1988	Sheerman and Bevan	Tuber disk		
1989	Ishida et al.	Tuber disks	pBI121	cv. Lemhi Russet and Russet Burbank
1991	Ishige et al.	Tuber disks	n/s	11 of 21 Japanese cultivars transformed
1991	Bludy et al.	Leaf disks	pBI101	Surface sterilized leaf disks
1992	Cardi et al.	Several explant sources	several binary vectors	*S. commersonii*
1994	Filho et al.	Leaf disks	pGV1040	1 of 3 Brazilian cultivars
1994	Garbarino and Belknap	Tuber disks	pCGN1547	Russet Burbank
1995	Vanengelen et al.	Stems	pBINPLUS	
1995	Liu et al.	Leaf	pBI121	LBA4404 vs. GV2260
1996	Sweetlove et al.	Leaf disks	pBIN19 deriv. pFW4101	cv. Prairie.
1997	Kirsch et al.	Leaf	n/s	Cold stress in cv. Desiree
1997	Duwenig et al.	Leaf disk	pBinAR derivative	Studying carb. metabolism
1997	Wallis et al.	Tuber disks	pKYLX7 derivative	Cold tolerance in cv. Russet Burbank
1998	Porsch et al.	n/s	pSR derivatives	Stabilizing effect of MAR
1998	Trethewey et al.	Leaf explant	pBinAR derivative	Starch metabolism in Desiree

somatic tissue sources (Shepard et al. 1980; Larkin and Scowcroft 1981; Sheerman and Bevan 1988; Hoekema et al. 1989).

2.3 Cultivar Dependence

Commercially viable cultivars vary in their physiological and agronomic characteristics and these differences appear to affect the efficiency of

Agrobacterium-mediated transformation. In our hands the effect has been dramatic. Generally speaking, round white varieties such as Atlantic and Lenape (Akeley et al. 1968) have been easier to transform with *Agrobacterium* than russeted varieties such as Russet Burbank, Lemhi Russet, and Ranger Russet. Wenzler et al. (1989), experienced a similar difference between Russet Burbank and a proprietary chipping variety. Early experiments into hormone ratios (Ishida et al. 1989) produced no dramatic difference in the effect of auxin to cytokinin ratio. Instead, sensitivity to the kanamycin used for selection was the key factor. While round white tuber disks withstand a minimal 100 µg/ml of kanamycin during regeneration, the russeted varieties often fail to thrive. By reducing the concentration of kanamycin in the early stages of transformation/regeneration to 25–50 µg/ml, the number of viable shoots at 200 µg/ml of kanamycin increases (data not shown). Similarly, Ishige et al. (1991), compared transformation efficiency of 21 Japanese cultivars and found that 11 could be transformed using binary vectors and *Agrobacterium*-mediated transformation but four lines were resistant to regeneration of whole plants after the procedure. Filho et al. (1994), showed that only one of three Brazilian varieties could be successfully transformed when using a binary vector and *Agrobacterium*-mediated transformation. As biotechnologists attempt to modify each of the commercially important varieties, they must be prepared to optimize performance of the targeted cultivar in the transformation procedure (Mitten et al. 1990; Liu et al. 1995).

2.4 Vectors and *Agrobacterium* Strains

Manipulation of transgenes is significantly simpler and more efficient in binary vector plasmids, which are capable of replication in both *Escherichia coli* and *Agrobacterium*. The ability to clone foreign DNA into small (10–15 kb) vectors capable of replication in both *E. coli* and *Agrobacterium* makes it possible to process transgenes rapidly. Most of the binary vectors also have convenient cloning sites that simplify introduction of engineered transgenes. Once the transgene is introduced into the binary vector, it is a simple matter to verify the presence of the construct using a variety of methods. The verified plasmid can then be introduced into the *Agrobacterium* vector by either freeze-thaw transformation (Hofgen and Willmitzer 1988) or by electroporation (Shen and Forde 1989). *Agrobacterium*-mediated transformation of explant tissue is the preferred method for introduction of new genetic information. Microprojectile bombardment, a widely used technique in monocotyledonous species, has not been reported as a successful means of transforming *Solanum* species. Hansen and Chilton (1996), however, have recently reported a new method – "agrolistic" transformation – that combines features of *Agrobacterium*-mediated transformation with a microprojectile DNA delivery system. Such a system could prove useful in the transformation of *Solanum* species.

2.5 Performance of Transgenic Clones

Several years of field evaluation are required to thoroughly characterize individual transgenic clones constructed using genetic transformation. Transgenic potato lines suitable for the marketplace must have transgene-dependent improved characteristics without compromising key agronomic or quality properties. In a previous study, Dale and McPartlan (1992) showed that a population of transgenic clones would be expected to contain a significant number of individuals with transformation induced variations severe enough to limit commercial viability. The stability of transgenic selections may also be dependent upon the propensity of the parental line to transformation induced variation. Jongedijk et al. (1992) field-tested selections of cultivars Bintji and Escort transformed to express the coat protein of potato virus X (PVX). They showed that 81.8% of the Escort and 17.9% of the Bintje selections maintained normal tuber yield and grading. The effect of such variations can be evaluated completely only when the transgenic plants are propagated under multiple plot field conditions.

Transformation induced variations were also noted when Belknap et al. evaluated the field performance of transgenic Russet Burbank and Lemhi Russet clones over two growing seasons (Belknap et al. 1994). The 57 independent clones, each containing one of three different transgenes were evaluated under both greenhouse and multiplot field conditions. Many of the clones examined showed reduced yield and increases in malformed and undersized tubers. Other characteristics, such as specific gravity and fry color, showed less variability. The majority of clones that failed to produce tubers in the field were those expressing the β-glucuronidase (GUS) transgene. The GUS clones were introduced into the field a year earlier than the other transgenic lines, but the propagation and growing conditions in the 2 years were similar (data not shown). Only four of the 57 clones tested maintained the agronomic and quality properties of the parental lines. The low frequency of the typical phenotype suggests that experiments designed to yield commercially-viable transgenic potatoes should be initiated with several hundred independent transgenic clones.

3 Applying the Methodology

In the following section, we will present a more defined example of the application of the technology – from transgene design to field and laboratory evaluation. The basic problem addressed is blackspot bruise – a post harvest problem in many commercial cultivars (Dean et al. 1993). In sensitive cultivars, dark spots develop in tissue after impact injury. The blackening results when melanin is formed by the enzymatic polymerization of phenolic compounds such as tyrosine with molecular oxygen. For this series of experiments, we attempt to down-regulate expression of polyphenol oxidase (PPO) (Hunt et al. 1993; Bachem et al. 1994) – a key enzyme in the formation of melanin – using antisense RNA transgenes (Izant and Weintraub 1984, 1985).

3.1 Promoters and Binary Plasmids

To be effective, a transgene must be under the control of an appropriate transcriptional element (promoter) to control expression of the cloned region. Two promoters were chosen for this series of experiments – CaMV 35S for generalized expression of the antisense RNA and granule bound starch synthase (GBSS; van der Liej et al. 1991) to limit expression to the tuber. The 35S promoter was derived from commercially available plasmids. The GBSS promoter was isolated from genomic DNA (*Solanum tuberosum* L. cv. Lemhi Russet) using polymerase chain reaction (PCR). Primers used in the PCR were based on the published sequence of van der Liej et al., but were modified slightly to add the AAGCTT hexamer at the 5′ end of the upstream primer to add a *Hind*III restriction site in the PCR reaction product (van der Liej et al. 1991). The two primers used were; 5′ end – AAG CTT TAA CGA GAT AGA AAA TTA and 3′ end – ATG TGT GGT CTA CAA AAA G. After PCR synthesis, primers and remaining *Taq* were removed by passing the mixture over a QIAquick Spin DNA Purification column (QIAGEN, Inc., Chatsworth, CA, USA). The fragment was then cloned into plasmid pCR2 using a TA Cloning Kit (Invitrogen, San Diego, CA, USA). The sequence of the isolated clone was verified using fmol DNA Sequencing Kit (Promega, Madison, WI, USA). Primer design was accomplished using Right Primer (BioDisk, San Francisco, CA, USA). Sequence analysis and comparison was accomplished using MacVector (Oxford Molecular Group, Palo Alto, CA, USA). The transcriptional activity of new promoters was verified by testing GBSS-GUS fusions in transgenic plants.

Both the GBSS and 35S promoters were then cloned into a binary plasmid, pCGN1547 vector with a 1.1-kb fragment of cDNA for potato tuber PPO (polyphenol oxidase) cloned in an inverse orientation. The PPO cDNA used in the construct was isolated from a wounded tuber cDNA library in λgt11 (Amersham Corp, Arlington Heights, IL, USA). The cDNA was isolated using a homologous probe from tomato leaf generously provided by John Steffens (Newman et al. 1993). The final construct was then verified by sequencing across border regions and/or by using PCR to simultaneously verify the presence of the promoter and the adjoining fragment.

3.2 Plant Material

Axenic stock plants are maintained by serial transfer in growth chambers at 23 °C with a 16-h photoperiod at 2000–3000 lux under cool-white fluorescent lights. Nodal stem cuttings are taken from established shoot cultures monthly and placed on shoot medium (Table 2). For in vitro propagation of microtubers, nodal cuttings are placed on medium containing 6% sucrose and kinetin.

Table 2. Media for potato culture

Ingredient	Shoot	Tuber	Co-Cult	Stage I	Stage II	Stage III
Salts, Type[a]	MSMO	MSMO	MSMO	MS	MSMO	MSMO
Sucrose, g/l	30	60	20	20	20	30
Inositol, mg/l	0	0	0	100	0	0
Vitamin Stock, ml/l	1.0[b]	0	1.0[b]	1.0[c]	1.0[c]	1.0[c]
IAA-Aspartic Acid, μM			0.3	0.3		
Zeatin Riboside μM			10	10	10	
Kinetin, mg/l		2.5				
Kanamycin, mg/l				25–100	100	200
Cefotaxime, mg/l				Optional	Optional	Optional
Carbenicillin, mg/l				500	500	500
Gibberellic Acid, 200 mg/l					10 (optional)	
Phytagel, g/l	2.0	2.0	2.0	2.0	2.0	2.0
pH	5.6	5.6	5.4	5.6	5.6	5.6

[a] MS and MSMO from Sigma Chemicals (Murashige and Skoog 1962).
[b] Gamborg's vitamin mix: thiamine HCl, 2.5 mg/l, pyridoxine HCl 12.5 mg/l, nicotinic acid 12.5 mg/l, glycine 50 mg/l, folic acid 0.5 mg/l and Biotin 0.05 mg/l. Filter sterilized. Store in aliquots at −20 °C.
[c] MSMO vitamin mix; nicotinic acid, 1 g/l; pyridoxine HCl, 1 g/l. Filter sterilize.

Microtuber cultures are incubated in the dark at 19 °C for 6 to 24 weeks before the progeny are used in transformation experiments. In all cases, cultures are grown in 50 ml of the appropriate medium in Magenta GA-7 vessels. Where the presence of ethylene could be a problem, e.g., while shoots are being induced or where senescent, non-transformed tissue is present, culture vessels are covered with closures that have a microporous polypropylene membrane to facilitate gas exchange.

3.3 *Agrobacterium* Culture Conditions and Bacterial Transformation

Agrobacterium tumefaciens strain LBA4404 with the disarmed Ti plasmid pAL4404, a derivative of the oncogenic pTiAch5, (Hoekema et al. 1983) is the bacterial host used in our laboratory for transgenic binary vectors. The binary vector pCGN1547 (McBride and Summerfelt 1990) uses an origin of replication derived from *A. rhizogenes* plasmid pRiHRI along with the ColE1 origin of replication for maintenance in *E. coli*. The transformation/regeneration protocols described in the following sections have resulted in successful transformation of several cultivars.

3.3.1 Agrobacterium *Transformation (Hofgen and Willmitzer 1988)*

Preparation of Frozen, Competent Agrobacterium cells. Inoculate 10 ml of LB with a single colony of the *Agrobacterium* strain (e.g., LBA4404) to be transformed. Grow 2 days at 28 °C. Transfer 5 ml to 50 ml of fresh L-Broth (10% v/v). Grow for 4 h. Centrifuge at $3000 \times g$ for 20 min. Wash the pellet with chilled TE pH 7.2. Resuspend in 5 ml of sterile L-Broth. Aliquot 500 µl portion into sterile tubes, freeze in liquid nitrogen, and store at −80 °C for up to 3 months.

Transformation of Frozen Competent Cells. Thaw a tube of frozen cells on ice. Mix with 1 µg of plasmid DNA. Incubate on ice for 5 min. Drop tube into liquid nitrogen and hold for 5 min. Transfer the tube to a water bath and incubate at 37 °C for 5 min. Add 1 ml of L-broth and incubate at 28–30 °C with shaking for 2–4 h. Plate 200 µl per plate on L-Agar (Difco) or YM Agar (Difco) with appropriate antibiotics and grow at 28 °C for 2 days. Select colonies and grow overnight in L-broth containing antibiotics. Verify the presence of the transformed plasmid by doing plasmid minipreps and restriction enzyme digests or PCR reactions. Many researchers now find that electroporation (Shen and Forde 1989) is preferable to freeze-thaw transformation. We, however, continue to depend on the simplicity and reliability of the older method.

Biochemical Confirmation of Agrobacterium. We have found it useful to be able to confirm the identity of the transformed *Agrobacterium*. By using the following method, it is possible to unambiguously confirm that the bacterial culture is *Agrobacterium*. The confirmation is accomplished by growth (28 °C for 48 h) on lactose agar (Bernaerts and De Ley 1963) followed by flooding the plate with Benedict's reagent. Benedict's reagent is prepared by mixing $CuSO_4 \cdot 5H_2O$ (1.73 g/10 ml water) with sodium citrate/sodium carbonate (17.3 g sodium citrate; 10 g sodium carbonate in 60–70 ml hot (55 °C) water; cool when dissolved). The final volume is 100 ml. This simple test for the conversion of the lactose to 3-ketolactose makes it possible to eliminate contaminating non-*Agrobacterium* that usually have no characteristic physiological or morphological properties. A brilliant yellow precipitate forms on plates inoculated with *Agrobacterium* (Fig. 1). Some *Agrobacterium* biovars are unable to carry out the conversion of lactose to the 3-ketolactose but can be differ-

A   B

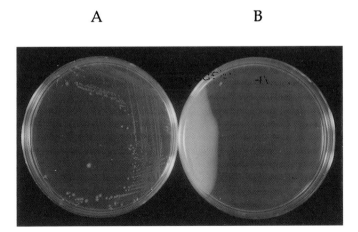

entiated on the basis of their motility, ability to produce acid, or presence of pectinolytic activity. (Bouzar et al. 1995)

3.4 Infecting Plant Tissue With *Agrobacterium*

The following procedure (Snyder and Belknap 1993) has been used for the potato varieties we have transformed, although with variable efficiency. First, inoculate 10 ml L-broth (LB) containing the appropriate antibiotics with the transformed *Agrobacterium* strain. Incubate the culture for 2 days at 28 °C on a gyratory shaker at 140 rpm. For optimum growth, use Erlenmeyer flasks with side baffles (Bellco Glass, Vineland, NJ, USA) to increase aeration and minimize clumping. To obtain a log-phase culture, transfer 10% (v/v) into a flask of fresh LB medium containing the appropriate antibiotic and 100 μM acetosyringone (3′,5′-dimethoxy-4′-hydroxyacetophenone). Grow for 4 h as above. Dilute the culture 1:20 in sterile MS, pH 5.3–5.5, containing 2% sucrose. Slice microtubers into 1–2 mm thick slices and place into the *Agrobacterium* solution. Incubate at room temperature for 10–15 min. After the incubation, decant the solution and add sterile water to the remaining tuber slices. Rinse briefly and decant again. Blot the slices gently on sterile filter paper and then place on co-cultivation medium with the smaller diameter end up. Incubate for 2–3 days at 23 °C under fluorescent lights with a 16-h photoperiod (8 h dark).

3.5 Regeneration of Transgenic Plants

After the bacterial treatment, the infected tissue must be maintained on the appropriate supporting medium. In stage 1, the tissue is allowed to form callus or dedifferentiated tissue as the transformation process proceeds. Then, shoot formation is induced in stage 2. Finally, nascent shoots are excised and placed on rooting medium (stage 3) to test the phenotypic expression of the selectable marker employed. At all times during the process, we strive to prevent the accumulation of excess moisture on the surface of sterilized media. Some workers (Liu et al. 1995) now find that liquid media increase the efficiency of transformation, but the practice has not been used in our laboratory. The following procedure, reported earlier (Snyder and Belknap 1993), with some minor modifications was used.

Day 1. Transfer the tuber slices to stage 1 medium in 15 × 100 mm Petri dishes.

Day 7. Transfer the tuber slices to fresh stage 1 medium. Repeat weekly for a total of 4 weeks.

Fig. 1A,B. *Agrobacterium* conversion of lactose to 3-ketolactose demonstrated by visualization with Benedict's reagent (Bernaerts and De Ley 1963). **A** *E. coli* produces no visible change in color. **B** *A. tumefaciens* produces a vivid yellow color change under the area containing heavy bacterial growth.

Day 28. Transfer to fresh stage 2 medium in 20×100 mm Petri dishes or Magenta vessels with covers that have microporous membranes. The first shoots that appear on the tuber disks usually arise from meristematic tissue and are not transgenic. These shoots are routinely excised and discarded. If, however, callus tissue is seen, it should be retained, as all actual transformations appear to arise from the dedifferentiated tissue. After a few more days, additional shoots may appear and callus should begin to form. If two, clearly separate, callus foci appear, the disk may be cut into two separate pieces. Otherwise, it is our practice to record only a single transformation event from an individual disk.

Day 42, and Every 14 Days Through Week 8. Transfer to fresh medium biweekly. As shoots appear and reach 1.5 to 2 cm in height, cut them off and transfer to stage 3 medium. Leave the disks on stage 2 to form more shoots. When shoots grow and form roots on stage 3, slice the shoot portion into segments with a node in each segment and transfer to stage 3 medium again. This increases the number of shoots and insures that the shoots are true transformants. Shoots that fail to form roots and/or thrive on the selective medium are not transgenic and may be discarded after 14 days. After rooting, plants may be "hardened" by a variety of methods. For example, gently remove the putative transformant and transfer to sterilized Vermiculite wetted with MS in a sealed Magenta vessel. Incubate in a growth room for 7–14 days to allow plants to form roots. Carefully transplant into potting soil. After 7 days in a protected (humid) environment away from direct sunlight, the plantlet can be subjected to normal glasshouse conditions. We routinely maintain an axenic culture of transgenic plants until a sufficient number of healthy plants are flourishing.

3.6 Greenhouse Conditions

After plants have been proven resistant to kanamycin, in vitro axenic specimens are transferred to 2-in. pots containing sterile Vermiculite wetted with sterile MS salts (Murashige and Skoog 1962). The young plants are maintained under greenhouse conditions except that they are kept in a shaded enclosure and are misted periodically to prevent wilt. In a few days, the root system is established and plants are gradually exposed to normal greenhouse illumination. As the plant grows, it is transferred to larger pots containing a commercial potting mix. At transplanting, each pot is treated with a slow release, complex fertilizer. Thereafter, supplemental fertilization is provided weekly during watering and a 20-20-20 fertilizer with micronutrients is employed. Plants are staked to prevent lodging and are allowed to grow 60 to 70 days from the date of transplantation. For proper tuberization, the photoperiod is supplemented with 1000 W halogen bulbs to maintain 16 h light and 8 h dark to avoid problems associated with a shortened photoperiod. Enstar (Sandoz Agro, Inc., Deplains, IL, USA) and other insecticidal sprays are applied as needed to control whiteflies.

3.7 Characterization of Clones, and Preparation for Fieldwork

CaMV 35S and GBSS antisense PPO transgenes (Fig. 2) were transformed into Lenape, Ranger Russet, Lemhi Russet, and Russet Burbank. The putative transformants were first used to prepare seed tubers in the greenhouse. Some transformants were then transferred to Aberdeen, Idaho, where they were tested for field performance as described previously (Belknap et al. 1994).

Fig. 2A,B. Antisense polyphenol constructs. Typical constructs for antisense down regulation in a "typical" binary vector (McBride and Summerfelt 1990). This vector, pCGN1547, provided by Calgene (Woodland, California) contains all the required functional elements to allow its propagation in either *E. coli* or *A. tumefaciens* hosts. *Left* and *right* borders (*LB* and *RB*) facilitate insertion of the transgene into the nuclear genome. Gentamycin and kanamycin resistance are conferred by genes expres-sed in prokaryotes (*Gm*) or eukaryotes (*npt*). Cloning in the vector is aided by the presence of the *lacZΔM15* fragment that confers blue-white selection in the appropriate host cell. **A** CaMV 35S antisense polyphenol oxi-dase. **B** GBSS antisense poly-phenol oxidase

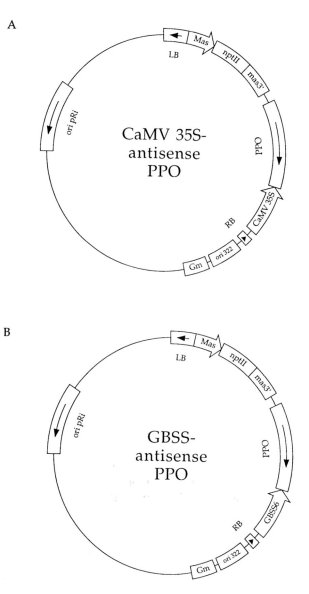

3.7.1 Northern and Southern Analysis

Hybridization of DNA probes to isolated RNA (Northern blots) represents an invaluable tool to eliminate "escapes" – non-transformed plants that grow in the presence of the selective agent. In our experience, PCR-based assays are not reliable as predictive tools when analyzing putative transgenic plants. In addition to identification of transformed clones, Northern blots also indicate the relative activity of a specific transgene at a transcriptional level. In our laboratory, RNA is routinely isolated using the small-scale procedure described by Verwoerd et al. (1989). This procedure has been modified slightly to shorten the time required for LiCl precipitation of the RNA. A minimum of 30 min at –20 °C is allowed for precipitation. Longer times do not result in significantly greater yields of RNA. One major advantage of the procedure is that it is not necessary to use reagents and glassware that have been baked and/or treated with diethyl pyrocarbonate (DEPC). The frozen material is treated with hot (80 °C) phenol plus extraction buffer directly. RNAase is destroyed and degradation of the target RNA is avoided. In addition, the procedure is done entirely in a microcentrifuge and results in high quality RNA in as little as 2 h. While we have had success with leaf tissue, yields of RNA from tuber tissue are much lower and may not yield sufficient quantities of RNA from microcentrifuge tube-scale extractions for (ca. 20 ng). Where larger quantities of RNA are needed, sufficient RNA can be obtained by increasing the quantity of tuber tissue extracted.

We routinely use two procedures to prepare probes for Northern blots. In most cases, radioactively labeled probes are prepared using an Oligonucleotide Labeling Kit (Pharmacia, Upsala). This procedure produces a heterogeneous probe consisting of regions of the template DNA molecule that are between about 300 base pairs and 1 kb in length.

Alternatively, full-length, homogeneous probes are prepared by PCR as described by Garbarino et al. (1992). In this protocol, 3–20 ng of template is mixed with 0.6 μM of each of two oligonucleotide primers flanking the probe and 50 μCi of (α-^{32}P) dCTP. The reaction mixture also contains 200 μM each of dATP, dGTP, and dTTP. The amount of unlabeled dCTP is normally decreased to 1–2 μM, depending on the length of the region to be labeled, to favor the incorporation of the labeled nucleotide. The PCR conditions used are; 1 min 94 °C, 1 min 45 °C, 30 s 72 °C, repeated 25 times. After the amplification, the probe is separated from unincorporated radionucleotides by fractionation through a 1-ml Sephadex G-50 spin column. Incorporation levels of 2×10^7 cpm/100 μl reaction are routinely obtained. Use of these full-length probes generally results in a higher signal to noise ratio.

To prepare Northern blots, total RNA (~8–20 μg/lane), is glyoxylated (Sambrook et al. 1989) and fractionated at 3–4 V/cm on a 0.8% agarose gel running in 0.1 M phosphate buffer, pH 7. The buffer is recirculated to the anode to prevent polarization of the buffer. The RNA is then transferred by capillary blot to charged nylon membranes (GeneScreen Plus, DuPont) as recommended by the manufacturer. Each membrane is placed in a sealed bag containing hybridization buffer (1% sodium dodecyl sulfate, 1 M NaCl, 10%

dextran sulfate, approximately 10 ml) for 15 to 60 min at 67 °C. To initiate the hybridization, ~3 × 10⁶ Cerenkov CPM of radiolabeled probe DNA and 3 mg of salmon testes DNA is heat denatured at 85 to 95 °C, and then added to the bag containing the filter. Hybridizations are incubated at 62 to 67 °C for 16 h. Filters are washed twice in 2 × SSC at room temperature for 10 min, twice in 2 × SSC, 1% SDS at the hybridization temperature for 30 min, and twice in 0.1 × SSC at room temperature for 30 min. After the wash, membranes are dried and exposed to suitable X-ray film. Using this procedure, it is possible to detect rare transcripts after autoradiography.

We have also employed the Genius System (Boehringer-Mannheim) to detect antisense transcripts using a single stranded, non radioactive probe. In this procedure, a digoxigenin (DIG) labeled probe is prepared using T7 or Sp6 RNA polymerase as recommended by the manufacturer. The resulting single-stranded probe is stable for many uses, if handled carefully. The non-radioactive method offers a useful alternative to the radioactive procedure but is not sufficiently sensitive to detect rare mRNAs. Autoradiographs from both methods are compared in Fig. 3. In this example, RNA extracted from transgenic plants containing a transgene for down-regulation of transaldolase is under the control of a CaMV 35S promoter. In the autoradiogram derived from the radioactive probe (Fig. 3A), signal intensity is higher, but the membrane was in contact with the X-ray film for 1 week. In Fig. 3B, the non-radioactive method clearly demonstrates the absence of the antisense RNA in sample 2.1 derived from a non-transformed "escape". In this particular example, the single-stranded RNA probe was prepared with SP6 RNA polymerase.

The Southern blot (DNA-DNA hybridization) is used to determine the copy number of a given DNA sequence. DNA is extracted from leaves as

Fig. 3A,B. Northern hybridizations comparing radioactive and non-radioactive probes. **A** CaMV 35S antisense transaldolase probed with radioactively labeled probe. **B** CaMV 35S Transaldolase RNA probed with single stranded probe prepared with Genius system

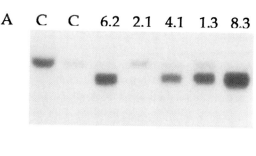

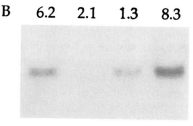

described by Draper and Scott (Draper and Scott 1988). After digestion with restriction endonuclease and electrophoretic fractionation, the DNA is transferred to charged nylon membranes by capillary transfer. The resulting blots are probed with labeled probes as described in Sambrook et al. (1989). The probes used for Southern analysis are best when a heterogeneous probe prepared by random-primed oligonucleotide synthesis is used. It is possible that the smaller, fragmented probe is able to form more stable hybrids than the longer homogeneous probes prepared by PCR (data not shown).

3.7.2 Western Blots and Oxygen Measurements

While Northern blots are useful in indicating transcript levels in transgenic clones, it many cases it is also necessary to determine changes in specific protein levels associated with the introduced transgene. In our laboratory, there have been occasions when high levels of antisense RNA could be demonstrated using a Northern blot, yet no concomitant change in the level of the target protein could be detected using a western blot (data not shown). One could argue that western blots are the most reliable of all tests designed to evaluate transgene efficiency. When a suitable antibody is available, early testing with a western blot is strongly recommended.

Oxygen uptake in tuber extracts was measured as described by Thygesen et al. (Thygesen et al. 1995) using a Diamond General Model no. 1231 Chemical Microsensor equipped with a Clark (Clark et al. 1953) oxygen electrode in a stirred oxygen uptake chamber.

4 Current Results

During the 1996 growing season, 27 independent Lenape clones containing the 35S antisense transgene for PPO were tested for changes in their tendency to produce blackspot bruise. After a normal growing season using greenhouse-grown seed, tubers from each clone were subject to an abrasive peel procedure as described by Pavek et al. (1985). Of these, four clones showed a decreased tendency to produce blackening after the peeling procedure. The same four clones showed a significant reduction in the amount of oxygen consumed in greenhouse grown tubers when analyzed as above (Fig. 4). When the clones in question were being tested for oxygen consumption, visual estimates of their propensity to bruise were recorded using an arbitrary scale of 0 to 3 where a 0 indicated no blackening of cut surfaces and 3 indicated severe blackening. Some of the more bruise resistant clones showed no blackening, even after extended incubation. Field-grown tubers containing these antisense constructs were recently examined by western blot. In Fig. 5 one can see that the PPO is greatly reduced in two of the transgenic clones – LSP016 and LSP031. Both clones showed reduced tendency to blacken in other, subjective,

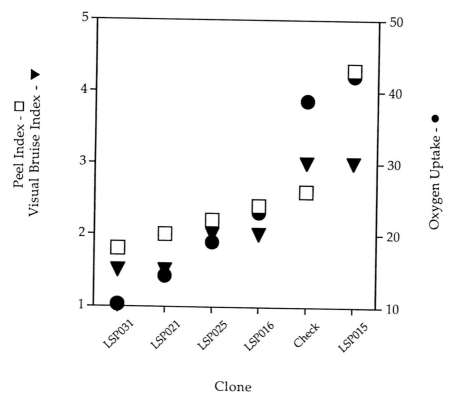

Fig. 4. Effect of different antisense polyphenol oxidase transgenes in potato visualized by abrasive peel (Pavek et al. 1985), visual bruise after cutting injury, and oxygen uptake (Hunt et al. 1993) in select Lenape plants

assays that are based on the degree of melanin formation on cut or abraded surfaces.

Transgenic potato selections suitable for the commercial market must have transgene-dependent improvements and be free of deleterious properties such as loss of yield and changes in agronomic or quality properties. Dale and McPartlan (1992) report that a population of transgenic clones would be expected to contain a significant number of individuals with variations severe enough to limit commercial viability. Indeed, we have noted that transgenic plants that perform well in greenhouse evaluations sometimes fail to either demonstrate the characteristic of the transgene or fail to maintain yield or quality characteristics of the parent cultivar (Belknap et al. 1994).

In recent years, field trials have been conducted with transgenic materials that show a general loss of parental phenotype. Typically, yield suffers or there

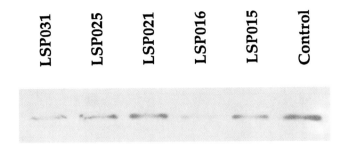

Fig. 5. Western blot (immunoblot) showing decreased polyphenol oxidase in CaMV 35S antisense polyphenol oxidase clones

is a loss of critically important quality properties such as French fry color. These changes may be due to somaclonal variation that occurs as the transformation proceeds. Since every successful transformation event includes a transition to callus before shoot and root initiation, it is possible that variations from the parental phenotype are brought about by a combination of the dedifferentiation of the tissue coupled with "position effects" of the inserted transgene.

5 Summary

There is a great deal of interest in the commercialization of transgenic selections that are insect resistant or that have improved handling properties. Around the world, researchers are developing transgenic plants with the goal of introducing resistance to a variety of diseases including those with bacterial, fungal and viral origin. Modern molecular genetics is being used successfully to aid in the study of carbohydrate assimilation and metabolism and the knowledge gained is being applied to the development of improved cultivars. Many laboratories are now searching for new regulatory elements to expand the capabilities of the molecular biologist. As genome sequencing projects expand the information base, it may soon be possible to use computational tools to predict the success of a given molecular genetics approach. In the end, the true test of any transgenic selection is its performance in field tests over several growing seasons.

Acknowledgments. The authors thank Dennis Corsini and his staff at Aberdeen, ID, USA, for field propagation of transgenic plants, subsequent analysis of the crop, and many helpful discussions. John Steffen generously provided polyclonal antibodies to polyphenol oxidase and the tomato cDNA clone for PPO. Thanks are also due to CalGene for the binary vector pCGN1547, and to Simon Robinson and Fieke van der Liej for helpful discussion. Wilson Chen, Grace Lu and Laura McGee provided invaluable technical support in the isolation and characterization of the GBSS promoter. Reference to a company and/or product by the USDA is only for purposes of information and does not imply approval or recommendation of the product to the exclusion of others that may also be suitable.

References

Akeley RV, Mills WR, Cunningham CE, Watts J (1968) Lenape: a new potato variety high in solids and chipping quality. Am Potato J 45:142–151

An G, Watson BD, Chiang CC (1986) Transformation of tobacco (*Nicotiana* spp.), tomato (*Lycopersicon esculentum*), potato (*Solanum tuberosum*) and *Arabidopsis thaliana* using a binary *Ti* vector system. Plant Physiol 81:301–305

Bachem CWB, Speckmann GJ, van der LInde PCG, Verheggen FTM, Hunt MD, Steffens JC, Zabeau M (1994) Antisense expression of polyphenol oxidase genes inhibits enzymatic browning in potato tubers. Bio/Technology 12:1101–1105

Beijersbergen A, Smith SJ, Hooykaas PJ (1994) Localization and topology of VirB proteins of *Agrobacterium tumefaciens*. Plasmid 32:212–218

Belknap WR, Corsini D, Pavek JJ, Snyder GW, Rockhold DR, Vayda ME (1994) Field performance of transgenic Russet Burbank and Lemhi Russet potatoes. Am Potato J 71:285–296

Bernaerts MJ, De Ley J (1963) A biochemical test for crown gall bacteria. Nature 167:406–407

Blundy KS, Blundy MAC, Carter D, Wilson F, Park WD, Burrell, MM (1991) The expression of class I patatin gene fusions in transgenic potato varies with both gene and cultivar. Plant Mol Biol 16:153–160

Bouzar H, Jones JB, Bishop AL (1995) Simple cultural tests for identification of *Agrobacterium* biovars. In: Gartland KMA, Davey MR (eds) *Agrobacterium* protocols, vol 44. Humana Press, Clifton, New Jersey

Cardi T, Iannamico V, D'Ambrosio F, Filippone E, Lurquin PF (1992) *Agrobacterium*-mediated genetic transformation of *Solanum commersonii* Dun. Plant Sci 87:178–189

Chung SH, Sims WS (1987) Transformation of potato (*Solanum tuberosum*) tuber cells with *Agrobacterium tumefaciens* and Ti plasmid DNA. Korean Biochem J 20:289–298

Citovsky V, Zupan J, Warnick D, Zambryski P (1992) Nuclear localization of *Agrobacterium* VirE2 protein in plant cells. Science 256:1802–1805

Clark LC Jr, Wold R, Granger D, Taylor Z (1953) Continuous recording of blood oxygen tension by polarography. J Appl Physiol 6:189–193

Croy RRD (ed) (1993) Plant molecular biology labfax. BIOS Scientific Publisher, Oxford, pp 1–382

Dale PJ, McPartlan HC (1992) Field performance of transgenic potato plants compared with controls regenerated from tuber discs and shoot cuttings. Theor Appl Genet 84:585–591

Dean BB, Jackowiak N, Nagle M, Pavek J, Corsini D (1993) Blackspot pigment development of resistant and susceptible *Solanum tuberosum* L genotypes at harvest and during storage measured by 3 methods of evaluation. Am Potato J 70:201–217

Deblaere R, Reynaerts A, Hofte H, Harnalsteens J-P, Leemans J, Van Montagu M (1987) Vectors for cloning in plant cells. In: Wu R, Grossman L (eds) Methods in enzymology, vol 153. Academic Press, San Diego, pp 277–292

De Block M (1988) Genotype-independent leaf disk transformation of potato (*Solanum tuberosum*) using *Agrobacterium tumefaciens*. Theor Appl Genet 76:767–774

Draper J, Scott R (1988) The isolation of plant nucleic acids. In: Draper J, Scott R, Armitage P, Walden R (eds) Plant genetic transformation and gene expression – a laboratory manual Blackwell, Oxford, pp 199–236

Duwenig E, Steup M, Willmitzer L, Kossmann J (1997) Antisense inhibition of cytosolic phosphorylase in potato plants (*Solanum tuberosum* L.) affects tuber sprouting and flower formation with only little impact on carbohydrate metabolism. Plant J 12:323–333

Filho FES, Figueiredo LFA, Monteneshich DC (1994) Transformation of potato (*Solanum tuberosum*) cv. Mantiqueira using *Agrobacterium tumefaciens* and evaluation of herbicide resistance. Plant Cell Rep 13:666–670

Fraley RT, Rogers SG, Horsch RB, Eichholtz DA, Flick JS, Fink CL, Hoffman NL, Sanders PR (1985) The sev system: a new disarmed Ti plasmid vector system for plant transformation. Bio/Technology 3:629–635

Garbarino JE, Belknap WR (1994) The use of ubiquitin promoters for transgene expression in potato. In: Belknap WR, Vayda ME, Park WD (eds) Molecular and cellular biology of the potato, 2nd edn. CAB International, Wallingford, pp 173–185

Garbarino JE, Rockhold DR, Belknap WR (1992) Expression of stress-responsive ubiquitin genes in potato tubers. Plant Mol Biol 20:235–244

Hanisch ten Cate CH, Ramulu S, Dijkhuis P, de Groot B (1987) Genetic stability of cultured hairy roots induced by *Agrobacterium rhizogenes* on tuber discs of potato cv. Bintje. Plant Sci 49:217–222

Hansen G, Chilton MD (1996) "Agrolistic" transformation of plant cells: integration of T-strands generated in planta. Proc Natl Acad Sci USA 93:14978–14983

Hoekema A, Hirsch PR, Hooykaas PJJ, Schilperoort RA (1983) A binary plant vector strategy based on separation of *vir-* and T-region of the *Agrobacterium tumefaciens* Ti-plasmid. Nature 303:179–181

Hoekema A, Huisman JJ, Molendijk L, van den Elzen PJM, Cornelissen BJC (1989) The genetic engineering of two commercial potato cultivars for resistance to potato virus X. Bio/Technology 7:273–278

Hofgen R, Willmitzer L (1988) Storage of competent cells for *Agrobacterium* transformation. Nucleic Acids Res 16:9877

Hooker WJ (ed) (1981) Compendium of potato diseases, Am Phytopathol Soc, St Paul, Michigan, pp 1–125

Hooykaas PJJ, Shilperoort RA (1992) *Agrobacterium* and plant genetic engineering. Plant Mol Biol 19:15–38

Hunt MD, Eannetta NT, Yu H, Newman SM, Steffens JC (1993) cDNA cloning and expression of potato polyphenol oxidase. Plant Mol Biol 21:59–68

Ishida BK, Snyder JGW, Belknap WR (1989) The use of in vitro-grown microtuber discs in *Agrobacterium*-mediated transformation of Russet Burbank and Lemhi Russet potatoes. Plant Cell Rep 8:325–328

Ishige T, Ohshima M, Ohashi Y (1991) Transformation of Japanese potato cultivars with the beta-glucuronidase gene fused with the promoter of the pathogenesis-related la-protein gene of tobacco. Plant Sci 73:167–174

Izant JG, Weintraub H (1984) Inhibition of thymidine kinase gene expression by anti-sense RNA: a molecular approach to genetic analysis. Cell 36:1007–1015

Izant JG, Weintraub H (1985) Constitutive and conditional suppression of exogenous and endogenous genes by anti-sense RNA. Science 229:345–352

Jongedijk E, de Schutter AAJM, Stolte T, van den Elzen PJM, Cornelissen BJC (1992) Increased resistance to potato virus X and preservation of cultivar properties in transgenic potato under field conditions. Bio/Technology 10:422–429

Kirch HH, van Berkel J, Glaczinski H, Salamini F, Gebhardt C (1997) Structural organization, expression and promoter activity of a cold-stress-inducible gene of potato (*Solanum tuberosum* L.). Plant Mol Biol 33:897–909

Kumar A (1995) *Agrobacterium*-mediated transformation of potato genotypes in *Agrobacterium* protocols. In: Gartland KMA, Davey MR (eds) Methods in molecular biology, vol 44. Humana Press, Totowa, New Jersey, pp 121–128

Laemmli UK (1970) Cleavage of structural proteins during the assembly of the head of bacteriophage T4. Nature 227:680–685

Larkin PJ, Scowcroft WR (1981) Somaclonal variation – a novel source of variability from cell cultures for plant improvement. Theor Appl Genet 60:197–214

Liu THA, Stephens LC, Hannapel DJ (1995) Transformation of *Solanum brevidens* using *Agrobacterium tumefaciens*. Plant Cell Rep 15:196–199

McBride KE, Summerfelt KR (1990) Improved binary vectors for *Agrobacterium*-mediated plant transformation. Plant Mol Biol 14:269–276

Mitten DH, Horm M, Burrell MM, Blundy KS (1990) Strategies for potato transfortion and regeneration. In: Vayda ME, Park WD (eds) The molecular and cellular biology of the potato. CAB International, Wallingford, pp 181–191

Murashige T, Skoog F (1962) A revised medium for rapid growth and bioassays with tobacco tissue cultures. Physiol Plant 15:473–497

Newman SM, Eannetta NT, Yu H, Prince JP, de Vicente MC, Tanksley SD, Steffens JC (1993) Organisation of the tomato polyphenol oxidase gene family. Plant Mol Biol 21:1035–1051

Ooms G, Karp A, Roberts J (1983) From tumor to tuber: tumor cell characteristics and chromosome numbers of crown gall derived tetraploid potato plants (*Solanum tuberosum* cv. Maris Bard). Theor Appl Genet 66:166–172

Ooms G, Burrell MM, Karp A, Bevan M, Hille J (1987) Genetic transformation in two potato cultivars with T-DNA from disarmed *Agrobacterium*. Theor Appl Genet 73:744–750

Pansegrau W, Schoumacher F, Hohn B, Lanka E (1993) Site-specific cleavage and joining of single-stranded DNA by VirD2 protein of *Agrobacterium tumefaciens* Ti plasmids: analogy to bacterial conjugation. Proc Natl Acad Sci USA 90:11538–11542

Pavek J, Corsini D, Nissley F (1985) A rapid method for determining blackspot susceptibility of potato clones. Am Potato J 62:511–517

Porsch P, Jahnke A, During K (1998) A plant transformation vector with a minimal T-DNA. II. Irregular integration patterns of the T-DNA in the plant genome. Plant Mol Biol 37:581–585

Sambrook J, Fritsch EF, Maniatus T (1989) Molecular cloning – a laboratory manual 2nd edn. Cold Spring Harbor Laboratory Press, Cold Spring Harbor, New York

Shahin EA, Simpson RB (1986) Gene transfer system for potato. Hort Sci 2:1199–1201

Sheerman S, Bevan MW (1988) A rapid transformation method for *Solanum tuberosum* using binary *Agrobacterium tumefaciens* vectors. Plant Cell Rep 7:13–16

Shen WJ, Forde BG (1989) Efficient transformation of *Agrobacterium* spp. by high voltage electroporation. Nucleic Acids Res 17:8385

Sheng J, Citovsky V (1996) *Agrobacterium*-plant cell DNA transport: have virulence proteins, will travel. Plant Cell 8:1699–1710

Shepard JF, Bidner D, Shahin E (1980) Potato protoplasts in crop improvement. Science 208: 17–24

Shirasu K, Koukolikova-Nicola Z, Hohn B, Kado CI (1994) An inner-membrane-associated virulence protein essential for T-DNA transfer from *Agrobacterium tumefaciens* to plants exhibits ATPase activity and similarities to conjugative transfer genes. Mol Microbiol 11:581–588

Snyder GW, Belknap WR (1993) A modified method for routine *Agrobacterium*-mediated transformation of in vitro grown potato microtubers. Plant Cell Rep 12:324–327

Stiekema WJ, Heidekamp F, Louwerse JD, Verhoeven HA, Dijkhuis P (1988) Introduction of foreign genes into potato cultivars Bintje and Desiree using an *Agrobacterium tumefaciens* binary vector. Plant Cell Rep 7:47–50

Sweetlove LJ, Burrell MM, Rees T (1996) Characterization of transgenic potato (*Solanum tuberosum*) tubers with increased ADPglucose phyrophosphorylase. Biochem J 320:487–492

Thygesen PW, Dry IB, Robinson SP (1995) Polyphenol oxidase in potato. A multigene family that exhibits differential expression patterns. Plant Physiol 109:525–531

Trethewey RN, Geigenberger P, Riedel K, Hajirezaei M, Sonnewald U, Stitt M, Riesmeier JW, Willmitzer L (1998) Combined expression of glucokinase and invertase in potato bubers leads to dramatic reduction in starch accumulation and a stimulation of glycolysis. Plant J 15:109–118

van der Lidj FR, Visser RGF, Ponstein AS, Jacombsen E, Feenstra WJ (1991) Sequence of the structural gene for granule-bound starch synthase of potato (*Solanum tuberosum* L.) and evidence for a single point deletion in the *amf* allele. Mol Gen Genet 228:240–248

Vanengelen FA, Molthoff JW, Conner AJ, Nap JP, Pereira A, Stiekema WJ (1995) pBinplus – an improved plant transformation vector based on pBin19. Transgenic Res 4:288–290

Verwoerd TC, Dekker BMM, Hoekema A (1989) A small-scale procedure for the rapid isolation of plant RNAs. Nucleic Acids Res 17:2362

Wallis JG, Wang H, Guerra DJ (1997) Expression of a synthetic antifreeze protein in potato reduces electrolyte release at freezing temperatures Plant Mol Biol 35:323–330

Wenzler H, Mignery G, May G, Park W (1989) A rapid and efficient transformation method for the production of large numbers of transgenic potato plants. Plant Sci 63:79–85

Zupan JR, Citovsky V, Zambryski P (1996) *Agrobacterium* VirE2 protein mediates nuclear uptake of single-stranded DNA in plant cells. Proc Natl Acad Sci USA 93:2392–2397

21 Transgenic Grapes (*Vitis* Species)

L. Martinelli[1] and G. Mandolino[2]

1 Introduction

1.1 Distribution and Importance of Grapes

Grape is a fruit crop with great cultural significance: tradition, habit, art and even religion have accompanied the development and the domestication of grape throughout human history (Unwin 1991; Martinelli 1997). This woody perennial plant belongs to the genus *Vitis* (family *Vitaceae*) (Willis 1973; Cronquist 1981), distinct in the *Euvitis* and *Muscadinia* subgenera, with diploid chromosome sets of 38 and 40, respectively. Chromosomes are very small (average length 1 μm; Shetty 1959) and the DNA content is approximately 1 pg/2C, in a genome of about 475 Mbp (Lodhi and Reisch 1995).

The systematics of *Vitis* is difficult and artificial. American *Vitis* species are cultivated for rootstocks or for industrial transformation (Colby and Meredith 1993). Within the Eurasian *Vitis vinifera*, the primary cultivated species, 5000 (Levadoux 1954; Dry and Gregory 1988) to 24000 (Viala and Vermorel 1909) cultivars have been described, but most of them are now considered not truly distinct genotypes (Alleweldt and Possingham 1988). The development of techniques for gaining knowledge of the germplasm has been strongly pursued (Martinelli 1997). Ampelography manuals have long been the only tools for the description of *Vitis* cultivars (Galet 1979), but recently, isozyme (Weeden et al. 1988) and DNA markers (Thomas et al. 1994; Bowers et al. 1996) have greatly enhanced the possibility of genotype characterization.

This versatile fruit crop is grown for wine, table grapes, raisin and juice production, and in 1996 the cultivated area was 7742000 ha, the grape harvest reached 58681200 tonnes and wine production was 272534000 hl (Dutruc-Rosset 1998).

1.2 Need for Genetic Transformation

Genetic improvement of grapes is carried out by clonal selection or breeding programmes. In wine grapes, this has been based on the occurrence of spon-

---

[1] Laboratorio Biotecnologie, Istituto Agrario, 38010 San Michele all'Adige (TN), Italy
[2] Istituto Sperimentale per le Colture Industriali, via di Corticella 133, 40129 Bologna, Italy

Biotechnology in Agriculture and Forestry, Vol. 47
Transgenic Crops II (ed. by Y.P.S. Bajaj)
© Springer-Verlag Berlin Heidelberg 2001

taneous mutations during vegetative propagation of traditional cultivars, while crossings have been used to improve table grape, raisin and particularly root-stock production, exploiting the huge genetic resources of grape germplasm. Crossings are relatively feasible within *Vitis* genotypes which are outbreeders; however, genetic drawbacks discourage the use of breeding, due to the high degree of heterozygosity, heavy loads of deleterious recessives and severe inbreeding depression, combined with a long generation time from seed to fruit (Colby and Meredith 1993). Furthermore, while dessert grape is a dynamic product, the wine market is very conservative and the acceptance of new varieties has been ruled by customs, law, and even traditional and emotional criteria (Alleweldt and Possingham 1988). In this framework, genetic transformation offers unique perspectives for the improvement of *Vitis*: most importantly to overcome disadvantages associated with conventional breeding, and to modify specific characteristics in genotypes already selected (Martinelli 1997). The most ambitious goal is the production of plants resistant to biotic stress and endowed with ecological and economical advantages. The introduction of viral constructs or their antisense sequences (Beachy 1990; Saldarelli et al. 1997; Minafra etal. 1997; Martinelli et al. 2000) into the grapevine genome could limit the diffusion of viral diseases; fungal resistance could be enhanced by introducing broad range resistance genes such as osmotin (Martinelli et al. 1996b) or lytic peptides (Jaynes et al. 1993); anthocyanin genes and the polyphenol oxidase gene have been cloned (Dry and Robinson 1994; Sparvoli et al. 1994) and open perspectives for the improvement of both chemical and sensorial properties of wine.

2 Genetic Transformation

2.1 Brief Review of Present Methods

Even though grape was one of the first plants employed in tissue culture studies (Morel 1944), genetic transformation in *Vitis* was reported only in the 1990s (Mullins et al. 1990). However, this technology is still far from routine, due to the low regeneration competence of grapevine which is the principal requirement for transformation strategies.

A summary of various studies on *Vitis* transformation is presented in Table 1 where it can be seen that transgenic plants have been obtained in only a few responding wild species, interspecific hybrids and rootstocks, as well as in some *V. vinifera* cultivars, since efficiency in grape regeneration was found to be strongly genotype dependent. Grapevines were found to be the least amenable to in vitro manipulation, although genes involved in biotic stress resistance have been successfully introduced into dessert and wine grapes.

Agrobacterium tumefaciens has been the most common vector employed for gene transfer while *A. rhizogenes* has been used less. Generally, the selection of transgenic cells during regeneration has been conducted in the

Table 1. Summary of various studies conducted on transformation of grapes

Reference	Genotype	Traits of interest/ objective	Exogenous genes	Explant/culture used	Vector/method used	Selection	Results
Baribault et al. (1989)	*V. vin.* Cab. Sauv.	Marker gene	NPTII	Cell cultures from pericarp calli	*A. tumefaciens*	Kan	Transgenic calli
Baribault et al. (1990)	*V. vin.* Cab. Sauv.	Marker gene	GUS	Fragmented shoot apices	*A. tumefaciens*	Kan	Chimeric shoots
Guellec et al. (1990)	*V. vin.* Grenache	Marker gene	NPTII	Whole micropropagated plants	*A. rhizogenes*	Kan	Root cultures, callogenesis
Mullins et al. (1990)	*V. rupestris, V. vin.* Cab. Sauv., Chardonnay	Marker gene	GUS	Anther somatic embryo sections	*A. tumefaciens*	Kan	Transgenic plants of *V. rupestris*
Colby et al. (1991)	*V. vin.* French Colombard, Thompson Seedl.	Marker gene	GUS	Leaves	*A. tumefaciens*	Kan	Chimeric shoots
Berres et al. (1992)	*V. vin.* Chardonnay, Gewürztram.; 41B; Kober 5BB; SO4	Marker gene	GUS	Stem pieces and leaf disks	*A. tumefaciens*	–	GUS(+) buds
Hébert et al. (1993)	*V. vin.* Chancellor	Marker gene	GUS	Embryogenic cell suspensions	Biolistic	Kan	Transformed calli
Le Gall et al. (1994)	110 Richter	Virus resistance	CP of GCMV	Anther embryogenic calli	*A. tumefaciens*	Hyg/Kan	ELISA(+) transgenic plants
Lupo et al. (1994)	Virus infected *V. vin.* cvs.; *V. rup.*; LN33; Kober 5BB	Virus purification	GUS	Internodes	*A. rhizogenes*	–	Root cultures for GFkV, GVA and GVB purification
Martinelli and Mandolino (1994, 1996)	*V. rupestris*	Marker gene	GUS	Petiole somatic embryos	*A. tumefaciens*	Kan	Transgenic plants

Table 1. *Continued*

Reference	Genotype	Traits of interest/ objective	Exogenous genes	Explant/culture used	Vector/method used	Selection	Results
Nakano et al. (1994)	*V. vin.* Koshusanjaku	Marker gene	GUS	Leaf embryogenic calli	*A. rhizogenes*	Kan	Plant regeneration
Gribaudo et al. (1995)	*V. vin.* Nebbiolo, Moscato, Barbera	Root line study	–	Micropropagated shoots	*A. rhizogenes*	–	Axenic root cultures
Krastanova et al. (1995)	*V. rupestris*; 110 Richter	Virus resistance	CP of GFLV	Embryogenic calli of anther and som. embryos	*A. tumefaciens*	Carb/Kan	ELISA(+) transgenic plants
Mauro et al. (1995)	*V. vin.* Chardonnay; 41B; SO4	Virus resistance	CP of GFLV	Anther embryogenic cell suspensions	*A. tumefaciens*	Kan	ELISA(+) transgenic plants
Scorza et al. (1995)	*V. vin.* cvs. (three table grapes)	Marker gene	GUS	Zygotic embryos	Biolistic + *A. tumefaciens*	Kan	Transgenic plants
Kikkert et al. (1996)	*V. vin.* Chancellor	Marker gene	GUS	Embryogenic cell suspensions	Biolistic	Kan	Transgenic plants
Martinelli et al. (1996b)	*V. rupestris*	Fungus resistance	Osmotin	Petiole somatic embryos	*A. tumefaciens*	Kan	Plant regeneration
Perl et al. (1996)	*V. vin.* Superior Seedl.	Herbicide resistance	Bar	Anther somatic embryos	*A. tumefaciens*	Basta/Hyg	Transgenic plants
Rombaldi et al. (1996)	*V. vin.* Gamay	Ethylene production	EFE	Protoplasts	Electroporation	–	Increased ethylene prod.

Scorza et al. (1996)	*V. vin.* Thompson Seedl.	Fungus resistance	Lytic peptide Shiva 1	Leaf somatic embryos	Biolistic + *A. tumefaciens*	Kan	Transgenic plants
Scorza et al. (1996)	*V. vin.* Thompson Seedl.	Virus resistance	CP of TomRSV	Leaf somatic embryos	Biolistic + *A. tumefaciens*	Kan	Transgenic plants
Gölles et al. (1998)	*V. vin.* Russalka; 110 Richter	Virus resistance	CP of GFLV, ArMV, GVA, GVB	Embryogenic tissues	*A. tumefaciens*	Kan	In vitro plant micrografting
Kikkert et al. (1998)	*V. vin.* Merlot, Chardonnay	*B. cin., U. enc.* resistance	*Trichoderma* endochitinase	Embryogenic cell suspensions	Biolistic	Kan	Disease resistance assays on plants
Krastanova et al. (1998)	Five rootstocks	Virus resistance	CP of GFLV, GLRaV	Anther embryogenic calli	*A. tumefaciens*	Kan	In vitro plant micrografting
Levenko and Rubtsova (1998)	*V. vin.* Podarok Magaracha	Herbicide resistance	Bar	Leaves and petioles	*A. tumefaciens*	Kan	In vivo Basta-resistant plants
Martinelli et al. (2000)	*V. vin.* Superior Seedl.; *V. rupestris*	Virus resistance	MP of GVA, GVB	Anther and petiole somatic embryos	*A. tumefaciens*	Kan	Transgenic plants
Mauro et al. (1998)	41B	Virus resistance	CP, polymerase, proteinase of GFLV	Cell cultures	*A. tumefaciens*	Kan	Experimental plots in vineyard

Abbreviations: ArMV, arabis mosaic virus; Carb, carbenicillin; CP, coat protein; EFE, ethylene-forming enzyme; GCMV, chrome mosaic nepovirus; GFkV, gnapevine fleck vinus; GFLV, grapevine fanleaf virus; GLRaV, grapevine leafroll associated closterovirus; GUS, β-glucuronidase; GVA, grapevine virus A; GVB, grapevine virus B; Hyg, hygromicin; Kan, kanamycin; TomRSV, tomato ringspot virus; MP, movement protein; NPTII, neomycin phosphotranspherase II.

presence of kanamycin, but herbicide and hygromycin resistance genes have been employed as well; the preferred marker for evaluating the fate of a foreign gene in transgenics has been β-glucuronidase gene expression.

Embryonic tissues (both zygotic and somatic) have proven to be the best cell source for transgenic plant regeneration and have given rise to homogeneous and stable gene insertion. Meristems have been less amenable because of the formation of chimeric tissues following adventitious bud formation in the presence of kanamycin as the selective agent. This fact hinders the use of an efficient regeneration system (direct organogenesis that occurs at suitable frequencies in several important genotypes within the genus *Vitis* (Martinelli et al. 1996a), while somatic embryogenesis is confined to a few genotypes at low efficiencies.

2.2 Methodology

Plant Material, Tissue Culture, and Genetic Transformation. Somatic embryos of *Vitis rupestris* Scheele have been employed for genetic transformation with *A. tumefaciens* strain LBA4404 harboring the plasmid pBI121 and containing the nopaline synthase and β-glucuronidase genes (Jefferson et al. 1987). The main steps of tissue culture and of transformation strategy, performed according to Martinelli et al. (1993) and Martinelli and Mandolino (1994, 1996), are summarized in Fig. 1. After co-culture, embryos were incubated for 3 days at 28 °C on secondary induction medium, then transferred to the same medium with 300 mg l^{-1} cefotaxime at 26 °C. Twenty days later, 100 mg l^{-1} kanamycin was also added and in the following subcultures only white tissues were selected on decreasing concentrations of both antibiotics (down to 15 mg l^{-1}), while further plant regeneration and micropropagation occurred on antibiotic-free media (Martinelli and Mandolino 1994). Finally, plantlets were transplanted

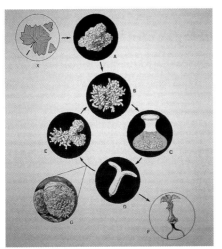

Fig. 1. Schematic representation of the main steps of somatic embryogenesis and genetic transformation in *Vitis rupestris*. From petiole- and leaf-(*X*) derived callus (*A*), embryogenic callus (*B*) has been induced. Further liquid culture (*C*) matured somatic embryos which were isolated (*D*) for respectively secondary embryogenesis (*E*) or plant germination (*F*) induction. Steps from (*D*) to (*C*) can be cyclically repeated over and over, and co-culture with *Agrobacterium* (*G*) occurs during secondary embryogenesis induction by submerging embryos for 5 min in an overnight culture. Drawing by Lucia Cardone (L. Martinelli 1997)

to in vivo conditions in a climate room after 1 month acclimation on a mixture of peat (80%) and sand (20%), at 25 °C and a 16h photoperiod of $100 \mu mol\, m^{-2}s^{-1}$ light.

Vitis DNA Isolation and Detection of Transgenes. Genomic DNA was extracted from somatic embryo cultures and from leaves of regenerated plantlets (0.1–0.5 g of fresh weight tissue) following the procedure described earlier Martinelli and Mandolino (1994, 1996). High molecular weight DNA, suitable for restriction analisis was obtained from embryos and in vitro grown grape plantlets.

In order to test for the presence of the inserted GUS gene, genomic DNA from different lines of transformed somatic embryo cultures and from eight different families of regenerated grapevine plantlets was digested with the enzymes EcoRI and HindIII, alone and in combination. The restriction fragments were separated and gels were blotted by capillary action onto nylon membranes. The hybridization conditions used in all experiments were as described in Martinelli and Mandolino (1994).

The probe used to test the effectiveness of the integration was the HindIII-EcoRI 3.1 kb fragment of the pBI121 plasmid, including the GUS coding sequence, the CaMV promoter and the Nos terminator. This fragment was gel-purified and radioactively labeled according to Feinberg and Vogelstein (1983).

Detection of GUS Gene Expression. Histochemical tests were conducted according to Jefferson (1987) on putatively transformed whole embryos, as well as on leaves and roots of regenerated plants, by incubation of histological sections in 5-bromo-4-chloro-3-indolyl glucuronide (X-Gluc; Martinelli and Mandolino 1994). The β-glucuronidase activity was evaluated in roots of plantlets by reading the 420nm absorbance, using p-nitrophenyl β-D-glucuronide (PNPG) as the substrate according to Naleway (1992).

3 Results and Discussion

3.1 Transgenic Embryo Production

Recurrent embryogenesis has frequently been described in *Vitis* (Krul and Worley 1977; Gray 1989; Matsuta and Hirabayashi 1989; Vilaplana and Mullins 1989). In our protocol (Martinelli et al. 1993), somatic embryogenesis develops in cycles that can be repeated again and again (Fig. 1). This morphogenic process proved to be particularly effective for *Agrobacterium tumefaciens*-mediated genetic transformation and is the basis of our protocol. Somatic embryos regenerate embryogenic callus within a 9-month culture period. Further isolation and maturation of individual embryos requires 40 days in liquid culture.

Grape tissues are extremely sensitive to kanamycin (Baribault et al. 1990; Colby and Meredith 1990). Consequently, the selection of transformed tissues has been performed mostly in the presence of low levels of this antibiotic (Baribault et al. 1990; Guellec et al. 1990; Mullins et al. 1990; Colby et al. 1991; Hébert et al. 1993). It is now generally accepted, however, that a rigorous selection is necessary to avoid chimeric tissues (Baribault et al. 1990; Colby et al. 1991; Berres et al. 1992). We observed that, out of 1204 cultured untransformed embryos, 23% generated secondary embryos during a 19-month culture period (unpubl.), while transformed embryos reached only a 4% efficiency. However, since further embryo production was abundant, low efficiency in the initial process was compensated. The adoption of a precocious selection at high levels of kanamycin is the other important part of our strategy.

The integration of the GUS gene was first tested in the embryo cultures deriving from single transformed embryos. This test was performed to investigate whether or not chimeric transformants occurred; in which case, it would be expected that tissues, after a certain number of subcultures, might lose the integrated gene. However, continued capability to proliferate in kanamycin-containing medium and the expression of β-glucuronidase suggests that transformation is stable. This was also confirmed at the molecular level since the DNA preparations from subsequent somatic embryo subcultures (at 12, 18 and 36 months) showed the same integration patterns for the GUS gene (Martinelli and Mandolino 1994).

3.2 Transgenic Plant Regeneration

Plant regeneration from distinct somatic embryos was obtained with agreeable efficiencies (13% within a 7-month culture period; Martinelli and Mandolino 1994), suggesting that the presence of foreign DNA in the genome does not significantly affect regeneration capability. The embryos first produced a green disorganized callus tissue and then many individual shoots regenerated via organogenesis from each embryo.

Eight transgenic somatic embryos developed into transformed cellular lines and were named A, B, F, J, K, M, N, and Z. These lines were competent for recurrent secondary somatic embryogenesis (Fig. 2) and for plant regeneration and, as a result, eight families of transgenic plants were obtained. Moreover, many sister plants were yielded within each family since every secondary somatic embryo regenerated numerous shoots via organogenesis (Martinelli and Mandolino 1996). When a sample of these regenerated plants was examined by Southern blotting analysis, it was found that out of 47 transformed plants tested, which belonged to eight families derived from eight different transformed embryos, only four gave no detectable hybridization signals; in the other cases, one to four bands were detectable (Fig. 3). This result confirms, at the level of a fairly wide population of transgenic plants, that the GUS gene is integrated into the genome. The most common pattern of integration observed consisted of one or two bands, and only in one case as an apparent multiple insertion (K5). The size and relative intensity of multiple

Fig. 2. Secondary embryogenesis
induced from an isolated somatic
embryo: embryogenesis begins on
the root/shoot transition zone and
embryo cotyledons can still be
recognized

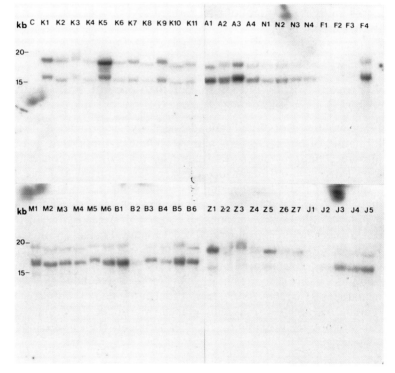

Fig. 3. Southern blot analysis of 47 transformed plants (belonging to the groups labelled as *K, A, N, F, M, B, Z, J*) and one untransformed plant (*lane C*). This figure was published in: Martinelli and Mandolino (1996) "Stability of the β-glucuronidase gene in a R0 population of grape (*Vitis rupestris* S.)." S Afr J Enol Viticult 17:27–30

bands in the same family of transformants shows some variation which is probably due to genomic rearrangements during the regeneration events (Martinelli and Mandolino 1996).

3.3 GUS Gene Expression

All cells of transformed secondary somatic embryos turned blue after a few hours of incubation with X-Gluc, and the same results have been obtained in regenerated shoots as well as in grown plantlets and in the roots. No endogenous GUS activity was detected in the control tissues (Martinelli and Mandolino 1994).

In a group of 33 plants that tested positive in the histochemical assay and of 7 untransformed plants, the average GUS activity was estimated to be 235.1 ± 16.4 and 1.7 ± 0.3 nmol of transformed PNPG $min^{-1} mg^{-1}$ of protein respectively (unpubl.).

4 Present Status and Field Performance of Transgenic Grapes

In the past few years, significant advances have been made in the genetic transformation of the genus *Vitis* and agronomically important genes have been introduced into rootstocks, table and vine grapes, mostly concerning virus and fungal diseases and herbicide resistance, as reported in Table 1. As a consequence of these breakthroughs, private companies are becoming increasingly involved in this research field. Recently, in the European Union, the first field tests with transgenic rootstocks were performed by the Institut National de la Recherche Agronomique of Colmar, France (notification number B/FR/94/11/04) and by the same institution with the French Moët et Chandon Company (notification number B/FR/96/03/14). The Californian Dry Creek Laboratories are carrying out a project for the protection of rootstocks against nematodes (Coghlan 1997), and the United States Tobacco Company is funding research on transgenic grapes (Martinelli 1997). Field performance data are not yet available for these experiments (Perl and Eshdat 1998).

5 Summary and Conclusions

Recently, promising results have been obtained in genetic transformation of the *Vitis* genus, even though this technology is still far from routine: transgenic plants have been regenerated in a few responding wild species, interspecific hybrids and rootstocks as well as in some *V. vinifera* cultivars. The most promising target seems to involve virus resistance, but examples of fungal resistance have also been reported.

The work reported here illustrates the possibility of stably transgenic grape regeneration. The described transformation strategy points out some of the potential advantages of using somatic embryos for foreign gene insertion. On the basis of our results and of similar reports of transgenic plant production (Table 1) it can be concluded that both zygotic and somatic embryos are very suitable material for stable insertion and high expression levels of exogenous genes, especially if severe criteria of selection are applied in order to restrict chimerism. Furthermore, if secondary embryogenesis is adopted, as in our strategy (Fig. 1), it is possible to propagate the transformed tissues from a single transformed somatic embryo and give rise to a population of transformed embryos capable of regenerating different stably transgenic plants. This system allowed us to test the insertion of genes involved in relevant traits (Martinelli et al. 1996b, 2000).

Acknowledgments. The authors wish to thank Prof. Maarten Koornneef and Mrs. Lucinda Smith for helpful suggestions.

References

Alleweldt G, Possingham JV (1988) Progress in grapevine breeding. Theor Appl Genet 75:669–673

Baribault TJ, Skene KGM, Scott NS (1989) Genetic transformation of grapevine cell. Plant Cell Rep 8:137–140

Baribault TJ, Skene KGM, Cain PA, Scott NS (1990) Transgenic grapevines: regeneration of shoots expressing β-glucuronidase. J Exp Bot 41:1045–1049

Beachy RN (1990) Coat protein-mediated resistance against virus infection. Annu Rev Phytopathol 28:451–474

Berres R, Tinland B, Malgarini-Clog E, Walter B (1992) Transformation of *Vitis* tissue by different strains of *Agrobacterium tumefaciens* containing the T-6b gene. Plant Cell Rep 11:192–195

Bowers JE, Dangl GS, Vignani R, Meredith CP (1996) Isolation and characterization of new polymorphic simple sequence repeat loci in grape (*Vitis vinifera* L.). Genome 39:628–633

Coghlan A (1997) Altered vines turn the worm. New Sci 153(2063):19

Colby SM, Meredith CP (1990) Kanamycin sensitivity of cultured tissues of *Vitis*. Plant Cell Rep 9:237–240

Colby SM, Meredith CP (1993) Transformation in grapevine (*Vitis* spp.). In: Bajaj YPS (ed) Biotechnology in agriculture and forestry, Vol 23, Plant protoplasts and genetic engineering IV, Springer, Berlin Heidelberg, New York, pp 375–385

Colby SM, Juncosa AM, Meredith CP (1991) Cellular differences in *Agrobacterium* susceptibility and regenerative capacity restrict the development of transgenic grapevines. J Am Soc Hortic Sci 116:356–361

Cronquist A (1981) An integrated system of classification of flowering plants. Columbia University Press, New York

Dry PR, Gregory GR (1988) Grapevine varieties. In: Coombe BG, Dry PR (eds) Viticulture. Australian Industrial Publ, Adelaide, pp 119–138

Dry IA, Robinson SP (1994) Molecular cloning and characterization of grape berry polyphenol oxidase. Plant Mol Biol 26:495–502

Dutruc-Rosset G (1998) The state of vitiviniculture in the world and the statistical information for 1996. Suppl Bull OIV 803–804:99–175

Feinberg AP, Vogelstein B (1983) A technique for radiolabelling DNA restriction endonuclease fragments to high specific activities. Anal Biochem 132:6–13

Galet P (1979) A practical ampelography: grape vine identification. Comstock Publ, Cornell University Press, Ithaca

Gölles R, da Câmara Machado A, Minafra A, Savino V, Saldarelli P, Martelli GP, Pühringer H, Katinger H, Laimer da Câmara Machado M (1998) Transgenic grapevines expressing coat protein gene sequences of grapevine fanleaf virus, arabis mosaic virus, grapevine virus A and grapevine virus B. Proc VIIth Int Symp on Grapevine Genetics and Breeding, Montpellier, France, Jul 6–10, 1998

Gray DJ (1989) Effects of dehydration and exogenous growth regulators on dormancy, quiescence and germination of grape somatic embryos. In vitro Cell Dev Biol 25:1173–1178

Gribaudo I, Schubert A, Camino C (1995) Establishment of grapevine axenic root lines by inoculation with *Agrobacterium rhizogenes*. Adv Hortic Sci 9:87–91

Guellec V, David C, Branchard M, Tempé J (1990) *Agrobacterium rhizogenes* mediated transformation of grapevine (*Vitis vinifera* L.). Plant Cell Tissue Organ Cult 20:211–215

Hébert d, Kikkert JR, Smith FD, Reisch BI (1993) Optimization of biolistic transformation of embryogenic grape cell suspensions. Plant Cell Rep 12:585–589

Jaynes JM, Nagpala P, Destefano-Beltran L, Huang JH, Kim JH, Denny T, Cetiner S (1993) Expression of a Cecropin B lytic peptide analog in transgenic tobacco confers enhanced resistance to bacterial wilt caused by *Pseudomonas solanacearum*. Plant Sci 89:43–53

Jefferson RA (1987) Assaying chimeric genes in plants: the GUS gene fusion system. Plant Mol Biol Rep 5:387–405

Jefferson RA, Kavanagh TA, Bevan MW (1987) GUS fusions: β-glucuronidase as a sensitive and versatile gene fusion marker in higher plants. EMBO J 6:3901–3907

Kikkert JR, Hébert-Soulé D, Wallace PG, Striem MJ, Reisch BI (1996) Transgenic plantlets of "Chancellor" grapevine (*Vitis* sp.) from biolistic transformation of embryogenic cell suspension. Plant Cell Rep 15:311–316

Kikkert JR, Reustle GM, Ali GS, Wallace PW, Reisch BI (1998) Expression of a fungal chitinase in *Vitis vinifera* L. "Merlot" and "Chardonnay" plants produced by biolistic transformation. Proc VIIth Int Symp on Grapevine Genetics and Breeding, Montpellier, France, Jul 6–10, 1998

Krastanova S, Perrin M, Barbier P, Demangeat G, Cornuet P, Bardonnet N, Otten L, Pinck L, Walter B (1995) Transformation of grapevine rootstocks with the coat protein gene of grapevine fanleaf nepovirus. Plant Cell Rep 14:550–554

Krastanova S, Ling KS, Zhu HY, Xue B, Burr TJ, Gonsalves D (1998) Development of transgenic grapevine rootstocks with genes from grapevine fanleaf virus and grapevine leafroll associated closteroviruses 2 and 3. Proc VIIth Int Symp on Grapevine Genetics and Breeding, Montpellier, France, Jul 6–10, 1998

Krul WR, Worley JF (1997) Formation of adventitious embryos in callus cultures of "Seyval", a French hybrid grape. J Am Soc Hortic Sci 102:360–363

Le Gall O, Torregrosa L, Danglot Y, Candresse T, Bouquet A (1994) *Agrobacterium*-mediated genetic transformation of grapevine somatic embryos and regeneration of transgenic plants expressing the coat protein of grapevine chrome mosaic nepovirus (GCMV). Plant Sci 102:161–170

Levadoux L (1954) La connaissance des cépages. Cah Vitivinicoles Rev Chambresgric (Suppl) 49:1–9

Levenko BA, Rubtsova MA (1998) Herbicide resistant transgenic plants of grapevine. Proc VIIth Int Symp on Grapevine Genetics and Breeding, Montpellier, France, Jul 6–10, 1998

Lodhi MA, Reisch BI (1995) Nuclear content of *Vitis* species, cultivars, and other genera of the *Vitaceae*. Theor Appl Genet 90:11–16

Lupo R, Martelli GP, Castellano MA, Boscia D, Savino V (1994) *Agrobacterium rhizogenes*-transformed plant roots as a source of grapevine viruses for purification. Plant Cell Tissue Organ Cult 36:291–301

Martinelli L (1997) Regeneration and genetic transformation in the *Vitis* genus. PhD dissertation, Agricultural University of Wageningen, Wageningen The Netherlands

Martinelli L, Mandolino G (1994) Genetic transformation and regeneration of transgenic plants in grapevine (*Vitis rupestris* S.). Theor Appl Genet 88:621–628

Martinelli L, Mandolino G (1996) Stability of the β-glucuronidase gene in a R0 population of grape (*Vivis rupestris* S.). S Afr J Enol Vitic 17:27–30

Martinelli L, Bragagna P, Poletti V, Scienza A (1993) Somatic embryogenesis from leaf- and peti-oladerived callus of *Vitis rupestris*. Plant Cell Rep 12:207–210

Martinelli L, Poletti V, Bragagna P, Poznanski E (1996a) A study on organogenic potential in the *Vitis* genus. Vitis 35:150–161

Martinelli L, Rugini E, Saccardo F (1996b) Genetic transformation for biotic stress resistance in horticultural plants. World Congr In vitro Biology, San Francisco, CA, June 22–27, 1996. In vitro 3:69A

Martinelli L, Buzkan N, Minafra A, Saldarelli P, Costa D, Poletti V, Festi S, Perl A, Martelli GP (2000) Genetic transformation of tobacco and grapevines for resistance to viruses related to the rugose wood disease complex. Proc VIIth Int Symp on Grapevine Genetics and Breeding, Montpellier, France, 6–10 July 1998, Acta Hortic, 528:321–327

Matsuta N, Hirabayashi T (1989) Embryogenic cell lines from somatic embryos of grape (*Vitis vinifera* L.). Plant Cell Rep 7:684–687

Mauro MC, Toutain S, Walter B, Pinck L, Otten L, Coutos-Thevenot P, Deloire A, Barbier P (1995) High efficiency regeneration of grapevine plants transformed with the GFLV coat protein gene. Plant Sci 112:97–106

Mauro MC, Walter B, Pinck L, Valat L, Barbier P, Boulay M, Coutos-Thevenot P (1998) Analysis of 41B grapevine rootstocks for grapevine fanleaf virus resistance. Proc VIIth Int Symp on Grapevine Genetics and Breeding, Montpellier, France, Jul 6–10, 1998

Minafra A, Saldarelli P, Martinelli L, Poznanski E, Costa D (1997) Genetic transformation for resistance to viruses related to the rugose wood disease complex. 1997 Congress on In Vitro Biology, Washington DC, June 14–18, 1997. In Vitro P-1033, 53A

Morel G (1944) Sur le développement de tissus de vigne cultivés in vitro. CR Soc Biol Paris 138:62

Mullins MG, Tang FCA, Facciotti D (1990) *Agrobacterium*-mediated genetic transformation of grapevines: transgenic plants of *Vitis rupestris* Scheele and buds of *Vitis vinifera* L. Bio/Technology 8:1041–1045

Nakano M, Hoshino Y, Mii M (1994) Regeneration of transgenic plants of grapevine (*Vitis vinifera* L.) via *Agrobacterium rhizogenes*-mediated transformation of embryogenic calli. J Exp Bot 45:649–656

Naleway JJ (1992) Histochemical, spectrophotometric and fluorometric GUS substrates. In: SR Gallagher (ed) GUS protocols: using the GUS gene as a reporter of gene expression. Academic Press, San Diego, New York, pp 61–76

Perl A, Eshdat Y (1989) DNA transfer and gene expression in transgenic grapes. In: Tombs MP (ed) Biotechnology and genetic engineering reviews, vol 15. Intercept, Andover, UK, pp 365–386

Perl A, Lotan O, Abu-Abied M, Holland D (1996) Establishment of an *Agrobacterium*-mediated transformation system for grape (*Vitis vinifera* L.): the role of antioxidants during grape-*Agrobacterium* interactions. Nat Biotechnol 14:624–628

Rombaldi CV, Girardi CL, Bilhalva AB, Ayub RA, Lelièvre JM, Latché A, Alibert G, Pech JC (1996) Expressão transitória de um clone de DNA da enzima formadora do etileno em protoplastos de uva. R Bras Fisiol Veg 8:201–207

Saldarelli P, Minafra A, Martinelli L, Costa D, Castellano MA, Poznanski E (1997) Putative movement proteins of grapevine viruses A and B: immunodetection in vivo and use for transformation of *Nicotiana* plants. Proc 12th Meeting Int Council for the study of Viruses and Virus-like diseases of Grapevine (ICGV), Lisbon, Portugal, Sept 28–Oct 2, 1997

Scorza R, Cordts JM, Ramming DW, Emershad RL (1995) Transformation of grape (*Vitis vinifera* L.) zygotic-derived somatic embryos and regeneration of transgenic plants. Plant Cell Rep 14:589–592

Scorza R, Cordts JM, Gray DJ, Gonsalves D, Emershad RL, Ramming DW (1996) Producing transgenic "Thompson Seedless" grape (*Vitis vinifera* L.) J Am Soc Hortic Sci 12:616–619

Shetty BV (1959) Cytotaxonomical studies in *Vitaceae*. Bibl Genet XVIII: 167–272

Sparvoli F, Martin C, Scienza A, Gavazzi G, Tonelli C (1994) Cloning and molecular analysis of structural genes involved in flavonoid and stilbene biosynthesis in grape (*Vitis vinifera* L.). Plant Mol Biol 24:743–755

Thomas MR, Cain P, Scott NS (1994) DNA typing of grapevines: a universal methodology and database for describing cultivars and evaluating genetic relatedness. Plant Mol Biol 25:939–949

Unwin PTH (1991) Wine and the vine: an historical geography of viticulture and the wine trade. Routledge, London

Viala P, Vermorel V (1909) Ampelographie. Masson, Paris

Vilaplana M, Mullins MG (1989) Regeneration of grapevines (*Vitis* spp) in vitro: formation of adventitious buds on hypocotyls and cotyledons of somatic embryos. J Plant Physiol 134:413–419

Weeden NF, Reisch BI, Martens ME (1988) Genetic analysis of isozyme polymorphism in grape. J Am Soc Hortic Sci 113:765–769

Willis JC (1973) A dictionary of the flowering plants and ferns, rev. Airy Shaw HK. 8th edn. Cambridge University Press, Cambridge

Subject Index

Printing (Computer to Film): Saladruck, Berlin
Binding: H. Stürtz AG, Würzburg